# Explaining Culture Scientifically

# EXPLAINING CULTURE SCIENTIFICALLY

EDITED BY

MELISSA J. BROWN

UNIVERSITY OF WASHINGTON PRESS | SEATTLE AND LONDON

University of Washington Press
P.O. Box 50096, Seattle, WA 98145 U.S.A.
www.washington.edu/uwpress


Library of Congress Cataloging-in-Publication Data

Explaining culture scientifically / edited by Melissa J. Brown.
    p.   cm.
Includes bibliographical references and index.
ISBN 978-0-295-98789-7 (pbk. : alk. paper)
1. Culture—Research.   2. Ethnology—Case studies.   I. Brown,
Melissa J.
GN357.E99   2008
306.07—dc22                                    2007048821

In memory of
the Department of Anthropological Sciences
at Stanford University
(1998–2007)
with hopes that what might have bloomed there
will find other soil in which to grow

# Contents

Acknowledgments | IX

Introduction: Developing a Scientific Paradigm
for Understanding Culture | 3
MELISSA J. BROWN

PART I / WHAT IS CULTURE? | 17

1 Some Kinds of Causal Powers That Make Up Culture | 19
ROY D'ANDRADE

2 Culture in Evolution: Toward an Integration of Chimpanzee
and Human Cultures | 37
CHRISTOPHE BOESCH

3 Dissent with Modification: Cultural Evolution and Social Niche
Construction | 55
MARCUS W. FELDMAN

PART II / MODELING-BASED CASE STUDIES | 73

4 Cultural Evolution: Accomplishments and Future Prospects | 75
PETER J. RICHERSON AND ROBERT BOYD

**5** Conditions for the Spread of Culturally Transmitted Costly
Punishment of Sib Mating | 100
KENICHI AOKI, YASUO IHARA, AND MARCUS W. FELDMAN

**6** Sexually Transmitted Infections as Biomarkers of Cultural
Behavior | 117
JAMES HOLLAND JONES

PART III / ETHNOGRAPHIC CASE STUDIES | 137

**7** When Culture Affects Behavior: A New Look at Kuru | 139
WILLIAM H. DURHAM

**8** When Culture Does Not Affect Behavior: The Structural Basis
of Ethnic Identity | 162
MELISSA J. BROWN

**9** A Cultural Species | 184
JOSEPH HENRICH

**10** Culture Matters: Inferences from Comparative Behavioral Experiments
and Evolutionary Models | 211
SAMUEL BOWLES AND HERBERT GINTIS

PART IV / CHALLENGES TO A SCIENCE OF CULTURE | 229

**11** Cultural Evolution and Uxorilocal Marriage in China:
A Second Opinion | 232
ARTHUR P. WOLF

**12** When Theory Is Data: Coming to Terms with "Culture"
as a Way of Life | 253
GREGORY STARRETT

**13** Studying "Culture" Scientifically Is an Oxymoron: The Interesting
Question Is Why People Don't Accept This | 275
ROBERT BOROFSKY

Epilogue: Future Considerations | 297
MELISSA J. BROWN

References | 307

Contributors | 359

Index | 361

# Acknowledgments

Several organizations at Stanford University helped to make this volume possible: the Department of Anthropological Sciences, the Morrison Institute for Population and Resource Studies, the Office of the Dean of Humanities and Sciences, the Institute for International Studies, the Humanities Center, and the Office of the Provost. By funding the January 2003 conference "Toward a Scientific Concept of Culture," they allowed me to bring the contributing authors together with additional scholars for discussion and debate. Contributors benefited from the comments of those additional scholars who presented papers at the conference—Rosaria Conte and Mario Paolucci, Lee Cronk, and Shannon A. Novak and Lars Rodseth—as well from those who served as discussants—Ronald L. Barrett, Terence Deacon, Hill Gates, Richard Klein, Kaiping Peng, Shripad Tuljapurkar, and Toshio Yamagishi. The Center for East Asian Studies at Stanford has also contributed by providing research assistants who helped with preparation of the manuscript.

I want to thank Elba Garcia, for helping with so much of the organizing of the conference; Helen Lee, for invaluable editorial and indexing assistance; and Bill Durham, Marc Feldman, and Arthur Wolf for encouraging me to undertake this project.

# Explaining Culture Scientifically

# Introduction

*Developing a Scientific Paradigm for Understanding Culture*

MELISSA J. BROWN

Anthropologists have been arguing for decades about what culture is, how useful it is as a concept, and how best to study it. The debate (see, e.g., Borofsky
1994; Fox and King 2002b) demonstrates that anthropology lacks a disciplinary paradigm, and thus still operates in a pre-science phase. The goal of this
book is to contribute to such a paradigm, and a scientific one at that, by taking
American anthropology's most important conceptual contribution—culture—
and exploring methods and epistemological underpinnings for scientific analyses of it. Most, but not all, of the authors in this volume not only see culture
as important to explanations of human behavior but also believe that a scientific study of culture is necessary for the development of a scientific paradigm for anthropology.

The exploration of scientific analyses of culture in this volume is empirical.
Each author examines empirical data about how culture influences behavior;
shows, rather than tells, how to approach culture scientifically; and demonstrates how empirical and theoretical material can be integrated. Not surprisingly, given the pre-science phase in which we find ourselves, contributors
approach their material from a range of viewpoints and methods across mathematical modeling and ethnographic empiricism.[1] However, there are important commonalities among the chapters regarding the importance of culture,
use of scientific methods, and some of the challenges posed to explaining culture scientifically.

This book is organized into parts that present views on how to think about
culture, modeling approaches to cultural influences on behavior, ethnographic

case studies addressing the question of culture's influence on behavior, and challenges to the possibility of a scientific approach to culture. Part introductions discuss the individual contributions, and the epilogue considers potential future research. Here, I introduce debates about culture and science in anthropology.

## IS CULTURE NECESSARY TO ANTHROPOLOGY?

Although twentieth-century American anthropology generally assumed the necessity and centrality of culture to anthropological explanation, other traditions have questioned its importance. Notwithstanding E. B. Tylor's (1871) often-quoted definition of culture and Malinowski's (1931) encyclopedia definition of it, British social anthropology largely did without the concept of culture, instead focusing on social structure via kinship, age grading, and other political economic organization (e.g., Malinowski 1922; Radcliffe-Brown 1952; Goody 1976; Gough 1981; Kuper 1982).[2] Marxist materialists also largely did without culture, except insofar as it constituted ideologies (see, e.g., White 1949; E. Wolf 1982; Mintz 1985; M. Harris 2001).[3] More recently, postmodernists questioned the usefulness of the concept of culture, suggesting that it is more productive to look at power and its discourses (e.g., Dirks et al. 1994; Ortner 1997, Fox and King 2002a; Trouillot 2002; R. Wilson 2002; see Starrett, this volume, for further discussion). Ironically, given that postmodernism disdains science as the power discourse most in need of deconstruction, postmodern anthropologists have arrived at the same position on culture—that it is unnecessary—taken by classic British social anthropologists and Marxist materialists, who saw themselves as conducting (social) science.[4]

We, in this volume, view culture as necessary to the development of a scientific paradigm for anthropology, a position based on evidence of culture's effects on behavior.[5] Not only do most contributors emphasize the potential for culture to influence and guide human actions, but many also extend culture's potential influence to the actions of our primate cousins and other social mammals (see especially the chapter by Boesch). Some authors discuss how such influence can lead to departures from expectations based on aggregated individual self-interest (see the chapters by D'Andrade, Feldman, Richerson and Boyd, Aoki et al., Henrich, and Bowles and Gintis). Some authors focus on culture's limitations in shaping behavior. Examining the relationship between individuals and the larger social group, these authors suggest that culture is mediated by other factors, most notably population structure and hierarchy (see the chapters by Jones, Durham, Brown, Wolf, Starrett, and Borofsky). Finally, some authors consider how specific ideas—the concepts of culture and science themselves—interfere with our ability to assess culture's influence (see the chapters by Starrett and by Borofsky). Most of the contributors find the concept of culture necessary to understanding behavior, though for different reasons and in different ways. Key to formulating this position on the necessity of culture is empirical evidence of its impact on behavior.

Most of the evidence, and indeed the staunchest support in recent years, for the ability of ideas or learned information to motivate behavior—an idealist position if ever there was one—comes from evolutionary anthropology and cognitive anthropology. Because of their ties to biology—genetics and ecology, and neuroscience, respectively—these two areas of anthropology have long been accused of materialist reductionism or biological determination by anthropologists working within the perspectives dominant in anthropological journals and organizations (e.g., Sahlins 1976b; Comaroff 1985: 53, 125–26; Rabinow and Sullivan 1987; see Roscoe 1995 for further discussion).[6] Most of the authors in this volume work in these or related research areas and thus provide evidence drawn from these purportedly biological areas of anthropology.

But what do we mean by culture? The contributors to this volume view culture as shared ideas or as socially learned and transmitted information.[7] These views are similar to the interpretive concept of culture that dominated American anthropology in the last decades of the twentieth century.[8] Interpretive anthropological theory (most famously articulated by Clifford Geertz [1973]) defined culture not as practices, which are behavior, but as the ideas shared across a society that can motivate practices.[9] The "interpretive" element in viewing culture as public meanings comes in fieldwork and analysis, when anthropologists must realize that actions constitute a text interpreted by different actors, each in a particular context, and thus different reports of a single event are all "true." It is an anthropologist's job to discover the ideas common to the various interpretations and thus to discover the cultural basis underlying events (Sperber 1996; Trigg 2001). This concept of culture as shared, abstract (i.e., ideational), meaning-laden, and public (i.e., external to individuals) has dominated American anthropology for decades. The contributors to this book do not emphasize the interpretive aspect of culture. We focus on the idealist aspects of culture—the potential for ideas to drive behavior—or on the quantifiable results of interpretation, such as variation in ideas across a population.[10]

The poststructuralist emphasis on power led some, in the 1990s, to criticize Geertz and his interpretive notion of culture as naïve in supposing that all members of a society share the same set of meaningful ideas (e.g., Abu-Lughod 1991; Dirks et al. 1994: 3, 22; Keesing 1994; Ortner 1997; Martin 1998: 24; as well as several of the contributions to Fox and King 2002b). Such dissatisfaction with viewing culture as monolithic is not new. Borofsky (this volume) discusses how ethnographers have always known of variation in the ideas held by different individuals in a society (see also Radin 1933: 42; Sapir 1938; Barth 2002). What is new, however, are recent arguments within American anthropology in favor of doing away with the concept of culture entirely (Brightman 1995 provides a good summary; see also Kuper 1999; A. Wolf 2001; Fox and King 2002b).

These desires to do away with culture derive, I think, from two ways in which the concept of culture—as understood in the broad interpretive consensus that has dominated American anthropology for decades—is unbounded.[11] First, although classical definitions of culture included the concept of sharing, many critics today mistakenly assume that those definitions required ideas to be shared by every single member of a society in order to be considered cultural (see especially Brightman 1995, but see also Borofsky, this volume; M. Brown 1997a, 1997b). Thus, in recognizing variation ("subaltern voices"), they think they must reject entirely the concept of a shared culture. By contrast, evolutionary anthropologists readily accept variation because they trace the transmission process from its origin at an idea held by a single individual to some distribution in the population (e.g., Cavalli-Sforza and Feldman 1981; Boyd and Richerson 1985; Durham 1991, 2002). Cognitive anthropologists also accept variation because they see meanings as internal to individuals, with variation deriving from the process of internalization from publicly received messages to internally coded schema (e.g., Strauss 1992a; D'Andrade 1995b, 1999; Strauss and Quinn 1997; Sperber 1996; see also Sapir 1938).

Like classical and interpretive anthropologists, the contributors to this book view sharing as a crucial aspect of culture. We also recognize—some based on empirical evidence, some based on mathematical modeling—that specific cultural ideas have some distribution or frequency across a population that is often not 100 percent. Some contributors focus on transmission processes as means for understanding the specific distribution in a population (see especially the chapters by Feldman, Richerson and Boyd, Aoki et al., and Jones, but see the chapter by Wolf for criticism), whereas others consider internalization or negotiation processes (e.g., D'Andrade, Durham, Brown, Starrett, Borofsky). Recognition that ideas have a distribution—and not necessarily the same distribution as practices (M. Brown, 2007)—raises the issue of the frequency at which an idea should be considered cultural. How many people need to share an idea for it to be considered a generally held—that is, cultural—idea? I suggest that this is an empirical question (M. Brown 1997a, 1997b, 2004:216, 2007), which needs further research (see the Epilogue to this volume for further discussion).[12]

The second way in which the broad consensus on the concept of culture is unbounded has largely gone unrecognized. Culture has become a catch-all explanation—if economics or genetics cannot account for behavior, then culture must be the cause. Unfortunately, in such accounts culture is also often seen as behavioral content (harking back to Tylor's definition of civilization). We are left, then, with a tautology that explains nothing: People do what they do because of culture, but culture is what people do (compare D'Andrade 1999).

What is particularly helpful in this book is that contributors explore specific circumstances under which culture can be said to cause behavior, and use the resulting insights to suggest bounds to what culture might be or bounds to when it might affect behavior. British social anthropologists, Marxists, poststructuralists, and others have been correct to criticize interpretive anthropology for assuming that, at root, culture explains the most important things—if not everything—about human behavior. That is not to say, however, that we should abandon the concept of culture as an explanation of human behavior. If we keep the concept of culture but limit its definition, we accept that culture may explain some human behavior under some conditions, but we also allow for the possibility that other factors may be explanatory as well. There are two main challenges here: first, to determine how often, and under what circumstances, cultural influence is subordinated or mediated by other factors versus

when culture is the driving influence in human behavior; and second, to identify which factors mediate culture and which may be driven by culture. Or, to
put it more simply, when does culture influence behavior, and when does it
not?

## IS SCIENCE POSSIBLE IN ANTHROPOLOGY?

All the contributors to this volume agree that explanations of cultural influences on behavior and societies must be empirically based, verifiable, and
critically evaluated; most of us consider an explanation that meets those criteria to be a scientific explanation. As with the concept of culture, however,
fundamental questions have been raised about whether it is possible to get at
the truth using particular scientific methods, which make the necessity and
even the possibility of science controversial.

Enlightenment scholars (Immanuel Kant, Henri de Saint Simon, Auguste
Comte, Thomas Hobbes, David Hume, Adolphe Quételet, Anne-Robert-
Jacques Turgot, the Marquis de Condorcet, and, to a lesser extent, John Stuart
Mill) thought that they and their intellectual heirs would rapidly attain a complete understanding of human societies and behavior which would allow
social engineering to "improve" human societies (see, e.g., Hankins 1985:
159; Hawthorn 1987: 83; Bendix 1988: 9, Trigg 2001: 222; see also Starrett, this
volume), but social evaluation and theory have proven much more difficult
than they anticipated. Attempts to apply a scientific method to sociological
studies have been thwarted not only by complexity (something natural scientists also faced) but also by the inability to conduct experiments and the
difficulties of adequately documenting existing societies at the individual level
in order to observe "naturally" occurring social experiments.[13]

Through the 1960s, anthropologists of many perspectives considered
anthropology a science. Franz Boas, often considered the father of American
anthropology and well known for his criticism of unilineal evolution and
social Darwinism, advocated anthropology as a science, though he argued that
much more empirical data were needed before the discipline could construct
a paradigm (e.g., Boas 1898: 107–8, 1887: 16; Benedict 1943: 61; Stocking
1968: 154–55; Bunzl 1996: 18, 62). American classical anthropologists, British
social anthropologists, French ethnologists, and German diffusionists col-

lected such data in detailed ethnographies that are still useful today.[14] Most did not pursue a deductive, hypothesis-testing form of science but rather an inductive, observation-driven form. They succeeded in documenting variation across societies and developing some important generalizations about universal attributes and common points of variation, for example, about marriage, kinship, incest taboos, and gift exchange.[15]

Foucault (e.g., 1965, 1973, 1977) criticized science and social science as tools for controlling populations and maintaining the status quo that, moreover, failed to recognize the inherent bias resulting from their creation by the system that they served. As a result, many anthropologists questioned whether science, or pursuit of truth at all, is possible in inquiries of human behavior. At the same time that interpretive anthropology was encouraging a shift to the examination of cultural meanings, criticism of science encouraged a retreat from the use of systematic, verifiable, and replicable methods as false objectivity (Roscoe 1995; Reyna 1994; Nettle 1997). Subsequently, much of American anthropology turned to increasingly postmodern forms of humanist approaches that largely avoided political involvement (Sangren 1995; see also Sangren 1988; Ensminger 1998, although there is a more recent trend toward political activism, e.g., Scheper-Hughes 1995.)

Perhaps not surprisingly, given their ties to biology, evolutionary anthropology and cognitive anthropology remained committed to anthropology as science (e.g., D'Andrade 1995a, 1995b; Bloch 1991; Nettle 1997; E. Wilson 1998; Laland and Brown 2002). Following its intellectual predecessor (ethnoscience), cognitive anthropology has focused on empirical studies (see, e.g., the contributions to D'Andrade and Strauss 1992 and to Hirschfeld and Gelman 1994). Evolutionary anthropology, on the other hand, has pursued two lines of contribution toward anthropology as science: empirical studies (see, e.g., the contributions to Cronk et al. 2000) and mathematical modeling (see, e.g., Cavalli-Sforza and Feldman 1981; Boyd and Richerson 1985; Odling-Smee et al. 2003).

Unfortunately, the lack of strong critical evaluation within evolutionary anthropology leaves the various efforts isolated. Modeling efforts used in evolutionary anthropology have generally created a model and then fit population data to that model—data variously derived from experiments, computer simulations, or limited historical, archaeological, or sociological sources (e.g.,

Cavalli-Sforza et al. 1982; Axelrod 1984; Richerson and Boyd 1989; Laland et al. 1995; Mackie 1996; Fehr and Gächter 2002; Henrich et al. 2004; Ihara and Feldman 2004). This method assumes that the model adequately captures the essential features of the social dynamics to be explained, and does not constitute critical evaluation or an empirical test. Empirical studies have generally relied on middle-range models from related fields (e.g., optimal foraging models drawn from ecology in Smith [1991], models of sex differences in parental investment drawn from sociobiology/evolutionary psychology in Daly and Wilson [1999]). They assume the grounding of these models in a cultural evolutionary modeling scheme, most commonly Boyd and Richerson's (1985), but again they do not provide critical evaluations. Moreover, works that explicitly discuss the links between human behavior and these larger evolutionary models, even when involving authors with substantial empirical experience (e.g., Smith and Winterhalder 1992), do not construct empirical tests of hypotheses derived from the theoretical framework. Instead, they assume the validity of the theoretical framework and argue that it provides new insights to the data. Other evolutionary empirical efforts have ignored entirely the specifics of the mathematical models (e.g., A. Wolf 1995, but see A. Wolf, this volume).

Reactions to empirical studies inspired by these models have been problematic. For example, Robert Aunger's dissertation (1992) was constructed as an empirical test of Boyd and Richerson's (1985) models, but he framed his thesis as an analysis of ethnographic methods, as most ethnographers would not consider sufficiently original his finding that people in the Ituri Forest learned food taboos via vertical transmission—that is, from their parents. Empirical studies supporting evolutionary models of culture (e.g., Durham 1991) are criticized for focusing too narrowly on one aspect of a society under change—often an aspect of society in which members of society easily accept change—or, at the other extreme, for trying to cover too broad a sweep of cultural phenomena (see Edmonds 2002, 2005, on the lack of empirical studies supporting one branch of evolutionary anthropology, memetics). Another kind of problematic reaction has been the response to empirical evidence challenging evolutionary models. For example, Dagg's (1998) finding, based on empirical data, that most lion cub mortality is due to maternal indifference rather than infanticide by adult males (a basic tenet of parental investment models) was considered wrong, preventing publication of her empirical results for ten years. Roughgarden's empirically derived questioning of standard models of sexual selection has at least been published (Roughgarden 2004; Roughgarden et al. 2006), though much of the response to it has been markedly emotional (see, e.g., the "Letters" section in the May 5, 2006, issue of *Science*, vol. 312).

Specialization and factionalism among science-oriented anthropologists— cognitive and evolutionary anthropologists, for example, have often been unaware of related work in each others' fields—has allowed the postmodern position on the impossibility of science to become canonical throughout much of the discipline.[16] There is a growing movement within the discipline to reestablish anthropology as a science, as evidenced by growing criticism of postmodernism (e.g., Sangren 1988, 1995; Reyna 1994; Roscoe 1995; D'Andrade 1995a; Nettle 1997) and by the development of the Society for Anthropological Sciences (SASci), born in 2002 in opposition to the postmodern dominance at the American Anthropological Association annual meeting.[17] This volume is a contribution to the anthropological sciences movement.

## IN SEARCH OF A PARADIGM

Even though all seem to agree that it is attention to methods that constitutes science, the factionalism of anthropological sciences indicates an underlying and often unspoken debate over which specific methods to use, or at least how to modify established methods of observation, hypothesis testing and replication for application to human societies (see O'Meara 1997 and Harris 1997 for one such exchange). Mathematical modeling and ethnographic empiricism focus on a different aspects of scientific method—mathematical modeling on the generation of theoretically derived hypotheses that can be empirically tested, ethnographic empiricism on accurately recording empirical cases that can yield comparative generalizations for theory building.[18]

Much of the disagreement stems from the lack of a paradigm. It is further complicated by those who advocate the scientific methods of "normal" science in striving to develop anthropological sciences. (A paradigm is a theoretical framework that is virtually universally agreed upon by scientists within a particular discipline [Kuhn 1962]. In normal science, a theoretical framework is already accepted as a paradigm, and science proceeds by testing hypotheses based on the established theory, even when these hypotheses push at the limits of the theory.)

Anthropology, and perhaps social science broadly, is still in a pre-science phase because there is no generally accepted theoretical framework (contra D'Andrade 1995b). Methods for the development of an initial paradigm in a pre-science context differ from normal scientific methods. Such development requires strong integration of theory and data: constant movement between using theory to inform empirical investigations and using empirical results to reformulate theory (Kuhn 1962). In order to achieve this integration, theoretical models must take into account factors and processes shown to be important in empirical cases, and empirical data must be gathered in such a way that they can inform theoretical models (for a similar position on integrating theory and data in ecology, see Kareiva 1989).

Moreover, because there is no paradigm, the need for critical appraisal of results and, where possible, independent replications, are even more important than under conditions of normal science (where a paradigm facilitates the identification of anomalous data). In a pre-science phase, the task of analyzing the theoretical relevance of specific empirical data can be daunting, if the data were not collected to test a particular theoretically informed hypothesis.

The contributors to this volume have begun the crucial task of integrating theory and data. Modelers here compare their mathematically derived expectations to specific empirical cases. They examine how modeling can address the range of empirical variation (e.g., the chapters by Richerson and Boyd; Jones; Bowles and Gintis) and how modeling can address the impact of historical contingencies (e.g., the chapters by Feldman; Aoki et al.). Ethnographers generalize from their empirical cases, offering theoretical propositions that expand some modeling assumptions (e.g., the chapters by D'Andrade; Boesch; Henrich) and challenge others (e.g., the chapters by Brown; Durham; Wolf; Starrett; Borofsky). The next step is to model such propositions, or hypotheses derived from them, and subsequently compare mathematical expectations with empirical data. The challenge of integration is to produce modeling and empirical results that make it easier for science researchers to evaluate, criticize and draw upon each others' work. Collaboration between modelers and empiricalists and across cognitive and evolutionary interests seems the most fruitful way to produce such mutually beneficial results. It will also, we hope, promote development of a broader scientific approach to human behavior, as modelers and empiricalists, cognitivists and evolutionists cooperate. The contributors to this volume present verifiable methods and analyses of cultural behavior and discuss implications for the importance of an anthropological concept of culture, thereby contributing to a science of culture.

## NOTES

I want to thank Monique Borgerhoff Mulder, Marc Feldman, Lorri Hagman, Jamie Jones, Jim Truncer, and Arthur Wolf for productive comments and/or discussion related to this introduction.

   1. By "empiricalism," I mean an approach where scholars come to theoretical conclusions based on empirical generalizations derived from field research. It is not empiricism, defined as an epistemological position which views truth as exclusively coming from empirical phenomena, nor simply generalization, defined as extrapolating from specific cases to more general rules. I want to emphasize the importance of the researcher's own empirical field work in contributing to the theoretical generalizations made. It contrasts with a modeling approach that derives theory and empirical studies from consideration of what is mathematically possible.

   2. "Culture or Civilization, taken in its wide ethnographic sense, is that complex whole which includes knowledge, belief, art, law, morals, custom, and any other capabilities and habits acquired by man as a member of society" (Tylor 1871: 1, italics added to indicate what is usually represented by an ellipsis). As Tylor himself indicated in his text, this definition refers to culture as a notion of civilization—it is not the interpretive notion of culture I discuss—although scholars no less authoritative than Alfred L. Kroeber and Clyde Kluckhohn (1952: 9) misleadingly credit Tylor's definition with establishing "culture" in its "modern technical or anthropological meaning" (see Stocking 1968: ch. 4, esp. 72–75). Malinowski's encyclopedia definition explicitly distinguishes "culture" from "civilization" and gives primacy to ideas (e.g., 1931: 621, 625), but elsewhere his treatment of culture (e.g., 1922: ch. 1, esp. 22–25; 1944: 36, 39–40), is closer to Tylor's, giving primacy to social structure. There are indicators that the concept of culture has invaded British social anthropology via cultural Marxism and some aspects of post-structuralism, which may have motivated Kuper's (1999) criticism of the concept, but I think it is fair to say that British social anthropology generally retained an interest in what is now fashionably called "power," even after American anthropology had ceded

this area of research to sociology. See Goody (1994) for a restatement of the British position against culture.

3. It is commonly said that nobody disagrees more than two Marxists, and certainly the impact of interpretivism on American Marxists in the 1980s (see, e.g., Taussig 1980, 1989) led to a revisionist Marxism that saw ideology as having the potential to drive society (on cultural Marxism, see Nelson and Grossberg 1988; Dworkin 1997). Despite prominent refutation by materialists Sidney Mintz and Eric Wolf (e.g., 1989), this cultural Marxism came to dominate American anthropology (e.g., Comaroff 1985; Donham 1990; Harrison 1997), which subsequently merged with poststructuralism. (For the persistence of a materialist Marxism, however, see Littlefield and Gates 1991; Gates 1996.)

4. Brumann (1999: S1) makes a similar point about the strange bedfellows opposed to culture in his argument in favor of retaining the concept of culture.

5. For most authors here, "behavior" refers to practices—behavior performed with some regularity. It sometimes specifically refers to cultural practices, which requires that the practice—or the idea behind it—be shared and meaningful across a population (see also M. Brown, 2007).

6. Such dismissals often portray anyone interested in biology or using methodological individualism as fascists and elitists. However, the origins of the idealist position in evolutionary anthropology (Cavalli-Sforza and Feldman 1973a, 1973b) came from geneticists providing mathematical proof that the social science data which Arthur Jensen (1969, 1970) spuriously used to claim African Americans as genetically inferior to Americans of European ancestry could be explained by the transmission of ideas with no genes whatsoever involved.

7. For discussion of the difference between ideas and beliefs, see Harrell (1977), Strauss (1992b), Sperber (1996), Brown (2007).

8. Some of this volume's contributors came to this definition via knowledge of interpretive anthropology, but others came to a very similar position by recognizing that socially learned information spreads through a population with a different dynamic than does genetic information.

9. Interpretive formulations of culture as public meanings were not derived in a vacuum. Geertz (1973: 5) explicitly drew on Weber's view of meanings and ideal motives (e.g., Weber 1922), but there is also much of Durkheim's (e.g., 1912: 228–29, 349) notion of collective representations and Wittgenstein's (1953: Part I, §§ 199–202, 208, 217, 219, 246, 378–86) stance on the public nature of rules to be found in the interpretive concept of culture (see also Trigg 2001: 26–30). Variation and debate existed among interpretivists in their views of culture (see, e.g., Geertz 1957, 1973, 1983; Sahlins 1976a; D. Schneider 1980, 1984). Indeed, Starrett (this volume) argues that this variation is sufficient to preclude a single concept of culture. However, in agreement with Kuper

(1999), I suggest that there existed a broad consensus on the basic aspects of an interpretive concept of culture. Moreover, Stocking's (1992: 4) reference to the emergence of a generalized "anthropological" concept of culture—which he defines as "pluralistic, holistic, non-hierarchical, relativistic, [and] behaviorally determinist" (see also Stocking 1968, 1974)—implies that he also sees a broad anthropological consensus on the concept of culture.

10. We do, however, grant the relativism of interpretation, just as physicists acknowledge Heisenberg's uncertainty principle but continue in their observations.

11. Goody (1994), Brightman (1995), Brumann (1999), Fox and King (2002a), and others criticize the "boundedness" of culture, by which they mean the presumption that one can specify where "a culture" stops—that is, they question whether one can draw a boundary between the people who share one culture and the people who share another (see also Borofsky, this volume). Indeed, the problem of where to draw such boundaries is central to Brumann's discussion. I am referring to a different aspect of culture as unbounded—the inability to define what culture is—what things are included in it and excluded from it. D'Andrade (1999) and Fox and King (2002a) raise this issue as well, though they do not refer to it as "unbounded."

12. Although Romney et al. (1986: 331–32) briefly consider the question of how much consensus is necessary to define something as cultural, it is not the main focus of their efforts, nor of subsequent debate about this method (Aunger 1999; Romney 1999).

13. Such difficulties include exorbitant cost (e.g., A. Wolf 1995: viii) and falsification of data about individuals, due to (legitimate) fears of individuals being documented, and/or local officials who represent them, that such documentation could be used against them by their governments (as indeed it could).

14. Authors of such rich ethnographies include: American scholars Paul Radin (1926), Alfred L. Kroeber (1925), Edward Sapir (1907, 1916), and Margaret Mead (1930); British scholars Bronislaw Malinowski (1922, 1929), Raymond Firth (1936), Edward E. Evans-Pritchard (1940), and Meyer Fortes (1945, 1949); French scholars Claude Lévi-Strauss (1955, 1964) and Maurice Godelier (1982); and German scholars Fritz Graebner (1909) and Wilhelm Schmidt (1931).

15. Although these institutions have cultural meanings attached to them, they also have strong structural components—that is, they serve a primary function of arranging network and hierarchical connections within a society. These structural elements raise the possibility that anthropologists made more systematic and cumulative—hence, scientific—contributions to these topics to the extent that they focused on their structural components. (See also Brown, this volume.)

16. The "culture wars" pit an often-shifting yet purportedly canonical thing called

"science" against a heroic or villainous—depending on your perspective—underdog challenger called "postmodernism." Postmodernism developed as an opposition perspective intended to critique, deconstruct, subvert. Its goal was to overturn the canon of science and promote subaltern voices. However, when postmodernism becomes the canon, as it has in anthropology and much of the social sciences and humanities over the last twenty years, what does it oppose but a straw man of what science used to be? Science also was originally an opposition perspective, developed as an argument against natural theology during the Enlightenment; it claimed that rigorous methods of empirical investigation subject to open debate would lead closer and closer to truth (i.e., a true understanding of reality). However, when a siege mentality labels critical examination as disloyal, as it has in reaction to the culture wars (see Laland and Brown 2002: 6–7), is science possible? For example, critics of Patrick Tierney's (2000a, 2000b) sensationalist accusations about Chagnon's work among the Yanomamo went beyond what was necessary to demonstrate Tierney's false claims, creating a climate in which Chagnon's work could not be questioned, even from within a scientific perspective. After Charles Mann (2001), reporting on the controversy, included Kim Hill's criticism of Chagnon's demographic analysis, Hill (2001) felt it necessary to write a reply to Mann denouncing Tierney as antiscientific and making it clear that he (Hill) supports Chagnon. See Borofsky et al. (2005) for a comparative view of the controversy.

17. The development of a scientific approach to culture could be an important contribution to the resolution of the culture wars because it unites a scientific approach with a concern for the transmission of ideas across societies and across time. For a history of SASci's beginnings, see http://anthrosciences.org/csac/SASci/rssite/salon.rss.

18. There is also epistemological debate about whether the fundamental scientific principle is falsification (e.g., Popper 1959) or verification (e.g., Lauden 1990). For debate within anthropology over what constitutes science, see, for example, the exchange between Aunger and his commentators, Layder, Sangren, and Hammersley (all in Aunger 1995).

# PART I  WHAT IS CULTURE?

How should we think about culture itself? In exploring this question, the chapters in this part address some of the most important current challenges to understanding the relationship(s) between culture and human behavior: how to determine the causal correspondence between culture and behavior; how to determine whether structure constitutes a system distinct from culture; how to assess the adaptiveness of culture and the applicability of a concept of culture to nonhumans; and how to define, detect, and quantify cultural units.

Roy D'Andrade, a cognitive anthropologist, focuses on causation, arguing that "the degree to which culture influences behavior depends primarily on what is included in or excluded from the definition of culture." Starting from an ideational concept of culture as necessary but not sufficient to motivate human behavior, D'Andrade wrestles with the issue of structure as a distinct system. Both at the philosophical level and in a detailed empirical case about a putative Southern-U.S. honor code, D'Andrade explores whether institutions belong in the definition of culture. Because institutions represent an intersection of ideational and structural systems, D'Andrade suggests that we should be asking about the causal powers of specific cultural forms, such as specific institutions, not about the causation of culture writ large.

Christophe Boesch, a primatologist, focuses on adaptiveness and cross-species comparison. On an empirical basis, he proposes uniting an adaptationist position on culture (that culture is adaptive, and thus natural selection gives culture causative power on behavior) with the contrary position (that culture rarely provides direct benefits to individuals and may well be genetically maladaptive) as "different levels in one continuum that embodies the complexity of different cultural systems." Most currently known cultural behaviors in primate and cetacean species—behavior patterns that are not genetically determined—appear adaptive. Moreover, there are structural contributions to culture via social learning. However, there is also ample evidence of cultural behaviors with no direct benefit to individual actors, including leaf-clipping among chimpanzees, which rely on shared meaning. Boesch's multilevel definition of culture shows continuities between humans and other animal species, but it also indicates distinctions that make us unique.

Marcus W. Feldman, a population geneticist, responds to criticism (such as that of Fraccia and Lewontin 1999) that culture must be viewed as history because of its contingent path and its variable units. Feldman argues that niche construction, which models the interactive effects of individuals and their ecological or social environments over time (e.g., Odling-Smee et al. 2003), effectively models environmental history—that is, the specific path taken, out of all the stochastic possibilities, which led to the current environmental structure. Moreover, "it is immaterial whether the units that are transmitted between members of a population are ideas, behaviors, or customs," as long as the model captures the social nature of transmission, although behavioral units are more easily identified and quantified than ideas. Feldman suggests that this formal, evolutionary approach to culture includes history in a quantitative manner, supporting the importance of cultural background and moving anthropology toward scientific explanations.

# 1

## Some Kinds of Causal Powers That Make Up Culture

ROY D'ANDRADE

In the introduction to this volume, Melissa Brown poses the question, *when does culture influence behavior and when does it not?* The answer to this question is not simple. Five different answers are presented below; each answer depends on what is included in the definition of culture and on the assumptions made about the distributions of motives and knowledge.

To start, let us provisionally agree to define *culture* as a polysemous term (for example, see Boesch's analysis, in this volume, of seven levels of culture) whose core meaning centers on shared cognition, sometimes described as ideas, information, or knowledge. These cognitive materials can be organized into schemas, models, taxonomies, prototypes, narratives, ideologies, discourses, and the like (D'Andrade 1995a).

With respect to Brown's question, the basic issue is, *when do ideas influence behavior and when do they not?* The standard answer from the point of models of the mind is that an idea will influence behavior only if the person who holds the idea has some want or need that motivates that person to do something, and only if the idea has some bearing on how to satisfy the relevant want or need (D'Andrade 1987). Max Weber's oft-quoted explication of this point was that ideas about salvation only influence those who want to be saved. Ideas tell us what is in the world, and by doing so they help us get what we want and avoid what we don't want, but they are not usually considered sufficient by themselves to cause behavior. It should be pointed out that without *any* ideas about what is in the world, most animals become completely unhinged, unable to do anything more than produce knee-jerks and sneezes. Ideas are like maps—they give information about what is in the world, but they don't make us go anywhere.

Thus, although ideas are not *sufficient* causes of behavior, they are *necessary* causes. This means that we can't predict what a group will do just because we know that the group has some shared idea. But we can predict that if a group does not have some idea of what an *atlatl* or a marriage is, then the group members won't use *atlatls* or get married. However, such negative predictions are usually not of much interest. Most controversies are about the causal effects of culture, not the causal effects of not having culture.

Depending on other assumptions, the proposition that ideas are necessary but not sufficient causes of behavior produces several different models about the relation between culture and behavior. Five of these models are laid out below.

## MODEL A

*Human goals (motives, desires, etc.) are random—different people want different things. Any cultural idea may influence one person but not another, because that cultural learning is connected in some relevant way to what a given person wants but not necessarily to anything other people want.*

In Model A, cultural ideas are not good predictors of behavior because human motives are so various. Most social scientists consider Model A inadequate because they generally agree that people everywhere have many similar goals and desires. Some goals, such as avoiding hunger and physical discomfort, are psychologically compelling to almost everyone. Other goals, such as earning money, are general prerequisites for achieving other goals, so that whatever a person wants, money may be helpful, and because of this almost everyone wants money.

## MODEL B

*Many human motives and goals are commonly shared. Cultural ideas that are connected to shared motives and goals can predict individual behavior because the necessary motivation can be assumed.*

Thus, if a group of hunters and gatherers knows how to make and use bows, but has no knowledge of other effective projectile technologies, this group will be strongly influenced by its cultural ideas concerning the manufacture and use of bows. Joseph Henrich, in this volume, describes Hadza and Kalahari conformity to their separate cultural traditions of bow making and poison extraction. Given that each technology works, and that the members of each society do not know a competitive alternative technology, each society's cultural ideas about how to make bows and extract poisons are almost perfect predictors of relevant behavior.

Of course, Model B doesn't say that cultural ideas *alone* influence behavior. But it does say that cultural ideas predict what people will do because the underlying motivations are shared. However, Model B is also considered by most social scientists to be too simple because it does not account for competition between cultural ideas. What if, in some group, ideas about making Hadza bows are in competition with ideas about making Kalahari bows? Which set of cultural ideas will affect their behavior? One would expect effort and efficacy to have strong effects on their choice.

## MODEL C

*Human motives and goals are shared to various degrees. Cultural ideas that are connected to shared motives and goals potentially influence individual behavior when the necessary motivation is present. But where competing cultural ideas are connected to the same goal, the cultural ideas that yield the more satisfactory outcomes will tend to be selected over cultural ideas that yield less satisfactory outcomes.*

The addition of the assumption of competition between cultural ideas diminishes the predictive power of cultural ideas. Given the presence of many competing cultural ideas, the situation reverts to that of Model A—there is not much to say about the relation of cultural ideas to behavior except that some cultural ideas are probably involved in whatever people do. The important predictive factors in Model C are not cultural ideas, but the differential *satisfactions* that come from holding one cultural idea or following one cultural practice rather than another. In this volume, Arthur Wolf's example of the economic and social considerations that influence the prevalence of uxorilocal marriage in China and Melissa Brown's example of the selection forces on social identity are cases in point.

The differential satisfactions in Model C can involve simple efficiency, such as which planting technique produces more corn. They can also involve direct testing of whether some cultural idea is true or not (e.g., does getting cold and wet cause one to catch cold?), or any other kind of criterion that is relevant to people. Thus the particular satisfactions may involve finding the idea that is most true, or most prestigious to hold, or most effective, or most beautiful, or least anxiety-producing. The kind of satisfaction is not the issue here. The point is that the major causal work is done by the processes that select cultural ideas—cultural ideas in themselves are not where the action is. Issues of self-interest, cost, prestige, social relations, adaptive fitness, and the like are where the causal action is.

Model C is probably close to being the standard model in the social sciences, and has been succinctly formulated by Sperber (1985). Despite Model C's status as standard social science, however, certain doubts persist. Some kinds of culture do not seem to be all that selectable. And perhaps some kinds of culture are not just ideas. For example, although the nation-state can be talked about as if it were basically just an idea—an "imagined community" (B. Anderson 1991)—it makes an odd sort of idea. States, unlike ideas of things, actual or imaginary, are corporate entities, and can go to war with each other, kill each other's citizenry, expand boundaries, and on and on. Such entities have an enormous range of causal powers. One might like to believe that the modern nation-state is basically just an idea, and so all we need to do is realize that the whole thing is a fantasy and it will lose its grip on us. But institutions don't work that way.

Are institutions culture? Here the extensive polysemy of the term culture creates confusions. At a high level of contrast, culture includes all kinds of institutions—few anthropologists would want to exclude institutions such as property or money or funeral ceremonies from the domain of culture. But, for a curious set of reasons, American cultural anthropologists like to talk about culture as if it were nothing but ideas. The Parsonian division between culture and social systems treats culture as shared symbols and idea systems. Parsons primarily conceived of institutions as role structures, and therefore as predominately part of the social system, leaving untheorized those aspects of culture that are institutionalized in systems of property, law, art, games, ceremonies, religious beliefs, and so on (see, e.g., Parsons 1961). Following Geertz and Schneider, most anthropologists today simply exclude the term *institution* from their vocabulary and theoretical framework. For example, in a debate about the concept of culture presented in *Current Anthropology* (Borofsky et al. 2001), none of the participants even mentions institutions. American cultural anthropologists, when they say anything at all about institutions, tend to treat them as if they were an epiphenomenon of ideas, ignoring the fact that the causal powers of institutions are strikingly different from the causal powers of ideas.

What, then, are institutions? Parsons defines institutions as "the integration of cultural-pattern elements of the motivational systems of individuals in such ways as to define and support structured systems of social interaction" (1961: 35). Sperber says that "an institution is the distribution of a set of representations which is governed by representations belonging to the set itself." (1985: 87). Furubotn and Richter define an institution as "a set of formal and informal rules, including their enforcement arrangements" (2000: 6). Many social-science definitions of institutions are equally vague and unhelpful, perhaps because institutions are such familiar objects of ordinary life that definitions seem unnecessary.

A detailed account of the nature of institutions has been presented by John Searle in *The Construction of Social Reality* (1995). Searle, a natural language philosopher, begins his discussion of institutions by pointing out that institutions are a distinctive kind of entity. Searle holds that institutions are real things, but a very special kind of real thing. Institutions are not facts that exist independent of people, like snow or electrons. Institutions could not exist if we did not believe they exist. But they are more than just ideas or meanings. Institutions are formed by the social agreement that some object, X, has a special symbolic status, Y, in some context, C, and because X is (counted as) Y, certain norms (deontic powers) apply to or are given to X. For example, certain objects (X), are given, by agreement, the special symbolic status of being someone's property (Y); because these objects are counted as someone's property, they constrain other people, who may not take someone else's property, and this enables the property holder to use, dispose of, or sell the object. Every role (e.g., father, president) is an institution in that some person in the proper context is given, by agreement, a special status of being a special entity (father or president), and because the person in this category is counted as this special entity, certain normative rights and duties are agreed upon for that person. Institutions can be corporate, such as the nation-state or the post office, or distributed, such as property or handshakes or chess. Institutions are ubiquitous, forming extensive hierarchies and interlocking chains of sub-institutions within themselves.

Much more can be said about institutions concerning their self-referential nature ("What is money?—whatever is counted as money"), the way speech acts can create certain institutional facts ("I hereby declare war"), the gradient between formal and informal institutions, and so on. Institutions are ordinary and common entities of the human world, and we all know a great deal about them. For example, we appreciate quite well the power of institutions. Policemen can arrest us. Debts have to be paid. Property cannot be taken from its rightful owner without repercussions. And so on for thousands of norms embedded in hundreds of institutions. *Institutions have great power over people because people have agreed to give their institutions these powers.* Institutions and their associated norms are the most obvious, ubiquitous, and immediate causal force in human life outside the biological and physical facts of existence. In fact, they are so immediate, obvious, and ubiquitous that many social scientists do not find them interesting to study—we already know too much about them. The more interesting questions concern what causes the institutions and their associated norms to be as they are, or to change, or to resist change.

Are institutions really culture? An institution is certainly more than just an idea, because it contains a social agreement that because X is Y, certain norms apply to X. And norms are a social agreement about what should be done. Does the power of an institution come from the idea or from the social agreement? Or should we say the power of an institution comes from the fusion of idea and social agreement, and is not separable into either? Certainly the melding of idea and social agreement into the seamless entity of an institution such as property is a fascinating phenomenon in its own right.

Perhaps some anthropologists and other social scientists see it as problematic that the distinction between the institutionalized and the uninstitutionalized cuts across the distinction between culture and social structure. Thus lineages, chiefdoms, and ethnic groups are institutionalized systems of roles—rules about how specially constructed categories of people should treat each other—and are considered part of the social structure, that is, the existing system of relationships and groups in a society, as Brown and Durham define it in this volume. Property, games, ceremonies, creeds, plays, honor codes, and the like, on the other hand, are institutionalized systems of rules which most social scientists consider parts of culture. The interesting fact is that humans can make institutions out of anything—objects (money, property), relationships (roles, groups), interactions (handshakes, kissing), time (calendars, holidays), competition (games, wars), mental defects (insanity), birth (citizenship, illegitimacy), beliefs (dogmas, axioms, totemic myths), and other possibilities limited only by human imagination.

Take away the institutions from a classic ethnography and there is not much left. Geertz held that it is the meanings that people give to their institutions that make up culture, not the institutions themselves. However, he had to describe the institution of the cockfight in detail before he could talk about its meaning (Geertz 1973). So one wonders: if it is the meaning of the cockfight that is culture, then what is the cockfight itself—its bets, ceremonial preparation of cocks, fighting ring, spurs, handlers, rules for winning, umpires, odds calling, and so on? The point here is not that one or the other way of defining culture is correct, for there can be theoretical reasons for defining culture in many ways. The point is that whether institutions are or are not included in the definition of culture makes a large difference in the causal powers that can be ascribed to culture.

A related issue is that selection processes do not seem to work the same way with institutions as they do with simple ideas. Some decades ago an article in *Scientific American* on timekeeping made the point that there is a very good calendar that any one can learn to use quickly and effectively. This calendar has thirteen months of four weeks each, making 364 days. The last day of the year is given its own special name. This calendar has various nice properties: Wednesdays, for example, always fall on the third, tenth, seventeenth and twenty-fourth of every month throughout the year. In an aside, the author said that despite the clear advantages of this calendar over any known alternative, it had no chance of being institutionalized in his lifetime, or probably ever. That sounds right. For one person's calendar to work, it must be the same as everyone else's calendar. How can anyone change if everyone has to change before any one person can do so?

With institutions, then, competition and selection on the basis of cost or reward do not always act directly or rapidly. Institutions create norms for intermeshed actions and systems of actions. The United States has a common-law system while France follows the Napoleonic code, and neither is likely to change soon, whatever the advantages of either system. Of course, institutions do change, sometimes by top-down authority, sometimes through complex bottom-up interactions, but the process seems different in many respects from the selection of independent and unconnected replicators (genes, memes, semes, etc.). Economic conditions, population pressure, new ideas, and many other factors do in the long run influence the creation and modification and elimination of institutions. New institutions are constantly arising and flourishing, while old institutions become otiose. Selective pressures are always present. But the timescale and the degree of complexity of interactions make institutional change different from the change of simple ideas.

It is interesting that under some conditions rapid change can take place with respect to institutions. Joel Robbins, who worked with the Urapmin of New Guinea, describes a case of culture change in which an entire Western institutional complex—Charismatic Christianity—was wholeheartedly absorbed into a relatively traditional tribal society (1998). The effects of the adoption of this institutional complex on other Urapmin institutions and ideas have been, from one perspective, pervasive and long reaching, but from another perspective it can be surprising that so much of the old culture (including institutions) remains in place and intact.

## MODEL D

*Culture includes both ideas simple and institutions. The relationship of cultural ideas to what people will do may be problematic because of the competition between ideas, but unlike ideas simple, institutions are relatively stable and powerful causal forces, acting as direct enablements and constraints on human behavior.*

This model is based on the assumption that ideas can sometimes be very compelling and powerful. Consider, for example, the power of prejudice. Racial and ethnic stereotypes have been found to be extremely difficult to change.

When not socially constrained, these stereotypes often lead to discrimination and abuse. It is unfortunately the case that racial and ethnic stereotypes are a highly shared part of many cultures.

Stereotypes and prejudice are just two among many types of affectively and motivationally compelling ideas. Religious convictions, political beliefs, and attitudes about food and sex are other examples. Here it seems that idea and motive have become one. Idea and affect are so fused together it becomes impossible to say whether the X hate the Y because they think the Y are monsters, or whether the X think the Y are monsters because they hate them so much. When ideas are held with such great conviction, they are said to be strongly internalized. In a discussion of internalization of cultural propositions, Melford Spiro (1997) points out that a variety of processes can bring about internalization. For example, a proposition may be internalized because it has strong empirical support, or because it provides effective guidelines for the accomplishment of some purpose, or because it fits previous cognitive or emotional experiences, or because strong motivations are satisfied by believing it. Durham, in this volume, gives the example of the Fore belief in witchcraft as an explanation of kuru, which the Fore still find more compelling than Western ideas concerning infectious diseases. Also in this volume, Bowles and Gintis find that choices in the ultimatum game are strongly influenced by norms that have been internalized; they follow Herbert Simon's interesting idea that once individuals have the machinery for internalizing personally fitness-enhancing norms, the machinery becomes available for the formation of more altruistic norms.

The phenomenon of internalization has other names. Philosophers, for example, speak of propositional attitudes. A propositional attitude is an attitude taken towards some proposition. If $p$ stands for some proposition, one can believe $p$, doubt $p$, deplore $p$, value $p$, intend $p$, want $p$, be enchanted by $p$, and so on. That is, one can be in almost any mental state with respect to a proposition. An idea that has been internalized is an idea about which someone has a strong propositional attitude—the idea is in some way important to that person and arouses attitudinal states.

To emphasize this point, I will use the phrase *affectively and motivationally charged ideas* to describe this phenomenon. The proposal that ideas and affects can combine to form a distinct and unified mental entity has support from research using functional MRI. A recent neuro-imaging study found that under certain conditions, cognitive and emotional brain activities converge and merge into a single brain state. In this brain state, differential aspects of cognitive and emotional brain processes are lost (Gray et al. 2002). It is notable that in cases such as, for example, strong prejudices, the prejudicial cognitive schema becomes isolated from normal cognitive capacities to reason and make distinctions, and at the same time the plasticity and context sensitivity of the regular affective-motivational system disappears.

Internalization is not the only way in which ideas can come to have exceptional causal effects. Ideas can also be collectivized. That is, a collectivity can come to some agreement that it, as a group, deplores X, or admires X, or fervently believes X, or takes whatever propositional attitude is collectively agreed upon to be appropriate (Gilbert 2002). Collectivizing belief further increases the causal power of an idea because it is not just that one learns to deplore X, one also learns that deploring X is what *we* do, bringing normative powers of obligation and conformity to bear on the deploring of X. For example, although most individual Americans probably fear and hate rapists, the present social understanding that this is how *we* feel probably adds to the force with which the sentiment and idea are expressed and acted upon.

## MODEL E

Culture is defined to include ideas, institutions, affectively and motivationally charged ideas, and collectivized ideas. Institutions, affectively and motivationally charged ideas, and collectivized ideas typically have considerable causal force. Cultural ideas alone, on the other hand, are not consistently related to what people will do, because of the competition between ideas and the necessity that ideas be linked to motives to influence behavior.

From the examination thus far of four different models, each of which accords culture a different amount of influence with respect to behavior, it seems clear that the degree to which culture influences behavior depends primarily on what is included in or excluded from the definition of culture. A secondary point is that there really isn't much disagreement about which kinds of cultural/ social/ psychological things have strong, medium, or weak influences on behavior. It is generally agreed that shared ideas, pure and simple, are not strongly predictive. However, processes of institutionalization, internalization, and collectivization meld ideas into strong causal forces.

Ideas simple, internalized ideas, collectivized ideas, and institutions do not exhaust the list of kinds of causal cultural entities. Examination of the effects of other cultural entities, such as the effects of collective symbolic representations (flags, plays, sacred texts, etc.) on behavior, or the special effects of ritual qua ritual, will not be attempted here. But consideration of these related cultural phenomena would reinforce the conclusion that there are many different kinds of cultural entities with different degrees of causal force.

## RESEARCH ISSUES

These philosophic problems tend to reappear in controversies generated by empirical research. Let us consider, as an example, research on violence and the culture of honor in the American South. In a series of publications, Nisbett and Cohen (1996, Cohen et al. 1998, Vandello and Cohen 2004) have shown that, when insulted, college undergraduates from the American South are more likely to have higher cortisol levels than undergraduates from the North. (Cortisol is a corticosteroid hormone produced by the adrenal and is considered a reliable indicator of stress; it increases blood pressure, blood sugar levels, may cause infertility in women, and suppresses the immune system). These Southern undergraduates are also more likely to show higher testosterone levels and marked physical signs of becoming aggressive in expressive behavior than undergraduates from the North. They explain this as resulting from the culture of the South, and specifically from a Southern code of honor; they call this the "Southern culture of honor." In a variety of survey and literature studies they have found that the cultural representations and social policies of the South, compared to the North, excuse and actively support the use of violence as a response to insult or threat.

### What Honor Code?

In contrast to their detailed documentation of reactions to insult and attitudes towards violence, Cohen and collaborators present little direct ethnographic data on the current code of honor in general in the American South.

Honor codes differ by society and by historical period, but they all share an important characteristic. To say a group has an honor code means not only that members of the group not only define certain things as honorable, but also that they treat an honorable person with respect and deference and treat those without honor with contempt or indifference. In a society with an honor code, social rank is strongly influenced by degree of honor, and degree of honor is considered to be an objective fact about a person. For members of a society just to admire strong, aggressive men or get angry at insults does not constitute an honor code. A true honor code is an institution, not just a set of shared attitudes.

Because an honor code is, by definition, an institution, it has a special kind of causal force that attitudes by themselves do not have. An honor code is a set of rules people are expected to follow, and must follow or suffer the consequences. Because honor entails having the *right* to respect from others, people who are honorable should be treated with respect, and people who are not honorable should not be so treated. This raises the question of the exact nature of the honor code in the current American South. The ethnographic material presented by Cohen, Nisbett, and collaborators is primarily based on eighteenth- and nineteenth-century historical accounts. The code they outline appears to be a version of the European gentleman's honor code. However, it is unlikely that gentlemanly honor is still the cultural code for Southern men; duels, ostentatious shows of good breeding, and a life of leisure as a way of garnering honor are things of the past. Even in the eighteenth and nineteenth centuries in the American South the European gentleman's honor code was probably adhered to by only a small percent of the population.

To investigate further the honor code of the American South, D'Andrade (2002b) developed a questionnaire asking respondents from the University of Tennessee, the University of Georgia, the State University of New York at New Paltz, the University of California at San Diego, and a high school in Long Beach, California, to rate how they would respond to fourteen insult scenarios. Examples of these scenarios are

> You are driving on the road and somebody cuts in front of you without sig-
> naling. You beep your horn and the person flips you off with an obscene
> gesture.

A homeless man approaches you and asks you for money. You say no and he calls you a "cheap bastard" in a loud voice.

You find out your lover has been cheating on you and lying about it.

You are with your Japanese-American family on a trip. You stop at a restaurant and ask if your daughter can use the rest room. The waitress tells you the rest room is "reserved for paying customers and red blooded Americans."

You are having an argument with someone. This person starts saying obscene things about your mother.

You are walking down the hallway when somebody bumps into you, knocking you aside. The person turns and calls you an asshole.

For each scenario, respondents recorded a series of ratings:

1  The anger aroused by the incident;

2  The amount of disrespectfulness shown by the other person in the incident;

3  The degree of insult involved in the incident;

4  The degree to which honor would be at stake in such an incident;

5  The strength of one's wish to do something back aroused by such an incident;

6  What one would do if this actually happened (nothing, walk away, etc.);

7  If this happened, what one thinks people would think if one did nothing back.

The expectation was that men, and, to a lesser degree, women, from the American South, in comparison to those from California or New York State, would give higher ratings for most of the scenarios, especially those having to do with sexual trespass, physical provocation, and slander, since these tend to be focal issues in most honor codes. This prediction was made on the basis of what it means to hold an honor code: If one has internalized an honor code one will invoke it frequently and saliently; breaches of the code will arouse anger and lead to physical violence; and one will expect to be judged on the basis of what one did back. It was expected that Southern women would show a profile of scores similar to the Southern men, but would be less involved in feeling that their honor was at stake in the various incidents and less concerned about what people would think of them if they did nothing.

Surprisingly, none of the scales showed significantly higher ratings for Southerners for any of the scenarios. No systematic differences were found among regions. The modal response to the various scenarios was to ask for an explanation (33 percent) or not respond (30 percent). Only 8 percent of the respondents indicated that they would retaliate physically for any insult.

Women, compared to men, responded with more anger and felt more disrespected in all scenarios except the date scenario (someone puts their hands on your date in a suggestive way). Women also felt more insulted in eleven of fourteen scenarios. One might speculate that the reason women show a pattern of greater anger along with stronger feelings of being insulted and disrespected is that women expect more courtesy than men and therefore find the same degree of violation more aggravating.

Overall, these results show that there is a strong common American code of civility (rather than of honor) with respect to the norms for response to insults. The norms of this civility code are quite similar across the country.

## How Come?

The most obvious objection to these findings is that questionnaires can't be trusted. However, Cohen, Vandello, and Rantilla (e.g., 1998) found the same results in a series of non-questionnaire experiments. In one of their studies Northern and Southern students were shown videos in which an aggressor bumps a college student as he walks down a hallway and calls him an asshole. In the videos the student either responded aggressively or shrugged off the incident and kept walking. Both Northerners and Southerners, asked to rate their preference for the two, preferred the less aggressive male, and no significant differences in preference were found between Northerners and Southerners.

In another study reported by Vandello and Cohen (2004), individual participants witnessed an event staged by three confederates. In this event, a male confederate stepped on a case supposed to hold the eyeglasses of a second male confederate, creating a loud cracking sound (actually dry pasta). When the victim complained and asked for an apology, the perpetrator was rude and confrontational. After the perpetrator left the room, the victim asked the subject a series of questions. For half of the subjects the victim asked questions about whether he should confront and attack the perpetrator. For the other half of the subjects the victim asked questions about whether he should go apologize. When the victim left the room, the third confederate, a woman, asked the subject open-ended questions about the confederate who had supposedly had his glasses broken. All responses of the subject were recorded and coded.

Strikingly, again there were no significant differences between Northern and Southern subjects in their answers to questions and suggestions about what the victim should do. The results of these experiments using live action and video presentations confirm the questionnaire findings reported above. It should also be noted that the studies found consistent gender differences with respect to perceptions of response to insult, with women making more negative judgments about aggressive responders.

This code widely shared by Americans can be described as one in which a person should be honest, respectable, and civil (E. Anderson 1999). If people are unpleasant, rude, or insulting, one should ignore them or tell them they are behaving improperly and ask them to explain or apologize, but one should not retaliate with verbal insult or physical force. Actually, the politeness aspect of this code is somewhat stronger in the South, as D. Cohen et al. (1996) report in a study in which Southern males were found to be more polite initially to an aggravating confederate than were Northerners.

Unlike honor codes, in which one needs to be concerned with how one is perceived because of the threat of loss of reputation, and in which one must destroy those who infringe on one's honor, the code of civility stresses not only civil treatment of others, but also relative unconcern about one's own reputation or getting even. Of course, decent individuals sometimes "lose it" in the face of provocations and resort to bad language, violence, and attempts to retaliate. However, when this happens, one has broken the code, and the normal response is shame, guilt, and apology. There is probably great variation in the degree to which individuals actually internalize this code of civility, but not in the understanding that this is how one should behave. Various aspects of the code are stronger or weaker in different areas of the country, but the basic elements appear to be understood almost everywhere, from the inner city of Long Beach to Tuscaloosa and New Paltz. One would guess that even rednecks and street warriors understand the civility code, although they may not hold to it.

The problem then, is to reconcile findings that on the surface appear contradictory. The South, it is argued here, has the same code of civility as the rest of the country. The findings of Cohen, Vandello, D'Andrade, and others are that Southerners are not different from Northerners in what they say about how they would respond to various insults, nor in the degree to which they believe their honor is at stake in these insults, nor in what they think they would do about these insults, nor in their preference for those who respond aggressively to insult compared to those who are not aggressive, nor in what they think people will think of them if they don't retaliate to insults. But they are different in that they become more physically aroused in the face of insult. Not only do Southerners show no evidence of a true honor code, they don't even show evidence of strong ideas about honor or honor-related attitudes.

Why then should undergraduates from the South show higher levels of cortisol in response to insults but not differ from non-Southerners in what they say about how angry such insults make them? Perhaps a deep sense of masculine pride is at work here. Something like this has been said to be characteristic of Southerners. Or it may be that greater use of physical punishment in early socialization has had an effect outside awareness on emotional responsiveness to any kind of aggression. Or it may be that just experiencing more physical violence sensitizes Southerners to violence, making them more physically aroused in situations that involve violence, and whether this is experienced as anger, or fun, or fear depends on context.

Another possible explanation involves differences in the cultural models for self-protection. Violence to protect oneself is endorsed, normatively supported, and represented in a positive light throughout the South, but less so in the Northeast. For example, the *New York Times* of March 12, 2001, reported that Charlie Condon, attorney general of South Carolina, had announced a policy that it was "open season" on home invaders—"invade a home and invite a bullet," he said. "I'm putting home invaders on notice that if an occupant chooses to use deadly force, there will be no prosecution." It has been speculated that Southern reliance on self-protecting violence is the result of a lack of reliable state mechanisms of social control during the pre- and post-bellum periods (Nisbett and Cohen 1996).

It may be that for some situations Southern norms involving the use of violence for self-protection, rather than the civility model, become the contextual system of meaning that steers the Southerners' emotional responses. According to this hypothesis, the Southerner has two relevant normative cultural models. The first is the civility model, the other is the self-protection-using-violence model. These two models are not necessarily in contradiction. Southern undergraduates did not punch or verbally abuse the insulters—they followed the American civility code because they were not under actual physical attack. But when a Southern male suffers an insult, there appears to be enough of a threat to invoke the model of self-protection by violence, which then triggers an expectation of having to fight. Southerners do not report experiencing more anger to insult than Northerners, but their bodies—as measured by cortisol response—are ready to fight.

*Implications*

The point to this example is that the culture-behavior model is much too simple to be of use in actual analysis. What is the culture of the South? Is it the cultural representations of the gentlemanly model found in advertisements and novels and movies and in artifacts like swords and dueling pistols? Is it the discourse of Southerners about honor—the talk and discussion that people carry on about their own actions? Is it what people really feel about things—not what they say, and not what is publicly represented in various media? Is it the institutional norms that they hold and follow about what one should do, like the code of civility, or different models about self-protection?

The research strategy used by Cohen, Nisbett, and their associates was to relate a difference in affective behavior—arousal in response to insult—to culture. They found newspaper stories that expressed the idea that it is excusable to commit murder in the face of extreme personal insult. They found crime statistics that show that Southerners are more likely than Northerners to murder someone in a personal dispute, but no more likely than Northerners to murder someone in a robbery. Was this Southern culture? There is no answer to this question, because the term *culture* includes too many kinds of different things. In including too much, it defines too little. But when the research is narrowed down to a question about the degree of difference between Southerners and other Americans with respect to something specific like the presence or absence of a institutionalized honor code, then it is possible to reach reasonable conclusions. Interesting problems remain about exactly what piece of Southern culture is causally active in the Cohen and Nisbett experiment, but not about whether the response is caused by an institutionalized Southern honor code. Southern honor, as a code, is gone.

This seems to be a typical research story. Some supposed fact about the human world is found. Some attempt is made to locate the possible cause of this fact. Initially, large abstract forces are invoked as explanations—culture, society, personality, biology, economics, etc. These categories turn out to include too many different kinds of things to make possible a determination of whether or not any one of them actually explains the fact. Further investigation turns out to show that specific things with particular causal force—such as institutions, historical events, social conditions, defense mechanisms, internalized cognitive models, and so on—are more probable causes of the fact, and discussion begins about exactly which of these particular things is the correct explanation. As Adam Kuper (1999) points out,

> Complex notions like culture . . . inhibit an analysis of the relationships among the variables they pack together. Even in sophisticated modern formulations, culture . . . tends to be represented as a single system, though one shot through with arguments and inconsistencies. However, to understand culture, we must first deconstruct it. Religious beliefs, rituals, knowledge, moral values, the arts, rhetorical genres, and so on should be separated out from each other rather than bound together into a single bundle labeled culture. (245)

But such complaints have not resulted in the categories of culture, society, personality, biology, etc. being replaced in our theories by these more specific kinds of causal entities. What happens is that, after a flurry of argument, everyone goes back to the same old debates about whether the really important causes in human life are culture, or society, or personality, or biology, in a rousing free-for-all with everyone using a different definition. But perhaps by explicitly posing the question and constructing theories about the casual powers of specific different cultural forms we can make something more like progress.

# 2

# Culture in Evolution

*Toward an Integration of Chimpanzee and Human Cultures*

## CHRISTOPHE BOESCH

Anthropologists and psychologists have traditionally viewed culture as a feature exclusive to human societies (Barnard 2000; Kuper 1999). Archeologists, following this assumption, have been searching through the records of our past for the first evidence showing the emergence of the ability to develop such a feature (Mithen 1996). In contrast with such unanimity, biologists have suggested that some animal species might also possess cultures (Bonner 1980; Kawai 1965; Goodall 1963, 1968; Maynard-Smith and Szathmáry 1995; E. Wilson 1975), a claim greeted with much skepticism by other disciplines (Heyes 1994; Sahlins 1976b). However, over the years, evidence has mounted documenting how different populations of chimpanzees have developed population-specific behavior patterns independent from any environmental factors, providing growing support to the notion of culture in chimpanzees (Goodall 1986; Nishida 1987; Boesch 1996, 2003; Whiten and Boesch 2001; Whiten et al. 1999, 2001).

The question of the presence of culture in other species is complicated by lack of agreement over the definition of culture—different people within the same discipline put different concepts under the same heading. I do not wish to review all the concepts of culture presented by all different schools of thought; it is important just to distinguish two main approaches. The first argues that cultural traits provide the individual with a direct benefit and that this explains why these traits spread within a population (Alexander 1979; Boyd and Richerson 1985; Cavalli-Sforza and Feldman 1981; Dawkins 1976; Durham 1991; Lumsden and Wilson 1981; E. Wilson 1975). The second approach argues that human cultural traits only rarely confer direct benefits, and that, by and large, culture provides no direct advantage to the individual, and might even be detrimental (Sahlins 1976b; Kuper 1999; Barnard 2000).

Researchers tend to follow the tradition of their respective fields in placing emphasis upon different aspects of culture. Psychologists, who are interested mainly in mechanisms, center their discussions of human culture on the different learning mechanisms that allow the acquisition of behavioral traits (Heyes 1994; Galef 1988; Tomasello 1990). Thus, they define culture in terms of the learning mechanisms necessary to acquire cultural traits. In so doing, some converge with anthropologists' view of culture as uniquely human, in the sense that they base their culture concept on learning mechanisms proposed to be unique to humans, such as imitation and teaching (Heyes and Galef 1996; Heyes 1998; Tomasello 1999a, b). Biologists, on the other hand, are attracted to the notion of culture because it represents an alternative to genetic evolution and allows for changes to be incorporated in a population in a much quicker and more flexible way than through genetic mutations (Bonner 1980; Maynard-Smith and Szathmáry 1995; E. Wilson 1975). Thus, for biologists, the decisive feature of a cultural behavior is that it is transmitted between generations on a nongenetic basis, allowing transmission of traits independent of reproduction and in many more directions than just the parent–offspring line.

This discrepancy of opinions about culture is not surprising, if one takes the point of view that "culture is not a thing but a set of processes" (Boesch and Tomasello 1998). Differences of opinion could therefore be viewed as representing different points on a continuum that embodies the complexity of different cultural systems, whether in humans or in other species. Some aspects of the cultural process may be shared between species, and others will be unique to some or a few of them. I propose here a seven-level definition of culture aimed at incorporating some of the complexity of this concept. I argue that recognizing the different aspects of culture will actually lead to an enrichment of the concept rather than to its dilution.

## MULTI-LEVEL DEFINITION OF CULTURE

Given that humans are the quintessentially cultural species, the proposed levels of culture include the highest sophistication seen in our species. In accordance with general consensus, I include in the present definition only behavior patterns that are not genetically determined. The hierarchical structure adopted here means that some levels of culture could not develop without the presence of the previous level. However, given our present knowledge of cultural complexity it is not possible to make a claim that this represents a sequential evolutionary scenario. Only tests making detailed interspecific comparisons will allow us to start to address this aspect, and we should at present view this with a lot of flexibility.

## Level 1

*Behavioral traits are acquired through a learning process influenced in part by observation of other group members.*

At this level, many different social learning processes might be involved, as long as they produce an alteration of the trait being acquired. Learning processes have been suggested to range from local enhancement or stimulus enhancement, to emulation learning and imitative learning. The classic example of blue tits in England learning to open milk bottles presents a nice illustration of this level of culture. Individual behavior was adapted to a novel environmental condition: the presence of milk bottles. This provided individuals with a net benefit in the form of access to a new food source, and the behavior was to some degree learned from the behavior of other individuals. Many more examples of this level have been observed in birds and other species (Bonner 1980).

## Level 2

*Behavioral traits are learned entirely through observation of other members of the immediate social group.*

Once individuals have started to live in close social units, a consequence is that social influences are generally restricted to individual members of the same social group rather than including all conspecifics encountered in more random associations. This often leads to the emergence of group-specific behavior patterns. For example, Caledonian crows have been observed to use three types of tools to gain access to new food sources, and different populations employ them with different frequencies (Hunt 1996; Hunt and Gray 2003).

## Level 3

*Individuals learn many of their behavior traits through observation of group members.*

Familiarity among group members will favor social influences during the phases of life in which learning is most important. In species where these phases are relatively longer, such as a life history with long infancy and juvenile periods, the opportunity for learning will be extended. Both factors will cause a higher proportion of behavior patterns to be learned, and the social influences will become apparent in many more traits. Thus, we move here from the level of simple tradition, in which only a single trait or a very limited number of traits are socially affected, to a level in which a whole set of traditions is socially learned. This effect is further enhanced when cognitive abilities for learning have attained the flexibility required to learn multiple or complex tasks. There is a growing body of evidence that capuchin monkeys possess such group-specific traditions concerning both feeding techniques and social traits, and that such traditions have life spans ranging from a few months to many years (Perry et al. 2003; Perry and Manson 2003).

## Level 4

*A "fidelity-copying" mechanism reinforces the precise learning of behavior traits.*

At this level, the reproduction of behavioral traits is more precise, and therefore more complex behavior patterns or longer sequences can be copied more faithfully. This may be the direct result either of individuals being able to use a more conservative learning mechanism, such as imitation, or of the social imposition mechanism that eliminates behaviors conflicting with a "social norm." This imposition can take a positive form as in teaching, where an expert guides the naïve individual in learning the correct behavior; or a negative form, where the naïve individual is punished for showing the incorrect behavior. Because innovation can also occur with learned behavior, cumulative changes of socially learned behavior may be observed in some of the previous cultural levels. However, at this level the cumulative effect is rendered more efficient by increased fidelity in the copying of innovative models. This is the lowest level at which psychologists classically recognize cultural processes, with some suggesting even that this level is uniquely human (Heyes 1994, 1998; Tomasello and Call 1997; Tomasello 1999a, b). This level corresponds as well to the main sense of the term "culture" as used by some biological anthropologists (Boyd and Richerson 1985, 1996; Cavalli-Sforza and Feldman 1981). Orangutans possess some very flexible cultural traits of this level based on a precise social learning mechanism (van Schaik et al. 2003; van Schaik et al. 1999).

*Level 5*

*Cultural traits have a socially determined value that makes them attractive to group members.*

Each group may develop for a given trait a specific meaning independent of the form of the behavior pattern itself. An effect of this is that cultural traits will become increasingly more arbitrary and some could potentially have shorter lifetimes. In addition, cultural traits will be less dependent upon adaptive value to survive and be disseminated within a social group. Here, a desire for "conformity" could increase the success of a trait, which then gains value purely from the fact that it is present in the majority of group members. The conformity effect endows the trait with a secondary adaptive value based on the trait's success within the social group. This is reminiscent of sexual selection, in which the success of some secondary sexual traits rests purely on their being preferred by females (Andersson 1994; Richerson and Boyd, this volume). A certain proportion of anthropologists recognize culture only at this level, in which shared meanings are an integral part of culture (Kuper 1999; McGrew 1992). Chimpanzees possess large cultural repertoires that vary across many different populations, and these repertoires are based on precise social learning mechanisms that include forms of teaching (Boesch 1991, 1996; Whiten et al. 1999, 2001). In addition, a certain number of these cultural traits are based on shared meanings that are specific to different social groups (Boesch 2003). To date, chimpanzees are the only nonhuman species in which such a level of cultural complexity has been documented (Boesch 2003).

## Level 6

*A generalization of abstract cultural traits, including shared ideas, beliefs, desire and knowledge is made possible through language.*

Language facilitates the development of this level of culture. It is not a requirement, however, as knowledge and beliefs can be inferred from others' behaviors with some accuracy and can then be the subject of the same valuation as in the previous level. Nevertheless, language accelerates the acquisition of shared meanings in large social groups, and permits much broader use of cultural traits at this level. Large parts of human culture involve ideas that are not visible but that mold the behavior of every one of us in very important ways: morality, prejudices, beliefs (including false ones), and religion. Obviously, without language, these aspects of our culture would be much less developed.

## Level 7

*Dissemination of cultural traits is accelerated by symbolic means, which not only allows the traits to be more complex, but permits their transmission to individuals from many different groups even without direct contact between individuals.*

At this level, language and writing are the crucial vehicles allowing for such dissemination. These are typical of modern humans and are responsible for the incredible expansion of cultural products throughout huge geographical areas and for the extremely rapid spread of fashions in our societies. At the same time, these means permit the survival of cultural traits over extended periods of time, and allow for the recovery of extinct cultural traits through written materials.

Anthropologists have distinguished additional levels of complexity in human cultures (see, e.g., Roy D'Andrade, this volume). As mentioned earlier, the order of the cultural levels presented above should not be viewed with too much rigidity. For example, the shared meaning of a trait could appear independently of the fact that a precise copying mechanism is in place, and could actually promote its appearance. On the other hand, we know that imitation is used also for behaviors that contain no meaning, such as the imitation of nut-cracking behavior or the opening of an artificial fruit (Boesch 1991; Whiten et al. 1996).

The first levels of culture are posited as consisting in large part of adaptive traits, because these traits are responses to problems an individual confronts in its environment on a daily basis. In contrast, the last levels consist predominantly of cultural traits that are independent of the environment and could be totally disconnected from it. Thus, the adaptive-nonadaptive dichotomy of the different approaches to culture is reconciled and made part of a general cultural process.

## CHIMPANZEE CULTURAL COMPLEXITY

Chimpanzees offer a rich example within which to discuss some aspects of this definition of culture, including how the two general types of cultural traits, adaptive and non-adaptive, are expressed. The above definition lends itself to the prediction that adaptive cultural traits were more common in early or simpler cultural processes; is this supported by the presently available data on chimpanzees and other species?

### Nut-cracking Technique, an Adaptive Cultural Behavior in Chimpanzees

The best-studied example of an adaptive cultural behavior in chimpanzees is the nut-cracking behavior commonly observed in the westernmost African populations of this species (Sugiyama and Koman 1979; Boesch and Boesch 1981, 1983, 1984; Boesch et al. 1994). Five species of nuts of different shapes and degrees of hardness are eaten by chimpanzees in the Taï forest. The most abundant one, *Coula edulis*, is the softest, *Panda oleosa* is the hardest, and the large, productive trees of *Parinari excelsa*, *Detarium senegalense* (irregularly cracked at our site), and *Sacoglottis gabonensis* (very rarely cracked at our site) produce nuts of intermediate hardness. Panda nuts require a force of 1600 kg to be cracked without pounding (Peters 1987). These nuts are rich in protein, sugar, and fat. Most of these trees are widely distributed throughout western and central Africa, where the majority of the remaining wild chimpanzee populations live. In addition, in forests such as those in Gabon or in Congo, *Coula* nuts are regularly eaten by bushpigs and forest duikers, so that chimpanzees regularly see partly eaten nuts left by the other species on the forest floor (personal observation). The distribution of the nut-cracking behavior, however, is much more limited than the tree distribution. Nut cracking occurs only in western Africa, and a survey of nut cracking showed that its distribution is surprisingly limited even within the west African forest region (Boesch et al. 1994), with the Sassandra River in the western part of Côte d'Ivoire marking the eastern border of this behavior. Comparison of chimpanzee density, tree density, and presence of nuts and hammers between sites east and west of the river revealed no difference in ecological factors between the two sides of the river. Only the chimpanzees on the west side pound these nuts, even though they are separated by just 30 km from east-side groups; this strongly suggests a cultural difference.

To crack nuts, the chimpanzees bring together three different objects: a hard tool to pound the nuts, the nuts, and a hard substrate as an anvil on which to place the nuts. These materials rarely occur together naturally, and the chimpanzees need to select and transport them to a given place. Anvils used by the chimpanzees to stabilize the nuts include emerging roots, the bases of large trees, outcrop rocks, and suitable branches in a tree. At Bossou, chimpanzees have been observed using mobile stones as anvils, a behavior not seen throughout Côte d'Ivoire. The most probable reason for this behavior is that Bossou chimpanzees pound palm fruits near the oil-palm trees within young secondary forests, and emerging roots in such forests are far too small to be suitable as anvils (Sugiyama 1994). Most hammers used to pound the nuts are fallen branches of various shapes, sizes, and degrees of hardness. Compared to branches, stones are rare in the forest, but the chimpanzees consistently use stones if they are available. A chimpanzee using a stone hammer on a soft *Coula* nut invests 30 percent less pounding energy than when using a wooden one. This efficiency gain increases to 43 percent when cracking the harder *Panda* nuts (Boesch and Boesch 1983). Chimpanzees content themselves with easily found wooden hammers for *Coula* nuts, but they more often seek out stones (a rarity in Taï forest) to crack the hard *Panda* nuts, which are difficult to open otherwise. Thus, transport of stone tools is frequent for utilization of the latter nut species. Our analysis showed that chimpanzees could memorize the locations of up to five stones and select the one nearest to a goal tree (Boesch and Boesch 1984; average transport distance was greater than 100 m). The selection is made anew for each different *Panda* tree at which they crack nuts and where no hammer is present. This requires rather elaborate planning that seems to be more complex than in other chimpanzee tool techniques (Goodall 1986).

Nut cracking in the Taï forest constitutes a very important aspect of the chimpanzees' life and diet. On average, during the four months of the *Coula* nut season, the chimpanzees crack nuts for 2 hours 15 minutes per day, which represents 270 nuts, or 700 grams of food. Chimpanzees obtain an average net gain of 3450 kcal per day from this activity. During the nut season, the chimpanzees obtain the large majority of their caloric intake and a large portion of their protein intake from nuts processed through tool use (Boesch and Boesch-Achermann 2000). Without tools, the Taï chimpanzees would have no access to this very rich food source, so we can say that they depend on tool use during the nut-cracking season. Nut sharing constitutes an important commitment of Taï chimpanzee mothers during the first eight years of the lives of their offspring. It represents a very rich food source for the youngsters: during the nut season, infants three to five years old receive from their mothers up to 1000 calories per day in nuts (Boesch and Boesch-Achermann 2000).

Nut cracking can be considered a typical cultural behavior with important beneficial payoff, and its evolution is thus easy to explain. This could provide the basis for our understanding of the emergence of cultural abilities, as any species that would have enough social learning skills would benefit by gaining access to rich food sources accessible only through special skills. Such beneficial payoffs have also been proposed as contributing to the appearance of possible cultural traits in birds and in some other primate species such as macaques (de Waal 2001).

*Adaptive Versus Nonadaptive Cultural Traits*

If the order of cultural levels presented above has some correspondence to the chronological order of their emergence in nature, we would expect culture to have first evolved as an adaptation to environmental conditions. We could use existing observations of chimpanzees and other primate species to make a first test of this proposition. In other words, is there evidence that the majority of cultural traits in chimpanzees and other primates produce a direct benefit to the performer? Is there a tendency for more complex primate cultural traits to be progressively less related to direct adaptive advantages?

An initial review of fifty cultural traits described in chimpanzees finds that 56 percent of them provide a high benefit to the individuals that perform them (table 2.1). But this is related to the context in which they occur. Hence, the majority of cultural traits performed in feeding contexts do provide high

46    **Christophe Boesch**

Table 2.1. Putative cultural traits in chimpanzees and their potential advantages

| | Behavior | Context[a] | Benefit[b] | Alternative[c] |
|---|---|---|---|---|
| 1 | Pestle-pound (mash palm crown with petiole) | Feed | High | No |
| 2 | Nut-hammer, stone hammer on stone anvil | Feed | High | No/Yes |
| 3 | Ant-dip-single (one-handed dipstick on ants) | Feed | Medium | Yes |
| 4 | Aimed-throw (throw object directionally) | Comm | Low | No |
| 5 | Food-pound onto wood (smash food) | Feed | High | Yes |
| 6 | Bee-probe (disable bees, flick with probe) | Feed | High | No |
| 7 | Index-hit (squash ectoparasite on arm) | Feed | Low | Yes |
| 8 | Nut-hammer, stone hammer on wood anvil | Feed | High | No/Yes |
| 9 | Nut-hammer, wood hammer on wood anvil | Feed | High | No/Yes |
| 10 | Nut-hammer, wood hammer on stone anvil | Feed | High | No/Yes |
| 11 | Nut-hammer, other (e.g., on ground) | Feed | High | No/Yes |
| 12 | Seat-vegetation (large leaves as seat) | Misc | Low | Yes |
| 13 | Marrow-pick (pick bone marrow out) | Feed | High | Yes |
| 14 | Food-pound onto other (e.g., stone) | Feed | High | Yes |
| 15 | Club (strike forcefully with stick) | Comm | Low | No |
| 16 | Ant-dip-wipe (manually wipe ants off wand) | Feed | High | Yes |
| 17 | Fluid-dip (use of probe to extract fluids) | Feed | High | No |
| 18 | Lever open (stick used to enlarge entrance) | Feed | High | No |
| 19 | Expel/stir (stick expels or stirs insects) | Feed | High | No |
| 20 | Self-tickle (tickle self using objects) | Misc | Low | Yes |
| 21 | Leaf-clip, fingers (rip single leaf with fingers) | Comm | Low | Yes |
| 22 | Leaf-squash (squash ectoparasite on leaf) | Misc | Low | Yes |
| 23 | Leaf-clip, mouth (rip parts off leaf with mouth) | Misc | Low | Yes |
| 24 | Knuckle-knock (knock to attract attention) | Comm | Low | Yes |
| 25 | Branch-slap (slap branch, for attention) | Comm | Low | Yes |
| 26 | Leaf-groom (intense "grooming" of leaves) | Misc | Low | Yes |
| 27 | Ant-fish (probe used to extract ants) | Feed | High | No |
| 28 | Hand-clasp (clasp arms overhead, groom) | Misc | Low | Yes |
| 29 | Shrub-bend (squash stems underfoot) | Misc | Low | Yes |
| 30 | Stem pull-through (pull stems noisily) | Misc | Low | Yes |
| 31 | Rain dance (slow display at start of rain) | Comm | Low | Yes |
| 32 | Termite-fish using leaf midrib | Feed | High | No/Yes |
| 33 | Termite-fish using non-leaf materials | Feed | High | No/Yes |
| 34 | Leaf-napkin (leaves used to clean body) | Misc | Low | No |

**Table 2.1.** *(continued)*

| Behavior | Context[a] | Benefit[b] | Alternative[c] |
|---|---|---|---|
| 35 Leaf-strip (rip leaves off stem, as threat) | Comm | Low | Yes |
| 36 Leaf-dab (leaf dabbed on wound, examined) | Misc | Low | Yes |
| 37 Fly-whisk (leafy stick used to fan flies) | Misc | Low | Yes |
| 38 Leaf-inspect (inspect ectoparasite placed on leaf) | Misc | Low | Yes |
| 39 Branch din (bend, release saplings to warn) | Misc | Low | Yes |
| 40 Deep-dig ants | Feed | High | Yes |
| 41 Surface-dig ants | Feed | Medium | Yes |
| 42 Termite-mound pound | Feed | High | No |
| 43 Days-nest rest | Misc | Low | Yes |
| 44 Days-nest play start | Comm | Low | Yes |
| 45 Days-nest courtship | Comm | Low | Yes |
| 46 Mature-pith eat | Feed | High | No |
| 47 Fresh-*Strychnos* eat | Feed | High | No |
| 48 Social scratch | Misc | Low | Yes |
| 49 Leaf swallow | Feed | High | No |
| 50 Little-stone hammers | Feed | Medium | Yes |
| | Feed=26 | Low=25 | Yes=30 |
| | Comm=9 | Medium=3 | Both=7 |
| | Misc=15 | High=22 | No=13 |

SOURCES: Boesch 2003; Whiten et al. 1999

[a] Indicates whether the trait is used in a feeding context ("Feed"), for communication purposes ("Comm"), or for a miscellaneous purpose ("Misc"; this refers mainly to personal comfort and grooming behavior).

[b] Indicates level of benefit of the behavior: Feeding traits provide a high benefit, by allowing access to difficult food sources, other contexts yield only low benefits either in terms of improved social communication or minor personal comfort, while medium indicates feeding techniques for which an alternative technique is more beneficial.

[c] Alternative is "Yes" when chimpanzees have been seen to achieve the same goal with another technique; "No/Yes" appears when alternatives are all cultural behaviors.

reward to the performer (twenty-two out of twenty-six feeding cultural traits equal 85 percent). At the same time, a large number of cultural traits are performed in other contexts, with much less obvious benefit and sometimes with known drawbacks. For examples, for three feeding-related cultural traits listed in table 2.1, alternative techniques exist that are more beneficial; these include techniques for dipping for ants (#3, Boesch and Boesch 1991), and different ant-gathering techniques (#16, 40, 41) and choice of hammers in Taï (#8, 9, 10, 11, Boesch 2003).

Note that in different species, cultural traits may tend to develop more in certain contexts than in others (see table 2.2). For example, in a recent list of twenty "very likely cultural variants" in orangutans (van Schaik et al. 2003), only four were performed in a feeding context, and only two (10 percent) in a communicatory context, whereas 14 (70 percent) were performed in contexts related to individual comfort and grooming. Thus, there might be species-specific inclinations in the development of cultural domains. Chimpanzees have been shown to be very manipulation-oriented (Savage-Rumbaugh et al. 1985), and this seems to be reflected in their cultural repertoire. Orangutans seem to be more self-oriented, as is apparent in their low level of sociability.

In examining table 2.2, it is necessary to keep in mind that biologists, in their search for possible cultural traits, have been using different methods and definitions in accordance with their varied motives. Those working with monkeys have been more interested in the search for precursors of culture, and include in their lists all behaviors that qualified as traditions—that is, behavioral practices that are relatively long-lasting, that are shared among members of a group in part through social learning, and that display variation between groups (Fragaszy and Perry 2003; Perry et al. 2003). For example, for capuchin monkeys (*Cebus apella*), similar processing techniques performed on different food species have been listed as separate traditions (Panger et al. 2002), which was not the case in the chimpanzee studies. For cetacean studies, the observational conditions are so much more difficult that less stringent criteria have often been used (Rendell and Whitehead 2001). In addition, all studies of cetacean culture have included vocal communication, which has not been considered in most studies of primates. Many recent field studies have started trying to document population differences, but this has generally come as a

**Table 2.2. Putative cultural traits in non-human animals and the context of their use**

|  | Feed | Communication | Miscellaneous | Total |
|---|---|---|---|---|
| Primate |  |  |  |  |
| Chimpanzee | 26 | 9 | 15 | 50 |
| Bonobo | 3 | 5 | 3 | 11 |
| Orangutan | 4 | 2 | 14 | 20 |
| Japanese macaque | 7 | 1 | 6 | 14 |
| White-faced capuchin | 17 | 5 | 1 | 23 |
| Brown capuchin | 1 | 0 | 0 | 1 |
| Vervet monkey | 2 | 0 | 0 | 2 |
| Cetacean |  |  |  |  |
| Bottlenose dolphin | 3 | 0 | 0 | 3 |
| Sperm whale | 1 | 1 | 0 | 2 |
| Killer whale | 3 | 2 | 0 | 5 |
|  | 60 | 22 | 39 | 121 |

SOURCES: Chimpanzee, see table 2.1; bonobo, Hohmann and Fruth 2003; orangutan, Van Schaik et al. 2003; macaque, capuchins, and vervet, Perry and Manson 2003; cetaceans, Rendell and Whitehead 2001.

byproduct of observations focusing on other topics. We do not yet have a first list of putative cultural differences among gorillas. Therefore, those lists have to be considered as provisional and under development.

Feeding-related cultural traits are more numerous in monkeys than in apes ($X^2$=7.67, df=1, p<0.01) (table 2.2). Because these are the cultural traits most likely to bring direct benefit to individuals, it would seem that cultural traits first developed mainly in the feeding domain. This is supported by the fact that most proposed traditions in other animal species are also performed in the feeding context. For example, tits in England access milk thanks to a newly acquired behavior, Caledonian crows extract more food with help of tools, and California sea otters are more efficient at eating oysters with the help of stones.

In apes we can already see a trend toward cultural traits being used more frequently in domains where they do not bring a clear direct benefit. This trend fits well with the multilevel definition of culture proposed above. Monkeys show traditions or culture patterns that fit with the first three levels, as they acquire behaviors under the influences of social partners and present group differences. In great apes, there is an apparent tendency to develop cultural traits related to grooming or communication interactions that bring no visible direct benefit.

## Nonadaptive Cultural Traits

What do nonadaptive cultural behaviors in chimpanzees look like? They may take on two different aspects, either by being less beneficial than other known alternatives or by being purely arbitrary. Ant dipping in wild chimpanzees is very instructive as a badly adaptive cultural behavior trait, the maintenance of which seems to rely on social norms prevailing in different social groups. Two different techniques used to dip for ants have been observed (Boesch and Boesch 1990; Goodall 1986). In Gombe, chimpanzees use one hand to hold a stick among the soldier ants guarding the nest entrance, and once the ants have swarmed about halfway up the tool, they withdraw the stick and sweep it through the closed fingers of the free hand; the mass of insects is then rapidly transferred to the mouth (McGrew 1974). Gombe tools are 66 cm long on average, and the dipping is performed 2.6 times per minute. McGrew (1974) estimated that 292 ants are gathered with each dipping movement. In Taï, the chimpanzees also use one hand to hold the stick among the soldier ants guarding the nest entrance, but only until the ants have swarmed about 10 cm up the tool. They then withdraw the stick, twist the hand holding it, and sweep off the ants directly with the lips. Taï chimpanzees use short sticks about 24 cm long, and perform the dipping movement about 12 times per minute (Boesch and Boesch 1990). We estimated that 15 ants are gathered with each dipping movement. In Gombe, the Taï dipping method has been observed being used only occasionally by just two individuals, McGregor and Pom (McGrew 1974). Each of the two ant-dipping techniques is essentially restricted to only one chimpanzee population. I have tested both techniques in each of the two sites and found no ecological factor that would prevent the use of either in both sites. The Gombe technique is four times more efficient than the one used in Taï (Gombe: 760 ants/minute, Taï: 180 ants/minute;

Boesch and Boesch 1990). So we can say that Taï chimpanzees restrict themselves to a suboptimal solution that must be maintained by a social norm that prevents individuals from testing other, more efficient possibilities.

Things become more complicated when we look at what Bossou chimpanzees (which live about 200 km from Taï) do when they dip for the same species of ants (*Dorylus* species) as do the Taï chimpanzees (Boesch and Boesch 1990; Humle and Matsuzawa 2002). Intriguingly, they regularly dip for these ants while the ants are moving in migrating columns, something that adult Taï chimpanzees never do. Furthermore, Bossou chimpanzees use only their hands to feed on the red *Dorylus* species, whereas Taï chimpanzees use only tools to feed on this ant species. Conversely, Taï chimpanzees feed on black *Dorylus* ants only with their hands, whereas Bossou chimpanzees use both tools and hands for this species. Another intriguing finding is that the Bossou animals are even less efficient with dipping tools (119 ants/minute) than are the Taï chimpanzees.

Arbitrariness in a cultural trait is most vividly shown in wild chimpanzees in the leaf-clipping behavior. In this behavior, a chimpanzee picks a single leaf, grasps the petiole with its hand, repeatedly pulls the petiole from side to side while removing the leaf blade with the incisors, and thus bites the leaf to pieces. A ripping sound is conspicuously and distinctly produced by removal of the leaf blade, and nothing of the leaf is eaten; Nishida (1987) describes this as a common behavior among chimpanzees in Mahale. The behavior has also been seen regularly at Bossou (Sugiyama 1981) and Taï (Boesch 1991), but only twice at Gombe (Goodall, pers. comm., cited in Nishida 1987). The fact that this behavior is present in three chimpanzee populations but practically absent in a fourth could be explained by an ecological difference, although none is known that would produce such an irregular distribution of that behavior. When present, this behavior serves a wide variety of functions. In Mahale, the chimpanzees most often use it as a herding or courtship behavior in sexual contexts (Nishida 1987). In Bossou, it occurs mostly in apparent frustration or in play (Sugiyama 1981). In Taï, leaf clipping occurs mainly as part of the drumming sequence of adult male display behavior (Boesch 1991). The shape of the behavior is the same in all populations, but it has acquired in each population a different and unique meaning. Group members respond to the signal according to the specific meaning it has for their group, even if they were not the intended recipients. The meaning of this behavior thus rests upon a meaning shared among group members (Boesch 2003).

## CULTURE COMPLEXITIES

The analyses traced above tend to indicate that cultural traits appeared in animals first in adaptive contexts where they provide individuals with a clear benefit. In addition, the number of proposed cultural traits increases when moving from the monkeys to the great apes, in a way predicted by the hierarchical definition of culture. Cognitive abilities related to learning and perception of other group members seem to develop in a parallel way in species ranging from monkeys to apes (Byrne 1995; Hauser 2000; Tomasello and Call 1997). For example, capuchin monkeys and chimpanzees are recognized as dextrous species that use many objects and tools both in the wild and in captivity (Savage-Rumbaugh et al. 1985; Goodall 1986; Chevalier-Skolnikoff 1988; Boesch and Boesch 1990). Therefore, it may be more than a coincidence that both species present the largest number of feeding-context cultural traits.

In great apes, we observe a tendency to acquire more cultural traits that do not present a direct adaptive value and that are not related to feeding context. These traits are much more independent from ecological conditions and have become more flexible, as well as having different meanings in different groups. Many anthropologists have posited this as a trend further exaggerated in humans, among whom most cultural traits occur in non-feeding contexts; however, extensive lists of cultural traits in the same format as those proposed here have yet to be drawn up for humans. Why do great apes present so many arbitrary nonadaptive cultural traits? One possible explanation implicates the greater demands of a complex social life. Humans, orangutans, and chimpanzees live in flexible social groups with a fission-fusion system, which is relatively rare in other primates (Dunbar 1988). In addition, humans and chimpanzees live in comparatively large social groups with accompanying propensities for conflict. Accordingly, we should expect a large difference between chimpanzees and the solitary orangutans in their numbers of cultural traits, and more of these traits will probably be found in the future. Another non-exclusive possibility is that some cognitive developments may be required in order for arbitrary signals to be used flexibly by numerous individuals of different sexes and ages within a social group.

Cultural traits relying on shared meaning have developed in chimpanzees in the communicatory domains. Gestural communication could be viewed as a precursor of language and can in the same way be based on shared meaning. Gesturing seems to have opened the way to language in our species. This is supported by the fact that some aspects of speech are still associated with gestures in different human languages, such as counting in African or Melanesian languages (Zaslavsky 1973; Saxe 1981).

The proposed level of cultural complexity in our multilevel definition seems to be well reflected in what has been observed in different monkey species. We must remember that the data are still quite fragmentary and very newly acquired, but nonetheless they encourage a multi-layered approach such as the one proposed above, which will allow us to show how complex culture can be. In addition, the multilayered definition will allow precise identification of what is unique or special in the cultural ability of a given species. Under the proposed scheme, humans share with many other species cultural abilities that are often very similar. Many of our daily cultural undertakings are on simple levels and are seen regularly in many other animal species. However, some of our cultural traits are quite unique and include two specific abilities. The first is our ability to share through language, in the absence of the appropriate situation and without having ever experienced the situation, a cultural trait and its meaning. Dissemination can thus occur out of context and between individuals that have never met. Second is our ability to use language as a vehicle to explain to social partners the meaning of abstract traits. This leads to the fact that human cultural elements based on shared meaning can become much more extensive than those in nonhuman primates.

## NOTE

I thank the Ministry of Scientific Research and the Ministry of Environment and National Parks of Côte d'Ivoire for supporting the Taï chimpanzee projects all these years, and especially the direction of the Taï National Park, the Centre Suisse de Recherches

Scientifiques and the Centre de Recherche en Ecologie in Abidjan, Côte d'Ivoire. The Swiss National Science Foundation and the Max Planck Society have supported this project financially. I thank the field assistants and students of the Taï chimpanzee project for constant help in the field, and Hedwige Boesch, William McGrew, William Durham, Linda Vigilant, and Andrew Whiten for helpful comments on this chapter.

# 3

# Dissent with Modification

## Cultural Evolution and Social Niche Construction

### MARCUS W. FELDMAN

Arguments against theories—especially quantitative theories—of cultural evolution claim either that there are no definable entities whose dynamics over time
can be evaluated, or that, even if such entities were to exist, there are no
detectable laws of transmission that apply to them. In this chapter, we claim that
models serve a number of purposes in the natural and social sciences and have
provided the impetus for vast numbers of experimental and observational studies. The meaning of *models* in anthropology is discussed with reference to
specific examples. These examples incorporate behavioral and demographic
data, which suggest models of cultural transmission that can be viewed as
either dynamical in making temporal predictions, or statistical in suggesting
causes of variation. The chapter also introduces the possibility of cultural and/or
behavioral niche construction that incorporates evolutionary feedback between
socially variable traits. This leads to a new theory of social niche construction
that might be widely applicable in anthropology, sociology, and economics.

The question of which of C. P. Snow's two cultures anthropology should
belong to continues to vex deans, and elicit numerous scholarly articles. At the
center of the debate is whether human culture or cultures should or can be
studied in an evolutionary framework or as part of history. Does the "problematic conflation of historical and evolutionary processes" (Fraccia and
Lewontin 1999: 60) preclude the formulation of any theory about culture and
therefore the acquisition of data relevant to that theory? In this chapter, I will
describe an evolutionary theory that extends the quantitative theoretical framework for cultural evolution that Luigi Cavalli-Sforza and I began in the early

1970s. This extension, developed over the past seven years in collaboration with Kevin Laland and John Odling-Smee, introduces the notion of "niche construction" and allows historical features of a population of organisms to be introduced into the continuing processes of evolution. The features may be biological or cultural in nature, and may result from actions or behaviors of the given population or other populations.

Niche construction occurs when the properties or behaviors of an organism or a population of organisms affect the environment experienced by the organism, its descendants, other species, and their descendants, in the current or future generations. The properties or behaviors may be biologically or culturally determined and transmitted. This dynamic symmetry between organisms and their environments entails a reconsideration of the term "adaptation." Because hominids have had tremendous power to affect their environments over the past 100,000 years, this extension of evolutionary theory is particularly important for the human sciences.

## EVOLUTION AND HISTORY

The process of evolution as understood in biology involves three components and an assumption about their interaction. The first component is the production of variation among the individuals in the population. In biology, this is mutation and occurs at the level of DNA. The second process is descent, or transmission. For biologists, this occurs from parent to child and is usually termed "vertical transmission," and what is transmitted is a set of genes. Third are differential survival and differential rates of reproduction of the different types of individuals. These constitute the process of natural selection. In a few well-studied cases, the process of transmission is itself subject to natural selection in the sense that some gametic variants are favored over others in the transmission process and Mendel's laws of random segregation are violated. However, for the overwhelming majority of genes, the replication-transmission process and the process of natural selection are independent of one another. Further, a fundamental assumption about the components in the process of descent with modification outlined above is that the production of variation is independent of what happens to that variation under the action of natural selection.

The entities that vary and are transmitted in biology are now known to be located in the DNA. Although in a number of microbes DNA may be transmitted laterally among individuals, this is thought to be rare among multicellular organisms; this generally vertical transmission produces expected values for statistical correlations between related individuals that were first elucidated by Fisher (1918) and Wright (1921). Earlier attempts by Francis Galton had foundered because he studied continuously varying traits that did not have the particulate transmission discovered by Mendel and that later produced the science of genetics (see, e.g., Bulmer 2003). The reverse process—that is, taking correlations between related individuals and attempting to infer the rules of transmission—remains extremely controversial. It makes heavy use of statistical models whose assumptions about the nature and role of the environment in contributing to the correlations strongly bias conclusions about the nature and role of genetics in forming the trait. In fact, it was this confounding factor that led Cavalli-Sforza and me to begin the quantitative analysis of cultural transmission and gene-culture coevolution in 1973 (Cavalli-Sforza and Feldman 1973a).

Despite the enormous variation among biological organisms, the laws of biological evolution described above have been shown to be general, and indeed it can be argued that the explosion of biological knowledge in the twentieth century would have been impossible if not for the generality of the rules of genetics and the laws of evolution.

Historical phenomena are defined by Fracchia and Lewontin (1999: 60) as "particulars embedded in particular socio-cultural forms, each with its own systemic properties and discrete logic of production and reproduction, its own dynamics of stasis and change." It follows from this definition that there can be no rules of variation, transmission, or change that apply to historical phenomena; furthermore, any formal modeling of such a phenomenon is rendered futile because any such model would have no relevance to any other.

In the context of culture, Fracchia and Lewontin (1999) argue that only the historical framework is legitimate. Their claim (1999: 64) is that "Theorists of cultural evolution, conscious of the need for a theory of inheritance, yet deprived of any compelling evidence for particular law-like mechanisms for the transgenerational passage of cultural change, are in a much more difficult position . . . because they do not even know whether an actor-to-actor, not to speak of parent-to-offspring, model of the passage of culture has any general applicability." They see the problem as parallel to the failure of Galton's theory of relationships, namely the lack of entities to be transmitted. They state: "The reason for this lack of generality is that no theory of cultural evolution has provided the elementary properties of these abstract units" (72).

Early opposition to an evolutionary approach to culture by Franz Boas, Alfred Kroeber, and others was based on the idea that evolutionary thinking entailed biological determinism and an overt or implied directionality that was value-laden and implied progress from simple to complex in a way that would justify racism (Boas 1938; Kroeber 1923). It can be argued that the expansion of human sociobiology and evolutionary psychology after the mid-1970s that produced biological interpretations of the causes of both variation and universality in human culture was made possible by the theoretical and conceptual vacuum left by the success of early opposition to evolutionary thinking in anthropology. The residue of the opposition remains today in the antipathy among many anthropologists toward quantitative models and quantitative analysis of ethnographic data. This is unfortunate, because in a number of cases quantitative analysis has shown that variation in cultural traits is most likely to have been the cause of the observed pattern of biological variation, reversing the usual hereditarian position of biological determinism.

This antipathy is strong, even to quantitative models of cultural evolution that do not include any biological assumptions. Most often it is expressed by forcing the definition of cultural entities into an ideational domain whose abstractness precludes measurements and statistical analysis. In this way, predictive and postdictive quantitative models are automatically deemed irrelevant because they concern entities that are viewed as not cultural. It is worth recalling the criticism by Ernst Mayr, the famous natural historian, of the work of Fisher, Wright, and Haldane, the founders of modern quantitative theory in biology, as "bean-bag genetics." This criticism is quite analogous to that made by Fracchia and Lewontin (1999: 74) of quantitative theory in cultural evolution: "Atomistic models based on the characteristics of individual humans or individual memes can be made, but they appear as formal structures with no possibility of testing their claim to reality."

I disagree, and believe rather that there are observable units of culture: they can be assessed at an individual and population level; useful statistical rules for their transmission can be formulated; where appropriate, environmental change that affects these rules can be incorporated; and from these rules useful predictions can be made about the future distribution of the units.

For quantitative theory, it is immaterial whether the units that are transmitted between members of a population are ideas, behaviors, or customs. It does matter that variation in these units is detectable and scorable on some scale. Behaviors and/or customs are much more valuable as markers of culture than ideas because there are limited data on ideas and their distribution within and between populations.

## MODELS FOR CULTURAL EVOLUTION

The first requirement for any evolutionary process is variation, and whether the cultural trait under study is discrete or continuously varying, the ultimate source of variation is innovation. My discussion here will concern discretely varying traits, such as behaviors; I regard these as cultural because an individual without a particular variant of the trait can acquire that variant by some form of learning or social transmission. An innovation may arise de novo in the population or be brought into the population by an immigrant.

The second factor in the evolutionary process is transmission. For cultural variants, Cavalli-Sforza and Feldman (1981) defined three modes of transmission: vertical, from parent to offspring; oblique, from nonparental members of one generation to individuals of a subsequent generation; and horizontal, among members of the same generation. A fourth kind of transmission was introduced by Cavalli-Sforza and Feldman (1973b) and termed "group transmission." It allows the trait value of any individual to come partly from a parent and partly from the average trait value in the parental generation. This special kind of oblique transmission has the effect of limiting variation within a population while allowing interpopulation variation to increase (see also Boyd and Richerson 1985: 287).

Our inclusion of nonvertical modes of transmission in the evolutionary paradigm for culture was intended to take into account dynamic processes that occur within a generation. Repeated horizontal transmission within a generation allows for evolution on an ecological or epidemiological time scale. Indeed, it may not even be necessary to include the vertical mode of transmission. In this case the process can be regarded as one of diffusion, and many examples of this kind of dynamical cultural process can be found in Rogers (1995). Of course, the superposition of vertical transmission between generations over horizontal transmission within generations increases the flexibility of the models. On the other hand, vertical cultural transmission may occur repeatedly between parents and offspring, unlike genetic transmission, which occurs just once. Repetition of the transmission, if it occurs in a large enough fraction of the population, may increase the rate of spread of an innovation, or if the transmitted variant is already frequent, it may act against the proliferation of less-common variants.

The third part of the evolutionary paradigm, selection, is the most contentious in the context of cultural evolution. If a variant of a trait is identified with the individual carrying (or performing) that variant, and if the different variants determine different survival or reproduction rates, then the selection process is equivalent to natural selection on a phenotype. In the human case, we might call this "demic" selection, by analogy with the term "demic diffusion" used by Ammerman and Cavalli-Sforza (1973) to distinguish the movement of people from the spread of ideas or behaviors without physical migration of their carriers. The difficulty arises if the variants of the cultural trait are acquired at different rates during individual development. This may occur as a result of variation in the transmission stage (called "awareness" by Cavalli-Sforza and Feldman 1981: 62) or in the rate of adoption by the receivers (see also Rogers 1995). In the transmission and natural selection of cultural variation, unlike those processes in biological variation, individual choice, intentionality, and goal recognition may affect innovation and transmission, and a selection process is possible without variation in birth and death rates among the carriers of the cultural variants. Cavalli-Sforza and Feldman (1981: 34) called this "cultural selection," and pointed out that the structure of a society may influence this kind of selection (353–55).

A fourth component of the evolutionary process is often omitted from discussions of cultural change. I refer to sampling effects. If a population is finite, and learning is not perfect, then at any point in time the variation in a population represents a sample of the variation at a previous time point. This sample may be random with respect to the variation, or it may be biased, as in the case of meiotic drive in genetics. The result of this sampling process is that

history can become important in determining the trajectory of cultural variation; there can be strong path dependence or founder effects that interact
with the transmission and selection processes. The overall effect can be called
"random cultural drift" (Binford 1963, Cavalli-Sforza and Feldman 1981).

I have purposely avoided the term "cultural inheritance" here in favor of
"cultural transmission" to stress that the vertical is not the only relevant mode
of transmission—that intragenerational processes may influence intergenerational dynamics, and that these transmission processes may have nothing to
do with biological inheritance.

## UNITS AND DATA

Hewlett et al. (2002) use the term "seme," originally proposed by Cavalli-
Sforza, for cultural units "because 'seme' comes from 'sign' and emphasizes
the symbolic nature of culture." These cultural units are in fact "schemas or
practices" whose presence or absence and frequency can be observed in the
demographic unit(s) under study. Included in the Hewlett et al. study of thirty-
six African societies are such properties as whether house walls are made of
wattle or mats, whether the house roof is gabled, or whether only men fish.
The study compared 109 such semes between all 630 pairs of societies, and of
these the distribution of 20 semes was consistent with demic diffusion, 12
fitted a cultural diffusion pattern, 4 had features of local adaptation, and nine
had features of all of these explanatory patterns. The patterns are based on
considerations of genetic, linguistic, and geographic distances between the
630 pairs of populations, and whether these pairs share semes. For example,
cultural diffusion would explain a shared seme if two populations are geographically close but not close linguistically or genetically. Earlier studies by
Hewlett and colleagues showed that semes that were consistent with demic
diffusion were less likely to be affected by local ecological factors (Guglielmino
et al. 1995), and that among the Aka pygmies there was vertical transmission
of many important aspects of their culture (Hewlett and Cavalli-Sforza 1986).
These studies are important examples of empirical research solidly grounded
in the theoretical frameworks of cultural transmission, population genetics,
and historical linguistics.

Recent studies of chimpanzees and orangutans have claimed that cultures exist in these primates. Whiten et al. (1999) synthesized years of research on chimpanzee behavior and found thirty-nine behaviors that are customary in some chimpanzee communities but absent in others, and for which the difference is not due to local ecologies. Van Schaik et al. (2003) carried out a similar analysis of thirty-six behaviors in orangutans, twenty-four of which were relatively common in at least one site but absent in at least one other ecologically similar site. Whiten et al. suggest that the chimpanzee behaviors are acquired by "a complex mix of imitation, other forms of social learning, and individual learning." In reviewing the chimpanzee study, de Waal (1999) concludes that "the evidence is overwhelming that chimpanzees have a remarkable ability to invent new customs and technologies and that they pass these on socially." In the same vein, van Schaik et al. state that "the size of the cultural repertoire at a given site is best predicted by the opportunities for oblique and horizontal social transmission during development."

In all three studies, the units subject to transmission are behaviors (see Boesch, this volume). In the human example, there may be an ideational component that resides behind each seme, but the primate researchers make no such suggestion. Nevertheless, underlying any quantitative statement about the likely historical reasons for the observed configuration of behaviors is a theory for the cultural transmission of the behaviors themselves. According to de Waal (1999), the political, economic, or psychological reasons for the behavior are "of secondary importance. In the same way that the definition of respiration doesn't specify whether the process takes place through skin, lungs, or gills, the concept of cultural propagation does not specify whether it rests on imitation, teaching or language." Of course, many cultural anthropologists would contest de Waal's claim.

## NICHE CONSTRUCTION

Many organisms from a wide array of taxonomic groups modify their local environments by their choice of resources, their protection of offspring, and their patterns of movement. Lewontin (1983) argued that organisms not only adapt to environments but also construct them. We call this the process of niche construction (Odling-Smee 1988; Laland et al. 1996, 1999), and have shown, using mathematical models, that the feedback from niche construction may substantially modify the action of natural selection. In this purely bio-logical context, niche construction can cause either evolutionary inertia or momentum; it can produce stable polymorphisms where none would exist without it, and it can eliminate what would otherwise be stable polymor-phisms. Niche construction introduces symmetry between organisms and environment by allowing changes in each to be functions of both.

The spectrum of possibilities for human niche construction is expanded by cultural transmission of information manifest in behaviors that can be socially learned. In some cases, environmental change brought about by culturally transmitted behaviors, that is, cultural niche construction, has had profound effects on the genetic evolution of some humans. For example, the culturally transmitted practice of consuming cow's milk affected the evolution of the genetically determined tolerance of lactose (Durham 1991; Feldman and Cavalli-Sforza 1989; Holden and Mace 1997; Tishkoff et al. 2007). Another less obvious example concerns the genetic consequences of patrilocal and matrilo-cal marriage, which are, of course, cultural traits. Among patrilocal commu-nities, because women move at marriage, intercommunity variation is expected to be less pronounced in mitochondrial DNA (mtDNA) than in Y-chromoso-mal DNA (Seielstad et al. 1998); the opposite relationship, which would be expected among matrilocal communities, has been observed in communities in northern Thailand (Oota et al. 2001). The environments in which the genetic patterns evolved were constructed by the cultural systems that are subject to cultural transmission.

The marriage-practices example cannot be regarded as a case of primary and secondary values (to use the terminology of Durham 1991: 200–201), because mtDNA and Y-chromosomal DNA obviously play no role in the deter-mination or transmission of marriage customs. In the lactose-tolerance case, if individual responses to digestion of milk led to decisions about dairying, then the lactose-tolerance gene(s) might be regarded as being responsible for primary value selection of the cultural trait of nonhuman-milk consumption. The evidence, however, suggests that this was not the case and that the dairy-ing-nondairying dichotomy preceded the spread of lactose tolerance (Holden and Mace 1997).

The first formal model of niche construction in the framework of evolu-tionary genetics took two biallelic loci **E** and **A** with gene **E** controlling niche

### Table 3.1. Two-locus fitnesses with gene-based niche construction

|     | EE | Ee | ee |
| --- | --- | --- | --- |
| AA | $w_{11} + f_{AA}(R)$ | $w_{12} + f_{AA}(R)$ | $w_{13} + f_{AA}(R)$ |
| Aa | $w_{21} + f_{Aa}(R)$ | $w_{22} + f_{Aa}(R)$ | $w_{23} + f_{Aa}(R)$ |
| aa | $w_{31} + f_{aa}(R)$ | $w_{32} + f_{aa}(R)$ | $w_{33} + f_{aa}(R)$ |

construction. This occurs via a resource R whose abundance (measured as a frequency) is given by R ($0 < R < 1$). Alleles at locus E are E and $e$, and the abundance of resource R is a function of the frequency of $x$ of allele E. The A locus genotypes AA, Aa, and aa have fitnesses that are functions of R and hence of $x$; in general these could be written $f_{AA}(x)$, $f_{Aa}(x)$, and $f_{aa}(x)$. The niche construction due to locus E therefore coevolves with the genotypes at the A locus whose fitness it affects. The two-locus fitnesses are shown in table 3.1, where the $w_{ij}$ values represent the usual two-locus fitnesses that would characterize a two-locus system in the absence of niche construction.

The properties of the functions $f_{AA}$, $f_{Aa}$, and $f_{aa}$ determine whether the effect of niche construction is to favor allele A or allele $a$. The argument of these functions is the resource frequency R, which at any generation $t$ might be affected by the properties of alleles E and $e$ in many previous generations, i.e., at generation $t$, R could be a function of $x_t, x_{t-1}, x_{t-2} \ldots x_{t-k}$, for example. Laland et al. (1996) show that many of the well-known properties of genetic evolution under the fitness matrix $||w_{ij}||$ are significantly changed by the inclusion of the niche construction terms. Stable polymorphisms may arise or be removed, fixation on one allele can be reversed, and depending on the $f$ functions, evolution can be retarded or accelerated. The spectrum of dynamics appears to be enriched considerably by the presence of niche construction.

For humans, niche construction may well occur as a consequence of some socially learned behavior (Laland et al. 2000). In this case, we consider a culturally transmitted trait E that occurs in one of two forms, E and $e$, and affects the amount R of the resource R in the environment. Once again, R may be a function of behaviors over a number of generations. In this case, the niche construction affects the evolution of the phenogenotypes AAE, Aae, AaE, Aae, aaE, and aae, where the genotypes AA, Aa, and aa obey the standard Mendelian

Table 3.2. Fitness of phenogenotypes in matrix form

|  | E | $e$ |
|---|---|---|
| AA | $\gamma_{11} + f_{AA}(R)$ | $\gamma_{12} + f_{AA}(R)$ |
| Aa | $\gamma_{21} + f_{Aa}(R)$ | $\gamma_{22} + f_{Aa}(R)$ |
| aa | $\gamma_{31} + f_{aa}(R)$ | $\gamma_{32} + f_{aa}(R)$ |

Table 3.3. Probabilities of E or $e$-offspring under cultural transmission

| | Offspring | |
|---|---|---|
| Matings | E | $e$ |
| E × E | $b_3$ | $1 - b_3$ |
| E × $e$ | $b_2$ | $1 - b_2$ |
| $e$ × E | $b_1$ | $1 - b_1$ |
| $e$ × $e$ | $b_0$ | $1 - b_0$ |

rules of transmission. Laland et al. (2000) used a phenogenotypic fitness array like that in table 3.2, where the $\gamma_{ij}$ represent fitnesses in the absence of niche construction. The cultural transmission of E and $e$ occurred via the vertical rules of Cavalli-Sforza and Feldman (1981), shown here in table 3.3.

Analysis of these models showed clearly that the evolution at locus **A** can be affected profoundly by biased cultural transmission of a niche-constructing behavior. Rates of change in the frequencies of E and $e$ are often faster than under natural selection alone. Thus, cultural niche construction can have a stronger effect on the resource R than niche construction resulting from genes as described above. The strength of selection on locus **A** can be amplified by bias in the cultural transmission of E and $e$.

This feedback between aspects of culture with their transmission and genetic evolution may be relevant to the claim by D'Andrade (2002a) that the development of culture modified the environment of humans, thereby selecting for the capacity to produce representative speech acts. Implicit here is the development of symmetry, with increased linguistic capability providing the appropriate environment for further cultural evolution.

The qualitative properties of the two models of niche construction described in tables 3.1, 3.2, and 3.3 lead to a number of predictions discussed in detail in Odling-Smee et al. (2003). For example, we might expect that strong niche constructors would show weaker signatures of genetic evolution in response to fluctuating environments than would weaker niche constructors. Thus, because hominids have a very high level of niche construction, they should show less evolution in morphology than other mammals. Bergmann's and Allen's rules suggest that organisms in warmer climates should have smaller bodies and longer extremities than those in cooler climates. We therefore expect that hominids should exhibit less correspondence with these rules than other organisms that are weaker niche constructors.

Now consider one of the arguments often raised in criticism of models for cultural evolution, namely that the transmission of variants of a cultural trait among individuals in a population cannot be treated in isolation from the cultural background of that population. The models of niche construction described above suggest a class of models of cultural transmission that do indeed consider the cultural background by incorporating cultural niche construction.

Suppose that there are two cultural traits $E$ and $H$ with variants $E/e$ and $H/h$, respectively, in a population. At time $t$, the frequencies of $E$ and $H$ are $x_t$ and $y_t$, respectively. Extending the semes of Hewlett et al. (2002) to more than one dimension, then, by analogy with the term "haplotype" from genetics, we call the four combinations $EH$, $Eh$, $eH$, and $eh$ "semotypes." Cultural niche construction will occur if the transmission of variants $H$ and $h$ is controlled to any extent by which variant of trait $E$ is present, $E$ or $e$. That is, $E$ is providing the context for the rule of transmission for $H$. If the cultural transmission is vertical, it can be quantified by the extension of table 3.3 shown in table 3.4, assuming that the transmission rate from a mating pair of semotypes to an offspring semotype is just the product of the entries in the first mating table by those in the second mating table. Thus, for example, mating $Eh \times eH$ gives offspring $Eh$ with probability $b_2 [1 - c_1 - g_1 (x)]$ and $eH$ with probability $(1 - b_2)[c_1 + g_1 (x)]$.

The idea behind table 3.4 is that the frequency of $E$ in the population affects the transmission of trait $H$. If all $g$ functions are the same, then the mating types for trait $H$ play no role in the functional dependence of the offspring's chance of becoming $H$ on the frequency $x$ of $E$. This would be the simplest case

Table 3.4. Cultural niche construction in vertical transmission rates

| Seme 1* | | | Seme 2 | | |
|---|---|---|---|---|---|
| Offspring | | | Offspring | | |
| Matings | E | e | Matings | H | h |
| E × E | $b_3$ | $1 - b_3$ | H × H | $c_3 + g_3(x)$ | $1 - c_3 - g_3(x)$ |
| E × e | $b_2$ | $1 - b_2$ | H × h | $c_2 + g_2(x)$ | $1 - c_2 - g_2(x)$ |
| E × E | $b_1$ | $1 - b_1$ | h × H | $c_1 + g_1(x)$ | $1 - c_1 - g_1(x)$ |
| E × e | $b_0$ | $1 - b_0$ | h × h | $c_0 + g_0(x)$ | $1 - c_0 - g_0(x)$ |

* E frequency is x.

of vertical cultural niche construction. A natural extension, analogous to those used by Laland et al. (1996, 2000), would allow dependence on the frequency of E in previous generations. Horizontal or oblique transmission can also be included, and it is straightforward to add a factor of conformity to the majority in the population (Ihara and Feldman 2004).

When the cultural background variation constructed by trait **E** is nonrandomly associated with the variation in the target trait **A**, we would find that, for example, semotype EH does not occur with frequency $xy$. We might define the quantity $\Delta = f_r$ (EH) $\times f_r$ (eh) $- f_r$ (Eh) $\times f_r$ (eH) as a semotypic phase imbalance, analogous to the standard linkage disequilibrium in genetics. Then $\Delta$ could measure the tightness of the control by the cultural background **E** on the target trait **A**.

An example of cultural traits that might be represented in terms of this model concerns the interactions of son preference and sex-selective abortion in China. The E/e dichotomy might represent son preference and no son preference, and H/h the dichotomy between the behavioral act of carrying out sex-selective abortion or not doing so. In those few Chinese communities where son preference appears to be weakest, the sex ratio at birth is the closest to normal (Li N., Feldman, and Tuljapurkar 1999; Li N., Feldman, and Li 1999, 2000). These studies strongly suggest that preference is culturally transmitted, although the availability of sex-selective abortion has not been wide for a long enough period to say much empirically about the vertical cultural transmission of the behavior of sex-selective abortion in favor of sons. There is probably strong horizontal transmission of information about availability of ultrasound B for fetal sex detection, but this would be difficult to study quantitatively. Also related to son preference is the rarity in rural China of uxorilocal marriage (S. Li et al. 2003). In those places where it does occur, it is related to the demographic contingency of a no-son family. By combining the life table features from the Chinese census with models of cultural transmission, the future sex ratio at birth (SRB) can be predicted (Li N., Feldman, and Li 2000). An SRB of 130 is a not-unlikely event within the next fifty years.

A second example, studied by Ihara and Feldman (2004), concerns level of education and the norm of fertility control. Here we might assume that the level of education is modeled by $E$ and fertility behavior by $A$. For this study, we assumed that both $E$ and $A$ were subject to standard vertical cultural transmission, but that during development, trait $A$ is subject to oblique transmission at rates that depend on the frequencies of variants of $E$. Two recent studies have shown the importance of social transmission in the diffusion of contraceptive use in Kenya (Behrman et al. 2002) and in Ghana (Montgomery et al. 2001), which suggests that we should also include horizontal transmission in such models of cultural niche construction.

A third example concerns the preference for cousin marriages that is prevalent among groups that until recently were primarily nomadic pastoralists. The high frequency of genetic abnormalities in a number of Middle Eastern countries is a result of such consanguineous marriages. As pointed out by Bittles and Erber (2005), the migration of significant numbers from these populations has resulted in pressures on public health systems in their adopted Western countries. Here we have the marriage customs in a putative cultural niche of nomadic pastoralism. On the other hand, specific matings might be prohibited in other environments (Aoki et al., this volume), where punishment was enforced or religious proscriptions had emerged. Again, the taboo could have emerged in an appropriate cultural niche.

Environmental change may also affect the rates of vertical and/or horizontal transmission. In their study of uxorilocal and virilocal marriages in Lueyang, China, Li S. et al. (2000) estimated the parameters of vertical transmission of marriage types. The results are shown in tables 3.5 and 3.6, where we see that after the economic reforms that were instituted in 1978, the rate at which husbands contributed to the vertical transmission of uxorilocal marriage

Table 3.5. Additive model of vertical transmission of uxorilocal marriage

| Husband's parents' marriage | Wife's parents' marriage | Transmission coefficient | Expected values of $b_i$ |
|---|---|---|---|
| Uxorilocal | Uxorilocal | $b_3$ | $a_0 + a_h + a_w$ |
| Uxorilocal | Nonuxorilocal | $b_2$ | $a_0 + a_h$ |
| Nonuxorilocal | Uxorilocal | $b_1$ | $a_0 + a_w$ |
| Nonuxorilocal | Nonuxorilocal | $b_0$ | $a_0$ |

Table 3.6. Maximum-likelihood estimates of additive models of vertical transmission of uxorilocal marriage

| Marriage cohort | $a_0$ | $a_h$ | $a_w$ | $\chi^2$ | $p^*$ | Number of cases |
|---|---|---|---|---|---|---|
| Before 1978 | 0.29 | 0.15 | 0.24 | 0.06 | > .80 | 572 |
| 1978 and after | 0.30 | -0.11 | 0.08 | 9.89 | < .01 | 891 |

*p is the probability of fit to the additive model.

changed substantially. It remains to be discovered what caused this change, but it should be noted that the Chinese national family-planning policy took hold at about that time. Indeed, there is very strong evidence that prior to the family-planning policy, the SRB was normal, even though son preference was ubiquitous. The manifestation of this preference through sex-selective abortion following the widespread availability of ultrasound B testing led to male bias in the SRB, a bias that continues to increase, despite improvements in living standards across the country (Li S. et al. 2003).

A further example of a form of cultural niche construction concerns a culturally transmitted behavioral dichotomy that affects the rate at which an infectious disease is transmitted. Tanaka et al. (2002) supposed that individuals in a population may be classified as susceptible (S) or infected (I) for their disease status. A second dichotomy defined the alternative forms careful (c) and risky (r) of a behavioral trait that affects the rate of disease transmission. The behavioral trait is horizontally transmitted within the population. Individuals are then specified as $S_c$, $I_c$, $S_r$, and $I_r$. The complete transmission system for disease and behavior is described in tables 3.7 and 3.8. In these tables, $\theta_1$ and $\theta_2$ spec-

Table 3.7. Disease transmission for each type of contact

|  | $I_c$ | $I_r$ |
|---|---|---|
| $S_c$ | $\beta_{cc} = \theta_1 b_{cc}$ | $\beta_{cr} = \theta_2 b_{cr}$ |
| $S_r$ | $\beta_{rc} = \theta_2 b_{rc}$ | $\beta_{rr} = \theta_1 b_{rr}$ |

NOTE: $\theta_1$ and $\theta_2$ are contact rates for individuals of the same or different behavior, respectively; $\beta_{ij}$ is the probability the disease is transmitted from an infected of behavior $j$ to a susceptible of behavior $i$ for $i,j = c,r$.

Table 3.8. Horizontal cultural transmission parameters for each type of contact

|  | Donor | | | |
|---|---|---|---|---|
| Receiver | $S_c$ | $I_c$ | $S_r$ | $I_r$ |
| $S_c$ |  |  | $\alpha_1 = \theta_2 a_{cr}$ | $\alpha_2 = \theta_2 a_{cr}\varepsilon$ |
| $I_c$ |  |  | $\alpha_3 = \theta_2 a_{cr}\delta$ | $\alpha_4 = \theta_2 a_{cr}\delta\varepsilon$ |
| $S_r$ | $\alpha_5 = \theta_2 a_{rc}$ | $\alpha_6 = \theta_2 a_{rc}\varepsilon$ |  |  |
| $I_r$ | $\alpha_7 = \theta_2 a_{rc}\delta$ | $\alpha_8 = \theta_2 a_{rc}\varepsilon\delta$ |  |  |

NOTE: Given that contact occurs, $a_{ij}$ is the probability of transmission of the behavior $j$ to an individual of the other behavior $i$ for $i,j = c,r$.

ify the rates at which individuals of the same or different behavioral type(s) come into contact; in table 3.8, $\delta$ and $\varepsilon$ modify the probabilities of acquiring a behavior, given that a contact occurs, relative to the rate for contacts between susceptibles. The cultural trait of careful or risky behavior is acquired by transmission on contact with an individual of the opposite behavior. In a sense, the cultural dichotomy is constructing the niche in which the epidemic proceeds. Tanaka et al. show that the culturally transmitted behavioral heterogeneity is important for the dynamics of the epidemic.

Recent work on the buildup of antibiotic resistance in bacteria in regimes of high antibiotic usage has also been shown to fit the paradigm of social niche construction. Here the antibiotic usage regime constructs the niche in which resistant and sensitive bacterial strains compete. The usage regime is, of course, a human behavior; antibiotic use could perhaps be reduced when the frequency of resistance in bacteria is high. This kind of behavior is shown by Boni and Feldman (2005) to produce a balance between resistant and sensitive strains that would not exist if the human hosts did not alter their behavior— i.e., construct the niche.

## CONCLUSION

I argue that behavioral variation is a useful measure and/or proxy for cultural variation, especially if the research goal is postdiction, prediction, classification, or comparison with, say, linguistic or genetic variation. Cultural transmission of behaviors can often be quantified theoretically and estimated empirically. In some cases, it is possible to model the structure of the cultural background that influences the transmission of a particular cultural variant. Although this approach is not likely to take care of every historical contingency that affects culture, it can be useful in some cases. Niche construction, formerly applied in the determination of fitness, can be usefully applied when several traits are under cultural transmission and the transmission of one set of traits—for example, attitudes and/or behaviors—is allowed to influence that of a second set of traits. The result can be a semotypic phase imbalance, analogous to linkage disequilibrium in genetics, which causes some semotypes to be non-random combinations of their separate semes. This is a quantitative expression of the importance of the cultural background. One can speculate that behaviors manifest in the game situations described by Bowles and Gintis (this volume) might be studied in this framework.

If we accept the notion that cultural variation over time can only be studied as "history," and that each case is unique and the search for principles futile, then among the social sciences anthropology would suffer the same fate as natural history among the biological sciences—it would become a compilation of curiosities. With more emphasis on the construction, analysis, and interpretation of models as well as on the fundamentals of evolutionary theory during the training of anthropologists, the seductiveness of misguided paradigms such as sociobiology and evolutionary psychology might be mitigated in the future.

# PART II   MODELING-BASED CASE STUDIES

The chapters in this part present case studies based on mathematical modeling, including examples of formal evolutionary approaches to culture. (Note that there are some differences in terminology among contributors: "gene-culture coevolutionary theory," "the Darwinian theory of cultural evolution," and "evolutionary culture theory" generally refer to the same set of evolutionary approaches.) Quantitative models clarify complicated relationships, which researchers may or may not be able to keep track of qualitatively. For example, models can reveal the potential consequences of repeated interactions in these relationships over time. To contribute to a paradigm, however, formal models need to be able to explain empirical data. These chapters (as well as Feldman's chapter in Part I) link formal models with empirical data to analyze the relationship between human behavior and culture.

Peter J. Richerson, an ecologist, and Robert Boyd, a biological anthropologist, outline contemporary evolutionary approaches to culture, including a basic introduction to formal modeling, making a case for their power and consilience. Richerson and Boyd examine a wide range of empirical studies of cultural change, which they suggest fit within an evolutionary perspective (regardless of whether the studies' authors intended to contribute to an evolutionary framework) and demonstrate the importance of culture in explaining human behavior. They highlight "the contributions that cultural evolutionary analysis has made and is prepared to make to a science of culture . . . quantitative, mechanistic answers to the questions of the contributions of genes, culture, and environment to human behavior."

Population geneticists Kenichi Aoki, Yasuo Ihara, and Marcus W. Feldman use formal coevolutionary modeling to demonstrate that the consequences of repeated social interactions do not meet qualitative expectations. They address the custom that anthropologists have long regarded as constituting a quintessential case of culture: incest taboos (see A. Wolf 1995, 2004 for reviews of the anthropological literature on incest taboos). In particular, they quantitatively assess Sir James Frazer's (1910) point, repeated by Sigmund Freud (e.g., 1918: 159–60), that laws only prohibit those behaviors that humans readily enact. Freud, of course, argued (e.g., 1918, 1920; see also chap. 29 in A. Wolf 1995) that incest taboos were necessary because humans have incestuous desires that must be repressed in order for societies to function. There exists a body of empirical data indicating that both Frazer and Freud were wrong (e.g., A. Wolf 1995; Wolf and Durham 2004). However, just as any empirical generalization is subject to dismissal, these data have been dismissed or reinterpreted by Freudians (e.g., Robin Fox 1962; Spain 1987). Aoki, Ihara, and Feldman's models show that Frazer's intuition, seized upon by Freud, is wrong at the theoretical level—it is not *possible* for an incest taboo to spread throughout a population when innate sexual preferences are strongly incestuous. Only where an innate incestuous preference occurs at low frequency in a population (as predicted by Westermarck [1891] and empirically corroborated by A. Wolf [1995]), and thus the cost of punishment is small "rendering the threat of punishment superfluous, will the incest taboo be accepted by all members of society as all become punishers."

James Holland Jones, a biological anthropologist, uses formal epidemiological models to suggest a potential answer to the empirical puzzle of the frequency of sexually transmitted infections (STIs) in the US varying by ethnic group. Focusing on assortative or nonrandom partnering (something epidemiologists have not tended to explore), Jones shows how ethnic labels affect the transmission of STIs by influencing preferences for sexual partners. Compartmental models predict that structured mixing—that is, ethnic preferences for sex partners—strongly affects dynamics of STI transmission and prevalence in the aggregated population. Not only do epidemiological patterns "provide information on the cultural norms that underlie transmission," they also show the importance of culture. Moreover, Jones argues by analogy that such formal epidemiological models can be used to explore cultural dynamics.

# 4

# Cultural Evolution

## Accomplishments and Future Prospects

### PETER J. RICHERSON AND ROBERT BOYD

Edward Wilson (1998) recently popularized William Whewell's nineteenth-century idea of "consilience." The idea was a favorite of Darwin's. It holds that seemingly disparate phenomena in the world are in fact connected. Nuclear physics is "remote" scientifically from the social sciences, yet nuclear reactions in the sun are the most important source of energy on earth, nuclear decay in the earth's interior drives seafloor spreading which in turn shapes terrestrial ecology, and nuclear weapons profoundly altered the shape of international politics in the twentieth century. Nothing is in principle irrelevant to the study of the human species—not nuclear reactions, not anything. Evolutionary theories apply to highly consilient phenomena.

Consilience appealed to Whewell, Darwin, and Wilson because the concept suggests that scientific theories have the desirable epistemological property of, on the one hand, explaining many things economically if they are correct, and, on the other, being horribly vulnerable to ambush from every direction if wrong. In this chapter, we argue that culture and cultural evolution are also consilient phenomena, and that the Darwinian theory of cultural evolution can take systematic advantage of the epistemological strengths of consilience. (In this volume, see Melissa Brown's Introduction for a definition of culture in terms of ideas acquired from others, and her and Roy D'Andrade's chapters for an analysis of some of the complexities that this definition causes. For more on the theory of cultural evolution, see Marcus Feldman's and Joseph Henrich's chapters.)

We wish to sketch the outlines of how the theory of cultural evolution can—and in some measure already does—underpin a vibrant empirical science. We will build our case on examples of theoretical and empirical investigations, which, in the spirit of consilience, range quite widely. Most of these examples consist of studies done by social scientists who would not consider themselves Darwinian evolutionary scientists. In fact, some might quite resent the sobriquet! From the point of view of Darwinians, anything bearing on the processes by which culture changes as a function of time and circumstance is a contribution to the project. Social scientists and historians certainly observe and comment upon such changes and upon the myriad consilient things that play a role in such change. What we can hope to do is convince such scientists that all the Darwinian perspective means to do is to draw together the many threads that, for lack of a proper theory, seem like unrelated domains of investigation. The Darwinian approach simply offers a set of tools designed to analyze the puzzles presented by a special sort of change over time—change involving populations with inheritance systems such as genes or culture. We can also expect to discover gaps and weaknesses in the present corpus of social-science investigations, areas where little or no good evolutionary work has yet been done.

We divide the totality of evolutionary studies into five broad realms: logical coherence, investigations of proximate mechanisms, microevolutionary studies, macroevolutionary studies, and patterns of adaptation and maladaptation. To make the discussion more concrete, we will focus on studies that could cause us to doubt that culture is uniquely important in the human species. By this we mean the proposition introduced by Edward Tylor (1871) and especially integral to the foundations of anthropology, that a great deal of variation in human behavior is acquired from others by teaching and imitation. Much of this variation is expressed in differences between societies, although not a little exists between individuals within societies, along the lines of gender, age, social role, occupational specialization, and the like. These cultural differences have arisen by processes of cultural evolution that are crudely similar to genetic evolution, though by no means identical. By contrast, Tylor and his successors argue, the biological differences between human individuals—especially the average biological differences between major subdivisions of the species—are small in comparison with the cultural differences. In modern language, the idea is that humans have both a genetic and cultural inheritance system, and that most of the differences between humans, particularly the differences between groups of humans, are cultural rather than genetic.

We have found it difficult to determine if any contemporary scholars actually doubt that culture is important in humans (see also D'Andrade, this volume). Some scholars, such as Tooby and Cosmides (1992) and Pinker (2002) argue that the standard picture of culture is typically wedded to an oversimplified blank-slate psychology, and that the direct effects of the environment on behavior may be underestimated. Some behavioral geneticists (e.g., Rushton 2000; Herrnstein and Murray 1994) seem to be arguing that genetic differences between major human groups are more important than the pure culture hypothesis allows. The economists' rational-choice theory, if taken literally, holds that perfectly rational actors with no information constraints adjust their behavior to the prevailing environment instantaneously and autonomously. This theory may cause economists to suspect that the role of culture relative to individual decision making is overemphasized by Tylorians. Because of its theoretical sophistication, rational-choice theory has been widely adopted as a theoretical foundation by other social scientists (e.g. Ostrom 1998; McCay 2002), who usually incorporate the culture concept or something like it into their formulation, using terms such as "bounded" or "situated" rational choice. Similarly, human behavioral ecologists (e.g. many of those represented in Smith and Winterhalder 1992), whose theories often formally amount to rational choice in light of fitness maximizing goals, are generally quite willing to believe that culture is proximally responsible for most behavioral adaptations. Arthur Wolf (this volume) notes that behavior is affected by social and ecological factors as well as culture, and that the former can easily be mistaken for the latter. His critique notwithstanding, we think that cultural evolutionists are quite sophisticated about the relationship between social organization and culture (see, e.g., Henrich, this volume).

Although hardly anyone seems to disagree with the importance of culture in humans, we think the exercise of systematically laying out the evidence to support the qualitative position we have outlined is useful. No one explicitly acknowledges how various and powerful the evidence is! Framing the case in terms of an evolutionarily based taxonomy highlights the contributions that cultural evolutionary analysis has made and is prepared to make to a science of culture. Ultimately, we seek quantitative, mechanistic answers to the questions of the contributions of genes, culture, and environment to human behavior. On this score, real ignorance and legitimate controversy litter the landscape we survey. Yet, across this huge field of scientific endeavor, we see prospects for progress everywhere.

## LOGICAL CONSISTENCY

Logical consistency is the proper domain of formal theoretical models. We do not emphasize this realm in this essay, but a brief characterization of its importance is in order. For an elementary treatment of cultural evolutionary models, see Richerson and Boyd (1992). For advanced treatments, see Cavalli-Sforza and Feldman (1981) and Boyd and Richerson (1985). See also the chapters by Feldman and by Aoki et al. in this volume.

Human minds aren't well adapted to thinking through the consequences of linked population-level processes. Although critics of mathematical models often recoil at their simplicity, simple models are an effective prosthesis for a brain that is rather poor at following intricate quantitative causal pathways. Model builders are often quite surprised at their own results, even when their intuitions are well-schooled by related models. The strategy of using simple models to study complex phenomena is well developed in many scientific disciplines, and should hardly be controversial (Richerson and Boyd 1987). We think of simple models as tactical reductionism, something entirely different from supposing that real human behavior is simple enough to be accurately described by such models. The models are just tools to help us think a little harder about complex problems. Not using these tools condemns the theorist to rely entirely upon verbal arguments and intuitions that are difficult to check for logical consistency, especially when phenomena obey quantitative rather than categorical causal rules.

Population genetics–style models of cultural evolution are well enough advanced to demonstrate that they can play exactly the same role in studies of cultural evolution as in studies of genetic evolution. One of the most important properties of models is that they are not constrained much by disciplinary history or by empirical difficulties. The logical exercise of making a complete evolutionary model, even a very simple one, typically forces the modeler to cut across traditional disciplines. The simplest sorts of recursion equation models of cultural evolution incorporate some sort of model of individual psychology and some sort of model of a population. Typically, the results of such models depend both upon the properties of individuals and upon the properties of the populations they make up. Psychologists study individuals, whereas cultural populations tend to fall under the disciplines of anthropology and sociology. The models show how the findings of each discipline are relevant to the other. Social scientists have had great difficulty theorizing about the relationship between individuals and societies; sociologists refer to the "macro-micro problem" (J. Alexander et al. 1987). When they start with considerations of individuals, sociologists find it hard to drive the properties of societies; when they start with "social facts," they find it hard to make a place for individuals. The evolutionist's recursion equation formalism that integrates over individuals within a generation and then iterates over time does what sociologists have otherwise found it hard to do.

Although the amount of modeling done to date is small relative to the immense complexity of the problems of cultural evolution, the diversity of problems tackled is certainly adequate to demonstrate the power of the technique. Studies range from the analysis of specific empirical cases to quite broad analyses of the general function of the cultural system (Aoki 2001). The very fact that such models can be made and that they behave in close accord with empirical cases may not seem like much, except that people have confidently asserted otherwise. Gregory Bateson (1972: 346–63) and David Hull (in Lamarck 1984: xl–lxvi) argue that systems with inheritance of acquired variation would be dysfunctional if they existed. This is not true; models linking learning to social transmission are easy to make and have quite interesting properties (Boyd and Richerson 1989). On the other hand, Bateson and Hull point to features of a system for the inheritance of acquired variation that is indeed dysfunctional under certain circumstances; more on this point below. To take another topical example, many people argue that unless culture comes in discrete units like genes, cultural evolution on the Darwinian plan is impossible (Durham and Weingart 1997). Actually, models show that Darwinian evolution is compatible with unitless cultural variation (Boyd and Richerson 1985: 70–79). Furthermore, near-perfect replication after the fashion of genes is not necessary to create patterns of heritable variation that respond to Darwinian evolutionary forces (Henrich and Boyd 2002).

The formal models thus demonstrate that culture *can* exist and have all the basic properties claimed for it by proponents of its importance.

## PROXIMAL MECHANISMS

If culture is important in humans, we must have a psychology that is adapted to acquiring information from other individuals by teaching or imitation. Classic experiments by Bandura and Walters (1963) and Rosenthal and Zimmerman (1978) showed that children are very rapid and accurate imitators, and that they use sophisticated rules to bias their imitations of people with different characteristics.

More recently, psychologists have begun to dissect these skills in greater detail. Developmental linguistics is perhaps the most advanced field in this regard. Tomasello (1999a) reviews evidence that human caregivers and children engage in a rather complex behavior called *joint attention*. Joint attention develops at around one year of age, and Tomasello infers that it is an innate capacity. Children follow the gaze of adults and gestures such as pointing. They can readily discern the focus of an adult's attention, which makes it easy for them to associate words with specific objects. Bloom (2000) discusses the many skills and strategies such as joint attention that children deploy to learn new words. Some of these are no doubt ancient and not at all specific to language learning. Babies already have an object-recognition system built into their visual system. Spelke (1994) shows that infants use rules such as *cohesion* to classify their visual experience into conceptions of objects: pieces that move together are taken to be objects. Thus, if an experimenter shows an infant a simple cartoon of a bicycle and rider appearing together on one side of screen, and reappearing still together on the other side of the screen, the infant will treat the combination of bicycle and rider as a single object. If after several repetitions of this pattern, the bicycle and rider suddenly reappear separately, the infant reacts with surprise, and quickly learns that people and bicycles are separate objects. Later, toddlers realize that the parts of people and bicycles do not move entirely together and that arms and wheels are semi-separate parts of people and bicycles. Native Americans are said to have mistaken Spanish cavalrymen for centaur-like humanoid beasts, suggesting that cohesion has a powerful influence even in adult cognition. Presumably all animals with good vision and reasonably advanced cognition use such strategies. Other tricks seem more directly tied to speech. For example in *fast mapping*, children use simple clues to deduce part of the meaning of a novel word. For example, if asked to bring the chromium tray when the choices are a blue tray (a color word the child already knows) and an olive one (a color word unknown to the child), children will assume that "chromium" refers not only to a color, because that is the only salient difference between the trays, but also specifically to olive. They retain this theory about "chromium" for some time on the basis of a single trial. The very rapid buildup of vocabulary in middle childhood probably depends crucially on such economical heuristic guesses of the meaning of new words. Presumably, similar skills underlie the imitation of motor skills (e.g., tying knots) and the learning of social rules and customs.

Baum (1994: 105, and personal communication) argues that humans differ from other animals in our high degree of sensitivity to social reinforcement. Interacting in a friendly way with other people is pleasurable, and being subject to verbal abuse, complaints, and the cold shoulder are quite unpleasant. Our responsiveness to social reinforcement makes humans relatively easy to teach. Some human skills are far too complex to learn by simple imitation. Polanich (1995) analyzed the case of diffusion of coiled basket construction from Yokuts to Mono peoples in the Southern Sierra region of California. Mono weavers by tradition made baskets by the twining technique. In coiled baskets, the main structural element of the basket spirals outwards from the center; in twined baskets the main structural elements radiate from the center like the spokes of a wheel. To learn the difficult Yokuts technique, Mono weavers had to work side by side with an experienced coiler and learn each stitch through careful demonstration and instructional feedback. Yokuts and Mono basket weavers became intimate enough friends for the student-teacher bond to become established, and thus for coiling to be transmitted to the Mono students. By contrast the decorative designs on baskets diffused much more easily because twiners could easily represent coiler's decorative patterns in twined baskets and vice versa. Henrich and Gil-White (2001) argue that human prestige systems are substantially built on the basis of rewards that imitators and learners provide to people they take to have superior knowledge in one domain or another.

Although many animals have rudimentary capacities for social learning

(Heyes and Galef 1996), humans are much more adept imitators and teachers than any species yet tested. Two independent research groups working at Yerkes Primate Center compared young children and adult human-reared chimpanzees on the same imitation tasks (Tomasello 1996; Whiten and Custance 1996). In these studies, children of three and four were considerably more accurate imitators than chimpanzees, although chimpanzees are in turn known to be better imitators than most other species. Similarly, anecdotal accounts of chimpanzees reared by humans as though they were children report that although chimpanzees respond to social reinforcement, they do so much less than children (Hayes 1951; Temerlin 1975).

Some doubt does remain about the precise gap between humans and chimpanzees as regards social learning. Boesch (this volume) notes that field researchers observe a considerable variety of what appear to be cultural behaviors in chimpanzees and other apes, more than sometimes seems consistent with controlled laboratory studies of social learning. The observations may be misleading, the experiments may contain artifacts, or much information may pass to young apes via relatively simple systems of social learning. Experimental evidence for more sophisticated social cognition in chimpanzees continues to be found in the laboratory (Hare et al. 2001). Nevertheless, we think it implausible that what Darwin called the "great gap" between living humans and our closest relatives will ever be closed. The sheer quantity of information that humans acquire by social learning outdistances that of our relatives by perhaps a couple of orders of magnitude.

Much work remains before we understand human social learning in detail. We and others have constructed mathematical models based on plausible assumptions about the social learning strategies that people might use. For example, theory suggests that people should use conformity to the majority under a wide variety of circumstances (Henrich and Boyd, 1998). Recently, we have begun an experimental program to dissect the way people (mostly undergraduate student volunteers, but also non-Western adults) deploy social learning strategies in laboratory tasks (McElreath et al. 2005; Efferson et al. 2007). This program is in its infancy, but it has already turned up surprises. Individual variation in strategies used is high and our participants often use less social learning and less conformity than would be optimal in the tasks they are given to solve.

## MICROEVOLUTION

At the heart of Darwinism is the close study of evolutionary processes on the generation-to-generation time scale, which is susceptible to precise observation and controlled experiment. What is the prevailing rate of change of gene frequencies, and why are they changing? Weiner's (1994) description of Peter and Rosemary Grant's famous study of the microevolution of the beaks of Galapagos finches is an accessible introduction to the genre.

Most cultural change is relatively gradual, and is apparently the result of modest innovations spreading by diffusion from their point of origin to other places. Cultural evolution is a population phenomenon. Individuals invent, and observe the behavior of others. Imitation by discriminating observers selectively retains and spreads innovations. Innovations accumulate and gradually build to complex technology and social organization. Darwin described such patterns of change as descent with modification. The theoretical and empirical tools designed by evolutionary biologists to study genes are well suited to describing cultural evolution, given suitable modification.

Anyone interested in human history is a student of cultural evolution. What Darwinians bring to the table is a commitment to quantitative study of the processes of cultural evolution. In order to gain leverage to compare across cases, Darwinians seek a taxonomy of forces that affect evolution. We (Boyd and Richerson 1985) favor a taxonomy that distinguishes three forces that are highly analogous to their counterparts in genetic evolution: *random variation, drift,* and *natural selection*. In addition to these are what we call the decision-making forces, which others call cultural selection (Cavalli-Sforza and Feldman 1981; see also Feldman, this volume). The decision-making forces can in turn be broken down into the effects of individual learning—*guided variation*— and the effects of biased choices of cultural parents or of the traits they carry—*biased transmission*. Biases depend upon the decision rules the learner or teacher uses, so several distinctive types exist, even at a rather general level. If the science of cultural evolution evolves as that of genetic evolution has, new forces will be discovered and old ones will be subdivided.

A study by Hewlett and Cavalli-Sforza (1986) on the transmission of hunting techniques by Aka Pygmies in the Congo illustrates the basic strategy of microevolutionary analysis. When asked how they had learned their hunting techniques, Aka men almost always answered that they learned them from their fathers and other male relatives with whom they hunted. The exception was the making and use of highly effective crossbows. Most men had learned this technique from one of the few men who had acquired their knowledge from outsiders. In fact, most other hunting techniques were long extant in the band and were known to all adult males. Fathers normally taught their sons to hunt. Although the lack of variety in hunting techniques resulted in a conservative pattern of vertical transmission for the most part, the Aka ability to use other modes of transmission and to employ biases is evident in the spread of the crossbow.

Only a handful of microevolutionary studies have yet been designed deliberately by Darwinians to study cultural evolution (Aunger 1994; Joseph Henrich, personal communication), but many studies designed for other purposes give a good picture of cultural microevolution in domains where social scientists have in effect reinvented Darwinism. An early and very self-consciously population-based approach is Foster's classic 1960 study of the effect of the Spanish Conquest upon Latin American culture. Latin Americans acquired a small sample of Iberian cultures, partly because immigrants were few and from selected areas, partly because church and state authorities deliberately biased the culture carried over to the New World (see Durham, this volume, on "imposition"), partly because the Native Americans were selective in what they borrowed, and partly because Spanish settlers selectively adopted Native American items. Scholars have conducted more than a thousand studies of the diffusion of innovations, predominantly technical innovations (Rogers 1995; Rogers and Shoemaker 1971). These studies usually use questionnaires and are designed much like Hewlett and Cavalli-Sforza's Aka study. They provide much useful evidence on the role of decision-making forces in cultural evolution. The study of language microevolution is also a well-developed field (Guy and Labov 1996; Labov 2001; Deutscher 2005). Dialect evolution is particularly well studied. Change is often appreciable on the generation-to-generation time scale; it is studied by sampling the speech of people of different ages. A number of forces have a significant role in driving language evolution, but many of them are prestige biases of one form or another. For example, speakers of stigmatized dialects often exaggerate (hypercorrect) features of the prestige dialect that they cannot manage to master exactly. When hypercorrecters are numerous relative speakers of the prestige dialect, hyper-correction actually drives the evolution of the target dialect. Local prestige is often the most potent force. Typically, the most advanced speakers of a changing dialect are upper-working-class or lower-middle-class women with high prestige in their communities. Because dialect is not normally altered after puberty, the inference is that popular pre-pubertal girls from the middling classes are the main drivers of language evolution. Language mavens, know it and weep!

A powerful method for studying cultural microevolution is to compare two communities derived from the same parent community, using quasi-experimental designs. Walter Goldschmidt designed such a study in East Africa, from which Robert Edgerton (1971) produced the most synthetic analysis. Goldschmidt chose four societies, each of which had some communities in the humid highlands specialized in horticulture, and some communities in the arid Rift Valley specialized in raising livestock. The exact degree and timing of the separation of the communities within a tribe was not known, but was surely not very great—perhaps a few generations. Because settled farming and migratory cattle-raising have huge implications for social organization as well as subsistence, communities at either extreme should be under strongly divergent evolutionary pressures. In fact, much more of the variation in the data Edgerton collected could be explained by common tribal affiliation than by economic mode. However, in a significant minority of cases the effect of economy was significant. Often the tribe by economic-mode interaction effect was also large; even when horticultural and herding communities were diverging, they were not necessarily converging on other communities with the same economic system. Goldschmidt and Edgerton's study demonstrates both that culture has significant "inertia" and that strong evolutionary pressures lead to quite measurable change in a few generations. These are the same general microevolutionary features we find in the case of genetic evolution and this design deserves much more widespread use to understand cultural evolution than it has received (McElreath 2004).

Two related questions about cultural evolutionary process are quite controversial. Most people seem to have the intuition that most directional cultural evolution is due to decision-making processes and that natural selection (on cultural variation) has little if any part to play. We think, to the contrary, that the microevolutionary evidence for the operation of natural selection is actually rather compelling. For example, Hout et al. (2001) conducted a careful and thorough analysis of the effects of fertility and of conversion on the growth of conservative Protestant denominations relative to mainline ones in the United States. Although decision-based movements between conservative and mainline denominations explain some of the difference, three quarters of the increase in conservatives was attributable to higher birthrates between 1903 and 1973, at which time conservative and mainline birthrates converged.

The second controversial question is whether group selection can play a role in human evolution. Darwin (1874: 178–79) clearly thought that humans group selected. Several prominent modern Darwinians (Hamilton 1975; E. Wilson 1975: 561–62; Eibl-Eibesfeldt 1982; R. Alexander 1987: 169) have also given serious consideration to group selection as a force in the special case of human ultra-sociality. They are impressed, as we are, by the organization of human populations into units that engage in sustained, lethal combat with other groups, not to mention other forms of cooperation. The trouble with a genetic group-selection hypothesis is our mating system. We do not build up concentrations of intra-demic relatedness, as do social insects, and few demic boundaries are without considerable intermarriage. By contrast, theoretical models show that group selection is a more plausible process if the variation selected is cultural (Boyd and Richerson 1982; Avital and Jablonka 2000). For example, if migrants are resocialized when they enter new groups, especially if their cultural variants are discriminated against by conformist transmission, then cultural group selection requires only the social, not the physical, extinction of groups. Soltis et al. (1995) reviewed the ethnography of warfare in simple societies in Highland New Guinea. The patterns of group extinction and new group formation in these cases conform well to the assumptions of the Boyd and Richerson (1982) model. The strength of group selection in Highland New Guinea was strong enough to cause the spread, in about 1,000 years, of a favorable new social institution through a region composed of many populations. Cases of group selection by demic expansion are quite well described, for example the spread of the Southern Sudanese Nuer at the expense of the Dinka (Kelly 1985); the expansion of the Marind-anim at the expense of their neighbors by means of large, well-organized head-hunting raids (including the capture and incorporation of women and children) at the expense of their neighbors (Knauft 1993: chap. 8); and the Spanish conquest of Latin America (Foster 1960).

Everyone agrees that human behavior is very diverse. To the extent that this variation is genetic, patterns of microevolution should conform to genetic patterns. To the extent that it is a direct product of individual interactions with environments, we should not see any microevolution of the Darwinian descent-with-modification sort at all. In fact, patterns of microevolution betraying the peculiar mix of decision-making, chance, and selection that culture theorists model are commonly observed. The diffusion of linguistic and technical innovations by biased horizontal and oblique transmission are fundamental to explaining these phenomena. Group selection on cultural variation is viewed with suspicion by many evolutionary social scientists, but we believe that this is the result of a prejudice acquired from the study of evolutionary biology, where the general case against group selection is more plausible.

## MACROEVOLUTION

Understanding what regulates the rate of evolution in different times and places is one of the main tasks of macroevolutionary studies. The large scale and comparative evidence suggest that cultural evolutionary theory, and the coevolution of genes in response to novel culturally constructed environments, have important roles to play in understanding the major events in human evolution. In favorable cases, detailed archaeological sequences can be investigated directly by fitting microevolutionary models. Bettinger and Eerkens have applied microevolutionary models to the analysis of stone point evolution in the Great Basin of North America, finding contrasting evolutionary dynamics in different times and places (Bettinger and Eerkens 1997; Eerkens and Lippo 2005; see also Shennan and Wilkinson 2001). Improved statistical techniques based on fitting models representing different hypotheses directly to data make this approach promising (Burnham and Anderson, 2002).

Two approaches to the study of macroevolution are more common than direct model fitting because they cope well with the sparse data normally available about deep historical events. One is to use comparative evidence from different species or different major cultural groups to try to understand how large-scale differences have arisen. The second is to closely examine major transitions and long-time-scale trajectories of species or societies, in an effort to disentangle the causal processes involved. Precise reconstructions of the past are seldom possible, but some things are well documented in the archaeological, paleoanthropological, and paleoenvironmental records.

The simplest and best-documented evidence is the rate of change of cultural (and gross biological) features of past populations. For example, the existence of decision-making forces in cultural evolution makes the cultural evolutionary process rather more rapid than ordinary genetic evolution. At least in the case of adaptive biases, these forces will act additively with each other and with natural selection to speed cultural evolution in adaptive directions. Rates of evolution of artifacts vary dramatically over the course of hominid evolution. Only in the late Pleistocene does cultural evolution become so rapid as to create temporally and spatially localized cultures that bear a strong resemblance to ethnographically known hunter-gatherers (Donald 1991; Klein 1999). Evidently, the cultural evolutionary system became fully modernized less than 250,000 years ago, perhaps as recently as 50,000 years ago (McBrearty and Brooks 2000).

At least one of the innovations that made the complex, rapidly evolving culture of modern humans possible is the ability to form social bonds with strangers, even across ethnic divides. The case of the exceedingly simple material culture of the Aboriginal Tasmanians is an interesting test of how large a culture area must be to maintain complex culture. Tasmania was isolated from Australia by the flooding of the Bass Strait in the early Holocene. Tasmanians started with a toolkit as complex as those anywhere else in Australia, but by the time of European contact, they had the simplest toolkit ever collected by European explorers (Diamond 1978). Henrich (2004b) gives an ingenious model of the loss. If the transmission of complex culture is fairly error-prone, transmission itself will result in selective degradation of more complex items in the repertoire. Rare, highly skilled craftspeople will tend to reinvent the otherwise degrading cutting edge of a technology. The 4,000 people on Tasmania likely included too few such experts to maintain a toolkit of ordinary complexity. If Henrich's hypothesis is correct, many more were necessary, and we do know that cross-cultural trade and other contacts linked peoples together across vast distances during the late Pleistocene, but probably not earlier (Klein 1999: 470–71, 544–45). In Upper Paleolithic Europe,

Moderns moved desirable raw materials hundreds of kilometers, compared with tens of kilometers for Neanderthals, and the settlement density of Moderns was greater as well. Gravettian Venus figurines were distributed over most of Europe. To judge from the size of their brains, Neanderthals were as intelligent as Moderns at the individual level, but if they lacked the social "instincts" to maintain wide-ranging social contacts, the complexity of their toolkit would have been limited by the "Tasmanian effect," as seems to be the case. Also consistent with the Tasmanian effect is the progressive increase in the complexity of toolkits and other aspects of culture in the last 10,000 years. Agriculture boosted population densities, transport technology was transformed by the use of inanimate and animate energy sources, and writing and mathematics substantially relieved memory limitations. In the last few centuries, innovations such as cheap printing, mass education, science, and industrial espionage have further revolutionized the number of experts available to each individual.

Recent discoveries of the nature of paleoclimates are revolutionizing our understanding of the environmental stage within which human macroevolution took place (Vrba et al. 1995; Potts 1996; Alley 2000; Richerson and Boyd 2000; Calvin 2002). During the Plio-Pleistocene eras the Earth's climates have changed dramatically and often very abruptly. Ice cores taken in Greenland in the late 1980s record proxies indicating climates with huge variation on time scales of a millennium or less during the last glacial—much more than during the preceding 11,500 years. Several similarly detailed core records from temperate and tropical latitudes, taken from anoxic lake and ocean sediments, show that very similar fluctuations occurred at these latitudes as well (e.g. Martrat et al. 2004). High-resolution data for the rest of the Plio-Pleistocene will likely be produced in the next few years, for example, from the sediments of Lake Malawi (Scholz 2005).

Already, some of the climate patterns are extremely suggestive of influences on cultural evolution. The transition from highly variable glacial to much more tranquil Holocene climates was quite abrupt, and coincides with the broadening of human diets to include high-processing-cost plant foods (Richerson et al. 2001). Many human populations began an evolutionary trajectory that sooner or later led to agricultural production. All but one of the known sequences start in the Holocene. The exception is the Natufian and neighboring cultures of the Near East. These cultures flourished in the Bølling-Allerød period of warm, calm climates just before the last cold, variable period (the Younger Dryas) that ended with the beginning of the Holocene. The Natufian plant-rich system substantially decayed during the Younger Dryas, but with the return of warm calm climates, the earliest domestications began almost immediately in the Fertile Crescent.

The climate change across the Pleistocene-Holocene boundary proves to be much more complex than a poleward shift of isotherms. Last glacial climates were lower in $CO_2$, on average much drier, and much more variable on time scales from decades to millennia than Holocene climates. In the Near East, the time required for societies to progress from an initial domestication to substantial reliance on crop plants was three or four thousand years (and, of course, increases in agricultural productivity continue right up to the present). Moves in the direction of agriculture all involved focusing on a handful of key proto-domesticate species that rewarded investments in labor and skill to collect (later, cultivate) and process. Considerable shifts in other aspects of subsistence followed from sedentary life, and from the nutritional implications of eating large amounts of one or two species of plants inevitably lacking in critical amino acids, vitamins, and minerals. In a world in which year-to-year climate variation probably exacerbated the risks of depending heavily upon one or two plants and in which environments were changing very rapidly relative to the cultural macroevolutionary time scale, the evolution of a complex, productive, but risk-prone subsistence strategy seems quite unlikely. By contrast, the demographic time scale is considerably shorter than the cultural evolutionary time scale. Populations can double each generation under favorable circumstances, and they probably did grow substantially at favorable times and places in Pleistocene, only to be cut back by climate deterioration. Had agriculture been merely a response to population pressure (M. Cohen 1977), people would have presumably shifted in and out of agriculture in the highly variable ice-age climates as good times for this subsistence system alternated with hard times.

The rapidly improving climate record is important not only for identifying environmental drivers of human evolution, but also for isolating other factors that regulate the long-term rate of cultural evolution. In the Holocene, societies everywhere in the world intensified their focus on plant resources, albeit at very different rates. The progress to agriculture, and to the spinoff effects of agriculture such as political sophistication, was most rapid in western Asia and China, but much slower in some other places. In Australia no societies practiced agriculture, and only a few did in western North America. The different regions of the world in the Holocene provide a useful natural experiment to test hypotheses about the rate-limiting processes affecting cultural evolution. Diamond (1997) takes on this project. He and others have proposed more than a half-dozen major types of explanations for the regulation of rates of cultural evolution (Richerson and Boyd 2001b).

The macroevolutionary record is highly consistent with culture playing a dominant role in human behavioral variation. On one hand, cultural evolution is distinctly less rate-limited than genetic evolution. The rates of behavioral evolution of modern humans are dramatic by the standards of genetic evolution. Given decision-making systems that have an adaptive tendency, humans can run up social and technological adaptations to variable environments much more quickly than could be accomplished by the evolution of new instincts and new anatomy. The success of modern humans in mastering the highly variable last glacial environment, in the course of the Out of Africa II migration, is testimony to the power of rapid adaptation to new environments. On the other hand, cultural evolution is very far from instantaneous. It follows the pattern of descent with modification familiar from genetic evolution. Holocene climates have varied very little, and for the past 11,500 years human societies have exhibited a progressive trend toward larger and more complex societies. This trend has run at different speeds in different parts of the world, but setbacks, perhaps sometimes in response to environmental changes, have been of modest depth and duration relative to advances. The telling fact, we believe, is that no other species in the last few hundred thousand years has had a macroevolutionary pattern at all similar to ours despite being subject to the same powerful climatic forces.

## PATTERNS OF ADAPTATION AND MALADAPTATION

Darwinians are often happy to be called adaptationists, and critics of Darwinism are often suspicious of adaptive explanations. We think that these stereotypes, to whatever extent they are true, are unfortunate. Adaptation is clearly important, especially in the human case. We are the Earth's dominant organism; perhaps no single species in the history of the Earth has ever been such a stunning adaptive success. On the other hand, people do many things that certainly do not appear to be adaptive. Perhaps because our adaptation is so dominating, we have more scope for doing ridiculous things than do species hedged in by tough competitors and dangerous predators.

Adaptations and maladaptations are interesting in their own right, but both are also interesting as clues to how evolutionary processes work (Boesch, this volume). In this regard, maladaptations are typically more revealing than adaptations. Adaptations are over-determined. Genes are selected to favor them. The cultural system is evolved to favor adaptations by the action of decision-making forces as well as by selection (Durham 1991). Individual learning favors adaptive behavior via reinforcement. Maladaptations, on the other hand, often arise through the operation of a more limited set of processes. Because clearly maladaptive behaviors are relatively rare, they are more likely to have arisen by one process, and hence to give direct insight into that process (see also Durham, this volume).

The social sciences have always had functionalist schools, and the recent rise of Darwinian social science has added evolutionary psychology (Barkow et al. 1992) and human behavioral ecology (Smith and Winterhalder 1992; Barrett et al. 2002) to this list. Humans have two closely related adaptive syndromes: technology and social organization (the two are related at least in the apparent need for large social networks to maintain complex technical adaptations using an error-prone cultural transmission system, as seen in the Tasmanian Effect). Both illustrate well the overdetermination of adaptation. Take social organization. People live in societies unusually large for a primate, and cooperate, or at least coordinate, with unusually distant relatives. Evolutionists have offered many explanations for human cooperation, including indirect reciprocity (R. Alexander 1987), group selection on genes (Hamilton 1975), sexually selected display (Smith and Bliege Bird 2000), innate algorithms for detecting rule violators (Cosmides and Tooby 1989), Machiavellian intelligence (Whiten and Byrne 1988), reputation effects (Nowak and Sigmund 1998), and the cultural group-selection process we describe above. None of these proposals can easily be ruled out. Likely many, if not all, play a role in the evolution of our social systems. Our own view, which combines the results of models with the broad empirical patterns, is rather baroque (Richerson and Boyd 2001a). We do think that the data do very strongly rule in a major role for cultural evolutionary processes, but perhaps the same can be said for competing explanations. A great many interesting if difficult puzzles lie in the problem of dissecting human adaptive systems.

The case is not quite so dire in the case of technology. We have so far found no colleagues willing to dispute that the skills to make and use tools are products of cultural evolution. However, many people have the intuition that new technology is rapidly mobilized to exploit new opportunities or adaptive pressures. Boserup (1981) famously argued that agricultural innovation so closely tracks population growth that Malthusian worries about demographic crises are unfounded. If so, cultural change must rest on such powerful decision-making forces as to obviate any role for natural selection on cultural variation and to make the decision rules in individual heads much more interesting than the population-level properties of culture as a system of traditions evolving, like organisms, by a relatively gradual process. Steven Pinker (2002) and other evolutionary psychologists sometimes seem to embrace a view like this, as do human behavioral ecologists when they pursue the phenotypic gambit. One might think that the slow but generally progressive trend toward more sophisticated technology during the Holocene compels one to take seriously the inertia of the cultural evolutionary system on time scales of millennia. Indeed, students of the evolution of technology typically stress how difficult a time inventors have had in developing such simple tools as paper clips and dinner forks (Petroski 1992). However, intuitions differ in this regard. A prominent human behavioral ecologist argued to one of us not long ago that environmental change since the Holocene-Pleistocene transition will likely turn out to explain the apparently progressive trajectory of cultural evolution over this span of time.

The typically shorter list of explanations for maladaptations often highlights the operation of evolutionary processes rather neatly. Perhaps the most important example is how asymmetries between two inheritance systems reveal the power of natural selection. The presence of genes on mitochondria that are transmitted only through the female line (or genes on y chromosomes only transmitted by males) can give rise to distorted sex ratios (Hamilton 1967). Mitochondria are transmitted only by females and y chromosomes only by males (in mammals). Mitochondrial genes sometimes arise that bias the sex ratio in favor of females, and y chromosome genes sometimes arise that bias the sex ratio in favor of males. Human culture is not transmitted completely symmetrically to genes, and highly analogous effects may occur. Kumm et al. (1994) argue that male-biased sex ratios in China and India could result from selective killing of high-cost female neonates, because when the cost of carrying a fetus to term is high enough, selection will favor reduced investment in the disfavored sex, rather than a compensating increase in the number of the disfavored sex, as would occur according to a simple Fisherian argument. Actually, Skinner's (1997) huge Chinese census sample is consistent with the hypothesis that the Chinese sex ratio is shaped directly by cultural processes, with a Fisherian primary sex ratio of around 106. According to Skinner's (1997) family systems theory, the Chinese desire a family with a ratio of 2 boys to one girl, and actively shape family configurations to meet this goal. Thus, in families with excess boys—say, two, three, or four boys and no girls—male infants will be at a progressively greater risk of infanticide as the family's sex ratio climbs beyond 2:1, just as girls are when it is below 2:1. In Skinner's large sample, rather rare family configurations are sufficiently numerous to test this hypothesis quite tightly. The data support the theory. Clearly some sort of cultural evolutionary process is acting here at variance to what selection on genes normally favors.

At least two cultural evolutionary mechanisms might lead to behavior that is maladaptive from the genetic point of view. First, prestige-biased cultural transmission is potentially prone to runaway dynamics, much as in models of sexual selection, either as a pure runaway or as a handicap signaling superior cultural variants (Boyd and Richerson 1985: chap. 7). Melissa Brown's (1995) description of the necessity for Taiwanese to bind the feet of women in order for them to achieve full status as ethnic Chinese strongly suggests this mechanism, probably among other processes (see also Brown's epilogue to this volume). Only other women had the skills necessary to bind feet, and in immigrant communities on Taiwan, which were made up mostly of men, ethnically Chinese families felt that they were still savages for lack of this practice. Many of the symbolic differences between human communities, ranging from dialect to ritual, are plausibly the result of runaway or quality-signaling dynamics.

The second source of cultural maladaptations is an analog of microbial pathogens. Culture is, to a greater or lesser extent, transmitted obliquely and horizontally as well as vertically (see also Jones, this volume). Selection can in theory act upon obliquely and horizontally transmitted cultural variation, selecting for "pathological" behaviors that parasitize effort devoted to biological reproduction (and vertically transmitted culture) in favor of effort devoted to horizontal and oblique cultural reproduction (Richerson and Boyd 2005: chap. 5). For example, individuals who delay marriage in favor of continuing their educations have a higher likelihood of entering the professions, including the teaching professions. Evidence suggests also that children from small families tend to do better in school. The advent of mass education in the last couple of centuries has had the effect of making oblique transmission relatively more important than previously, giving greater scope for the process just described. Blake (1989) argues the case for small family size increasing educational attainment. Kasarda et al. (1986) detail the case for the tendency of educated women to acquire professional aspirations that conflict with their reproductive careers. The progressive decline in fertility over time in the developed West, and increasingly elsewhere, is due not so much to this conflict becoming more severe as it is to ever more women being exposed to education at a level that triggers career aspirations. Anabaptists are a rare example of a group in the developed world that takes very great care to inculcate traditional family roles for both men and women and very great care to limit children's exposure to modern prestige norms (Peter 1987; Kraybill and Olshan 1994). For example, Anabaptist groups usually restrict education to the first eight grades, and resort to parochial schools when public schools introduce movies and other forms of modern mass communication into their curriculum. Anabaptist reproductive rates remain high.

Oddly, the existence of maladaptive and neutral variation has been used to contest the Darwinian approach to social science (Sahlins 1976a, 1976b), when in fact explanations for maladaptations are the strongest arrows in our quiver! Special creation is a good account of adaptations, one that most pre-Darwinian scientists, including the youthful Darwin, subscribed to. Darwin's account was superior because it could account for maladaptation—individual variation, vestigial organs, uselessly exaggerated ornaments, and the like. To take a modern example, Hamilton's (1964) theory of inclusive fitness is basically a theory of why cooperation is drastically undersupplied relative to what would optimize fitness. A benevolent God could have designed a much more harmonious world. Interesting special cases such as human societies demonstrate the rule. Humans are a hugely successful species in large part because we can organize cooperation on a scale that no other vertebrate can accomplish. We are not especially good large-scale cooperators on an absolute scale (relative to an imaginary society of saints, say), but we are so much better than any close competitors that we succeed abundantly. Darwinians should take care not to appear to be naïve adaptationists in their quite legitimate enthusiasm for discovering exotic and interesting new adaptations.

Cultural maladaptations follow a familiar pattern. When elements of inheritance are transmitted asymmetrically, evolutionary conflicts of interest easily arise. The modern world, with its emphasis on mobility, exposure to formal education, and work outside the home, has reduced the role of parents and other family members in enculturation relative to modern roles such as teachers, coworkers, supervisors. The mass media have become important sources of influence upon children and adults. These changes have made it much easier for maladaptive cultural elements, such as those favoring low fertility and wasteful consumption of prestige goods, to spread. Parents notice how children bring colds, flus, and bad habits home from school. Only groups like the Anabaptists that seal their culture off from the influences of modernity have resisted cultural changes like the demographic transition to small families (Newson et al. 2007). We expect that many cases of evolved maladaptations will be discovered when the data are better explored. Take the prevalence of female-preferential relative to male-preferential infanticide in the ethnographic record. The human sexes have gendered subcultures, more extreme in some cultures than others. If the general theory is correct, such subcultures, especially when most highly differentiated, should lead to sex-ratio distortion. Given that males generally manage to dominate females politically and culturally, the result should be expressed as female-preferential infanticide, as in the case of the Chinese family system. The female subsystem should normally be coadapted to the dominant system by the process that Durham (this volume) terms *hegemony*. The case of Chinese women being proximally responsible for footbinding is an example of this process.

## CONCLUSION

Critics of the Darwinian approach to cultural evolution typically make categorical arguments against it (Marks and Staski 1988; Sober 1991; Fracchia and Lewontin 1999; Lewontin 2005). Their argument boils down to the claim that because culture has some properties different from genes, borrowing concepts and methods from evolutionary biology to study culture is an error. We think that the evidence we have sketched here is perfectly sufficient to answer any such critics (see also Feldman, this volume). Recursion models are a comfortable formalism to capture cultural evolutionary processes, and related empirical techniques are equally apt. The critics are certainly right that cultural evolution differs in very important ways from genetic evolution, but these differences are easy to model. Indeed, the entire theoretical project initiated by Cavalli-Sforza and Feldman's (1973a, 1973b) pioneering papers has been to explore the *differences* between cultural and genetic evolution. If the two systems were the same, the theoretical project would be trivial!

Are culture and cultural evolution massively important in the human case? At least some evidence can be supplied in all five of our consilient domains—logical coherence, investigations of proximate mechanisms, microevolutionary studies, macroevolutionary studies, and patterns of adaptation and maladaptation. The math works. The psychology of humans appears to be designed to acquire and manage a cultural repertoire. No other known animal is nearly as well equipped as humans to do this. The microevolutionary foundations of the program find support. The macroevolutionary data rather clearly show that cultural evolution is gradual, but faster—sometimes much faster—than genetic evolution. Culture is pretty clearly an adaptation-generating system that arose in response to the onset of high-frequency climate change in the Pleistocene. It is prone to a characteristic suite of maladaptations that follow from its evolutionary properties.

Such a very general conclusion about the importance of culture seems to us to be as safe as scientific generalizations ever get. The importance of culture to human behavior is about as likely to be disproved as the inverse square law. That said, the work of understanding cultural evolution is in its infancy. Newton started a revolution in physics; he didn't end it. It wasn't until the

1930s that biologists finally concluded that Darwin's nineteenth-century picture of evolution, wedded to the genetic theory of inheritance, was basically correct. Rather than ending the field of evolutionary biology, this realization set off its golden age. The same ought to be true for cultural evolution. Once we understand that cultural evolution is basically a Darwinian phenomenon, myriad questions spring to mind. Major gaps still exist in the modeling effort. For example, as far as we know, only one model exists in which cultural change is driven by a collective process (Roemer 2002). Boehm (1996), using simple societies as his examples, and Turner (1995: 16–18), with complex societies in mind, both point to the fact that people use village meetings, committees, councils, legislative bodies, constitutional conventions, academies of sciences, and other such institutions to make collective decisions, many of which constitute or cause cultural change. As regards to proximal mechanisms, no consensus yet exists even for language learning, where the most effort has been expended (Thomason 2001). Our knowledge of cultural variation is largely qualitative. No extant study of cultural microevolution yet meets standards that are routine in evolutionary biology. Descriptively, we know more about human macroevolution than about most other aspects of our evolution. In terms of understanding the processes that control the macroevolutionary record, however, we stand almost at the beginning of the field.

An interesting small problem in cultural evolution is why a Darwinian field did not emerge a century ago in social science as it did in biology. The mythical history, believed by many credulous social scientists, is that a Darwinian social science did emerge, but got entangled in genetic reductionism and racism, and suffered the deserved fate of extinction along with the horrific political regimes it patronized (Hofstadter 1945). Bannister (1979) pointed out the cartoonish nature of Hofstadter's analysis, which applied accurately to only a few Darwinians, such as Ernst Haeckel. More recently Richards (1987), focusing on psychology, and Hodgson (2004), looking at economics, have told a much different story. Darwin's own *Descent of Man* (1874) was much more sophisticated as regards culture than critics give it credit for being. Followers were active in the late nineteenth and early twentieth centuries. In psychology Mark Baldwin and in economics Thorstein Veblen successfully showed how the main problem in Darwin's own thinking, the confounding of cultural and genetic inheritance, could easily be rectified.

Neither Baldwin nor Veblen was a genetic reductionist in any sense nor a racist. Veblen is remembered mainly for his biting satire of consumer capitalism; he was much influenced by Marx. Both men made a number of other conceptual and methodological contributions. They were influential figures in their day, and yet they were, in the end, largely written out of the emerging social-science disciplines of their day. The reasons Richards and Hodgson give for their being ignored are contingent historical ones. For example, both men became entangled in sexual scandals that complicated their professional lives. In the small scientific community of the day, too few hands existed to pursue all reasonable paths, and, fateful choice by fateful choice, Darwinian social scientists lost effective proponents in every discipline, leaving the social sciences without a collection of essential tools to do their business. We must not let it happen again!

# 5

## Conditions for the Spread of Culturally Transmitted Costly Punishment of Sib Mating

KENICHI AOKI, YASUO IHARA, AND MARCUS W. FELDMAN

James George Frazer believed that the "universal" occurrence of the incest taboo vitiated the Westermarck hypothesis, for "the law only forbids men to do what their instincts incline them to do; what nature itself prohibits and punishes, it would be *superfluous* for the law to prohibit and punish" (1910, vol. 4: 97, italics added). But Frazer may have had it wrong. In this chapter we focus on brother-sister mating and describe three gene-culture coevolutionary models showing that culturally transmitted costly punishment of sibling couples is more likely to spread through an egalitarian society precisely when such punishment is superfluous. In particular, this will be true when incestuous sex is avoided due to an "innate aversion" as postulated by Westermarck (1891).

### DEFINING THE QUESTION

Westermarck (1891: 320) proposed that "there is an innate aversion to sexual intercourse between persons living very closely together from early youth, . . . as such persons are in most cases related, this feeling displays itself chiefly as a horror of intercourse between near kin." Because sibs usually grow up in the same household, this psychological effect would function to prevent brother-sister mating. However, aversion is also predicted when unrelated boys and girls are reared together, and studies on the marriage patterns of Israelis socialized in kibbutzim (Spiro 1958; Talmon 1964; Shepher 1971, 1983) and on the demographic correlates of the Chinese custom of *sim-pua* ("minor marriage"; A. Wolf 1966, 1968, 1970, 1995) seem to support this prediction.

Westermarck further proposes that this "aversion to sexual relations with one another displays itself in custom and law as a prohibition of intercourse between near kin" (1926: 80), in other words, the incest taboo mirrors, rather than conflicts with, human nature. In support of Westermarck, Lieberman et al. (2003) have recently shown that the duration of childhood co-residence with an opposite-sex individual predicts the strength of opposition to third-party sibling incest. Fessler and Navarette (2004) have replicated this finding, noting that this effect is stronger in females. Arguably, aversion might give "meaning" to (Durham, this volume) and generate "want or need which motivates" (D'Andrade, this volume) prohibition and punishment. However, the occurrence of the incest taboo is more often invoked to refute the existence of an innate aversion. Hence the memorable words of Frazer quoted above.

Murdock (1949) claimed that sex and marriage among first-degree relatives were universally prohibited. The issue of universality is something of a red herring, but much has been made of it (e.g., Lévi-Strauss 1969), so it is worth noting that well-documented exceptions are now known, e.g., full brother-sister marriages in Roman Egypt (Hopkins 1980; Bagnall and Frier 1994). With reference to the earlier studies, Minturn and Lapporte (1985: 159–60) remark "the designation of nuclear family incest as a universal taboo is based on the absence of marriages between members of nuclear families." In fact, in their sample of fifty-two societies, they found that forty explicitly prohibited sex between brother and sister. Thornhill (1991), an evolutionary psychologist who takes Frazer's criticism of Westermarck seriously, suggests that the majority of societies in the ethnographic present *do not* have rules against nuclear family incest.

In simple societies that "lack recognizable leadership roles and status differentials among adult men . . . aggressively self-interested persons may be killed with the consent or active collaboration of the community at large" (Knauft 1991: 400; see also Boehm 2000). Violators of the incest taboo are also punished, although the severity of punishment is quite variable across societies (Fox 1980). Death (including suicide) and ostracism are among the extreme secular sanctions, whereas in the more lenient societies the guilty parties may simply be ridiculed.

However, and more importantly, such disciplinary actions presumably entail a cost to the punishers as well. If, as we assume, there is no external authority to impose or enforce the taboo, this cost falls squarely on the members of society who feel obliged to punish the incestuous pair. Gates (2004: 148) describes a case among the Trobriand Islanders (Malinowski 1926) in which the male offender commits suicide and the "punisher" suffers injury. "A boy had an affair with his mother's sister's daughter. . . . Her previous boyfriend, jealous, accused him in public of incest, shaming him beyond endurance. . . . The male exogamy-breaker . . . climbed a coconut tree. . . . Accusing his enemy of driving him to death, . . . [he] leaped, and died. . . . Taking up his cause, his relatives wounded the rival." Here, the motivation for punishing is not entirely disinterested, but the incident does illustrate the possible cost of such behavior.

This suggests that punishers will be better off the fewer are the occasions requiring punishment—in other words, when incest is for the most part avoided, either because of an innate aversion or in response to the threat of punishment. We can thus say that the incest taboo is more likely to spread among equals when it, or rather the punishment, is superfluous. In this chapter we focus on brother-sister mating and describe three gene-culture coevolutionary models that formalize the above intuitive arguments. The exposition in the main text is entirely verbal except for the occasional inequality. Readers interested in the precise quantitative models are invited to consult this chapter's appendices.

## SOME PRELIMINARY CONSIDERATIONS

Before dealing with the incest taboo, we briefly review some recent data on close inbreeding with special reference to sib mating. First, it should be noted that the propensity to inbreed varies widely among species (Thornhill 1993). Close inbreeding occurs routinely in some plants and insects, but is generally rare in birds and mammals including the African apes (Pusey 1980; Goodall 1986; Stewart and Harcourt 1987). How about humans?

The prevalence of sex between members of the nuclear family has been estimated in various modern Western societies. For example, 42 of 930 (4.5 percent) adult women interviewed in San Francisco had been sexually abused by their fathers (Russell 1984; although a Finnish study, Sariola and Uutela 1996, suggests a lower incidence of father-daughter incest). In a survey of 796 New England undergraduates, 13 percent reported some type of sexual experience involving a sibling, including thirty-eight cases (4.8 percent) of attempted or actual intercourse (Finkelhor 1980). According to Bevc and Silverman (1993), 5.2 percent of Canadian undergraduates surveyed had experienced sibling intercourse.

It is well known that brother-sister marriages were common in Roman Egypt (Hopkins 1980; Bagnall and Frier 1994). The evidence comprises about 300 census returns preserved on papyri, from which pedigrees have been reconstructed. Of 121 marriages attested in these documents, 20 are between full sibs, where the female spouse is described as "my wife and sister of the same father and of the same mother" (Hopkins 1980: 320). Given the low life expectancy and the preference for younger wives, the incidence of brother-sister marriages perhaps approached the demographically possible and socially acceptable maximum.

If we accept the innate avoidance mechanism postulated by Westermarck, then we expect a low incidence of sib mating under normal conditions. Moreover, when for whatever reason(s) societal rules encourage or mandate such unions, marital dissatisfaction should result. Most ethnographic descriptions (Murdock 1949; Fox 1980), the above figures on sibling incest in modern Western societies, and Arthur Wolf's (1966, 1968, 1970, 1995) studies on minor marriage are all consistent with this prediction.

On the other hand, it is not clear how to account for the Roman Egyptian brother-sister marriages, where there is no indication that fertility was compromised. Perhaps one solution is to invoke a vertically transmitted cultural component to sexual preference, such that the offspring of sibling couples are more likely to be themselves incestuous (Aoki and Feldman 1997). According to this view, the Roman Egyptian institution of brother-sister marriage is a transient phenomenon, initiated by as-yet-unidentified socioeconomic forces. From our standpoint, the detailed dynamics of the disappearance of this custom after several centuries (Scheidel 1995) would be highly interesting.

In what follows, however, we ignore the difficulties introduced by the Roman Egyptian evidence and assume for simplicity that all humans share the same level of propensity to mate with a sib. Then, according to Westermarck (1891) and prevailing evolutionary theory, this propensity is by nature weak, and the observed low incidence of sib mating reflects an "error" rate. That is, although natural selection favors complete avoidance, a proximate mechanism to implement this ideal has not evolved. Note that Westermarck makes no claim that the mechanism he proposes is foolproof. See Aoki (2005) for an alternative view.

The reason why inbreeding in general and sib mating in particular can be disadvantageous is that the offspring produced are less viable or fertile. The decrease in fitness relative to the fitness of an outbred individual is called the inbreeding depression (Lande and Schemske 1985) and is estimated in practice by comparing survival to some arbitrary age. In humans, a direct estimate of the inbreeding depression in eighteen cases of nuclear family incest has been calculated at 29 percent (Adams and Neel 1967), although this figure is inflated by the inclusion of morbidity.

Understandably, better data are available for the inbreeding effect of first-cousin marriages. In a comparative study of thirty-eight populations, Bittles and Neel (1994) find an excess mortality of 4.4 percent in the children of first cousins relative to the children of non-consanguineous couples, across a wide range (between 3.1 percent and 39.5 percent) of pre-adult mortality values for the latter. This implies that the inbreeding depression due to first-cousin marriage in humans increases as the population mortality (presumably reflecting environmental stress) increases. In fact, it ranges between 4.5 percent and 7.3 percent. A simple method of extrapolating to the case of offspring of sibs is to multiply the above figures by four, yielding an inbreeding depression between 18 percent and 29 percent. Alternatively, by applying the method of Morton et al. (1956), we obtain an estimate between 17 percent and 26 percent.

To complete the picture, it is necessary to consider the possibility that humans in fact experience a strong sexual attraction to sibs that is held in check by the threat of punishment. This we do at the end of the chapter.

## MODEL 1: VERTICAL TRANSMISSION OF PUNISHMENT WITH EXOGENOUS MATING PREFERENCE

The basic assumptions of Model 1 are as follows:

1 The children of sibs suffer a higher mortality than do outbred children (inbreeding depression, $d$).

**2**  Mate choice is exercised by the females, and all females prefer to mate with their brothers with the same constant probability $\alpha$; a positive value of $\alpha$ implies that aversion sometimes fails to develop.

**3**  The realized frequency of sib matings depends on the availability of surviving brothers, and on whether the incestuously inclined females whose brothers die will accept unrelated males (with probability h).

**4**  The alternative memes for punishment (P) or nonpunishment (NP) of sibling couples are vertically transmitted from mother to daughter (Cavalli-Sforza and Feldman 1981). When the mother has the P meme, her daughter acquires the P meme with probability $b_p$ and the NP meme with probability $1-b_p$. Similarly, when the mother has the NP meme, her daughter acquires the NP meme with probability $b_n$ and the P meme with probability $1-b_n$. Husbands adopt the meme possessed by their wives. (Dawkins [1976: 206] defines the "meme" as a "unit of cultural transmission" that "propagate[s] itself, spreading from brain to brain." We use the term as a convenient label for any unit of information that is transmitted by social learning.)

**5**  Couples with the P meme are punishers, punishing all sibling couples of the same generation.

**6**  The fertility of sibling couples is reduced in proportion to the frequency of punishers (cost of being punished). The proportionality constant is $\pi$, where $0 < \pi < 1$.

**7**  The fertility of punishers is reduced in proportion to the frequency of sibling couples that need to be punished (cost of punishing). The proportionality constant is $\gamma$, where $0 < \gamma < 1$.

Further details, including the mathematical recursions and the additional assumptions necessary to write them down, can be found in Appendix 1.

Cultural dynamics were investigated after setting $b_n < b_p = 1$. Clearly, if there is no transmission bias ($b_n = b_p$) or there is a transmission bias in favor of the NP meme ($b_p < b_n$), the cost of punishing ($\gamma > 0$) ensures that the P meme will be disadvantaged. The P meme can achieve a majority only if $b_n < b_p$. It is a reasonable inference that the existence of an innate aversion might entail a cultural transmission bias in favor of punishment. Hence, the assumption $b_n < b_p$ may be justified based on the Westermarck hypothesis.

An additional simplifying condition, $b_p = 1$, is imposed partly for mathematical convenience, so that an equilibrium in which all individuals have the P meme (the frequency of the P meme is one) can exist. At this equilibrium, which we call $E_p$, there is a consensus that sibling couples should be punished. Thus, a society at or near $E_p$ satisfies a precondition for morality, which is "common agreement as to which behaviours are unacceptable" (Boehm 2000: 80).

Let us now look at the predictions of the model. First, mathematical analysis shows that the punishment equilibrium, $E_p$, will be locally stable if

$$1\text{-}b_n > \gamma \hat{S}_p, \tag{1}$$

where $\hat{S}_p$ is the equilibrium frequency of sib matings. Roughly speaking, $E_p$ is locally stable if the frequency of the P meme increases in the vicinity of $E_p$, where the frequency of the P meme is already high. In other words, trajectories that begin near $E_p$ converge to $E_p$. On the other hand, numerical work exemplified by fig. 5.1 and summarized in fig. 5.2 suggests that $E_p$ will also be globally stable as long as (1) holds. Global stability of $E_p$ means that all trajectories converge to $E_p$, regardless of where they begin (fig. 5.1). In this case the P meme will increase in frequency and completely replace the NP meme, no matter how low its initial frequency.

The meaning of (1) is clear. The left-hand side is the transmission bias in favor of the P meme, $b_p\text{-}b_n = 1\text{-}b_n$. The right-hand side is the cost of punishing at $E_p$, as $\gamma$ is the cost incurred per sibling couple, and $\hat{S}_p$ is the equilibrium frequency of sib matings. The punishment equilibrium, $E_p$, will therefore be stable provided the transmission bias exceeds the cost of punishing. Moreover, because $\gamma < 1$, the bias $1\text{-}b_n$ need not exceed $\hat{S}_p$.

The analysis we present here is based on the assumption that female mate choice is determined exogenously—specifically, that it is innate. According to the Westermarck hypothesis, their preference for brothers, $\alpha$, is small. In this case, $\hat{S}_p$ will be determined almost entirely by $\alpha$, and punishment will have only a negligible effect. In fact, $\hat{S}_p \approx \alpha$. Because the incidence of sib matings will be low, costly punishment need only be administered on rare occasions. Thus, (1) predicts that a small transmission bias will suffice. On the other hand, if there

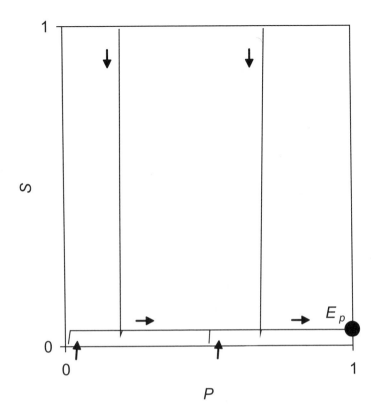

5.1. Cultural dynamics of Model 1 (exogenous preference and biased vertical transmission) viewed in the two-dimensional (P, S)-variable space. The horizontal and vertical axes represent, respectively, the frequency of the P meme (punishers), $P = S_p + R_p$, and the frequency of sibling couples, $S = S_p + S_n$. Here, as in figs. 5.3 and 5.5, we use P to denote the frequency of the P meme. At the punishment equilibrium, $E_p$, which is indicated by the solid circle, $\hat{P} = 1$ and $\hat{S} = \hat{S}_p \approx \alpha$. Parameter values are $d = 0.3$, $h = 0.5$, $\pi = 0.5$, $\gamma = 0.1$, $\alpha = 0.05$, $b_n = 0.99$. Note that in this case the trajectories, which were obtained by iteration of equations (A1) and (A2), all converge to $E_p$, which is globally stable.

is a natural instinct in favor of incest, as Frazer (1910) suggests, $\alpha$ is large. As shown in fig. 5.2, $E_p$ can be globally stable for large values of $\alpha$, but relatively large transmission biases are necessary for this to be true.

No information is available on the cultural transmission rates of punishment and nonpunishment memes, either in general or with regard to the

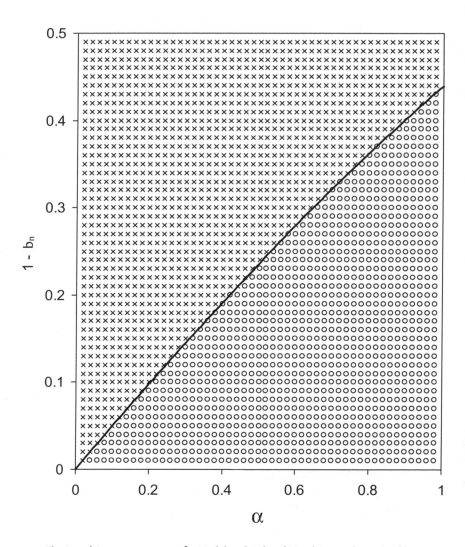

5.2. The $(\alpha, 1-b_n)$-parameter space for Model 1, showing the regions, as determined by numerical work, where $E_p$ is globally stable (crosses) and unstable (circles). The boundary of the two regions is $\gamma \cdot \hat{S}_p$, where $\hat{S}_p \approx \alpha$ when $\alpha$ is small. When $E_p$ is unstable, there exists a stable polymorphic equilibrium. Other parameter values are $d = 0.3$, $h = 0.5$, $\pi = 0.5$, $\gamma = 0.5$.

specific case of sib mating. Neither are there any quantitative estimates of the cost of punishing. We therefore do not know whether the inequality (1) is satisfied. We can only claim that a small cost of punishing, by placing fewer restrictions on the transmission rates, facilitates the spread of the punishment meme. Nevertheless, we feel justified in suggesting that the cultural origins of the taboo against sibling incest can profitably be sought in the biological lack of inclination (small $\alpha$) postulated by Westermarck (1891), as this implies a small cost of punishing.

## MODEL 2: HORIZONTAL CONFORMIST TRANSMISSION

The preceding arguments are based on the assumption that the memes are transmitted vertically from mother to daughter (with bias), and then horizontally from wife to husband (without bias). In traditional societies, the cultural transmission of established skills is usually vertical, but innovations may spread horizontally (Cavalli-Sforza and Feldman 1981; Hewlett and Cavalli-Sforza 1986; Shennan and Steele 1999). Henrich (2001) argues that the asymmetrical S-shaped adoption curves often described in the literature on diffusion of innovations (e.g., Rogers 1995) are consistent with horizontal conformist transmission. Conformity is a type of transmission bias in which the majority (frequency greater than one-half) meme is preferentially copied.

Model 1 can be rewritten with horizontal conformist transmission replacing biased vertical transmission. A new parameter, $c$, is defined (see Appendix 5.2) that measures the amount of attention paid to frequency-dependent cues in choosing between alternative memes. The resulting Model 2 predicts two equilibria: the punishment equilibrium, $E_p$ (in which the frequency of the P meme is one), and the nonpunishment equilibrium, $E_n$ (in which the frequency of the P meme is zero). Here, in parallel with (1), and assuming a small exogenous mating preference, $\alpha$, we find that $E_p$ will be locally stable if

$$c > \gamma\alpha. \tag{2}$$

Clearly, (2) has the same form as (1), with the horizontal conformist bias, $c$, replacing the vertical bias, $1-b_n$, and with the right-hand side representing the cost of punishing. However, in contrast to Model 1 the second equilibrium, $E_n$,

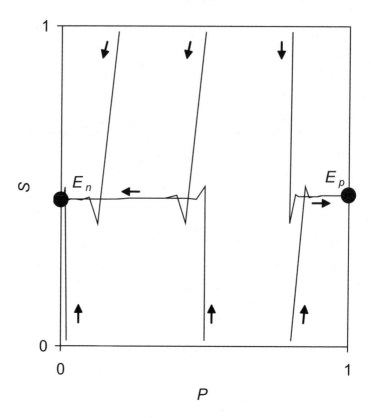

5.3. Cultural dynamics of Model 2 (exogenous preference and horizontal conformist transmission) viewed in the two-dimensional (P, S)-variable space. Parameter values are $d = 0.3$, $h = 0.5$, $\pi = 0.5$, $\gamma = 0.2$, $\alpha = 0.5$, $c = 0.5$. In this case the trajectories, obtained by iteration of equations (A4), (A5), and (A2), converge to two alternative locally stable equilibria, $E_p$ or $E_n$, depending on where they start. In particular, $E_p$ cannot be reached from small initial values of P.

will always be locally stable. Note that there can be two locally stable equilibria, because the conformist bias works both ways, to either increase or decrease the frequency of the P meme (punishers); trajectories may converge either to $E_p$ or to $E_n$, depending on the initial frequency of the P meme (fig. 5.3). As a result, the NP meme will be maintained if its frequency is sufficiently high, and an initially rare P meme will not be able to spread. Numerical work suggests that the condition for local stability of $E_p$ becomes more stringent as $\alpha$ increases.

However, if we step outside Model 2, whose dynamics are entirely deterministic, and permit random fluctuations in meme frequencies to occur, we recognize the possibility, albeit small, that the P meme may achieve an appreciable frequency simply by chance. Once the random fluctuations carry the frequency of the P meme over a certain threshold, the deterministic cultural dynamics as defined by Model 2 will, when $E_p$ is locally stable, cause a further increase in frequency until $E_p$ is reached. Hence, given an array of partially isolated human societies, horizontal conformist transmission entails that in a small minority of these societies the frequency of the P meme will approach $E_p$, whereas in the great majority it will remain at or near $E_n$.

If societies that punish sibling couples do better than those that tolerate them (because of lower mortality or higher fertility of their members), cultural group selection may cause a gradual increase in the relative number of societies that punish sibling couples (societies at or near $E_p$). As a result, the taboo against sib matings may be established in the majority of societies and may even become a cultural universal. Readers will have noticed that here we are applying Boyd and Richerson's (1982, 1985) argument for the evolution of cooperation. Their model includes a cooperative equilibrium and a non-cooperative equilibrium, both of which are locally stable. Boyd and Richerson make the reasonable assumption that societies at or near the cooperative equilibrium have a competitive advantage over those at or near the noncooperative equilibrium. However, unlike the case of cooperation, it is unclear how a society would benefit from the regulation of sib matings (see, e.g., Hammel et al. 1979).

## MODEL 3: THREAT OF PUNISHMENT

We have so far assumed that female mate choice is unaffected by the threat of punishment. Where sexual attraction exists, even the possibility of premature death may not be a sufficient deterrent, as the current AIDS epidemic would seem to attest. On the other hand, the Roman Egyptian evidence suggests that sexual preferences are malleable in the face of social pressures (see also Graziano et al. 1993). (In a similar vein, the threat of punishment appears to be immediately effective in inducing cooperation [Fehr and Gächter 2002].) This possibility is certainly worth investigating, especially as there is a simple way to incorporate the threat of punishment into Model 1: we assume that the preference for brothers, $\alpha$, is reduced in direct proportion to the frequency of punishers—see equation (A3) in Appendix 1. Then $\alpha$ becomes an endogenous variable that decreases monotonically with the frequency of punishers and that attains its maximum value, $\alpha_0$, in the absence of punishers.

When modified in this way, the model indicates that sib matings are absent at the punishment equilibrium, $E_p$, where the frequency of punishers is one. This result is contrary to fact—the data discussed above indicate a low but non-negligible incidence of sibling sex in the United States and Canada despite its prohibition (Glendon 1989). Nevertheless, the model may be acceptable as a first approximation. We also find that $E_p$ will be locally stable provided $b_n < 1$, which is true by assumption. The reason why $E_p$ is always locally stable is that at this equilibrium the threat of punishment has obviated the need for actual punishment.

Numerical work also suggests that $E_p$ will be globally stable if

$$4(1-b_n) > \gamma\alpha_0. \tag{3}$$

When (3) is reversed, the same numerical work suggests that there will be a culturally polymorphic equilibrium, $E_{pn}$, at which punishers and nonpunishers coexist and which is also locally stable (fig. 5.4). In this case, the punishment equilibrium, $E_p$, will no longer be globally stable, and trajectories may converge to $E_{pn}$ instead of to $E_p$.

In Model 3 it is possible to regard the parameter $\alpha_0$ as measuring an innate female predisposition to mate with a brother. If $\alpha_0$ were large, this would indicate that the incest taboo does forbid "[a crime] which many men have a natural propensity to commit" (Frazer 1910, 4: 97). Sib matings are avoided not because of an innate aversion, but rather because of fear of punishment. We know that even in this case the punishment equilibrium, $E_p$, is locally stable. However, if $\alpha_0$ is large enough relative to the transmission bias that inequality (3) is reversed, an initially rare P meme cannot completely replace the NP meme, because the cultural dynamics will drive the frequency of the P meme to the locally stable polymorphism, $E_{pn}$ (fig. 5.4). Only when $\alpha_0$ is small enough that inequality (3) is satisfied, rendering the threat of punishment superfluous, will the incest taboo be accepted by all members of society, as all become punishers. Once again, we note the linkage between the cultural origins of the taboo and the biological lack of inclination (small $\alpha_0$).

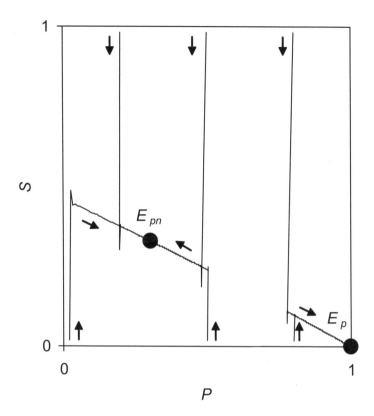

Figure 5.4. Cultural dynamics of Model 3 (threat of punishment and biased vertical transmission) viewed in the two-dimensional (P, S)-variable space. Parameter values are $d = 0.3$, $h = 0.5$, $\pi = 0.5$, $\gamma = 0.1$, $\alpha_0 = 0.5$, $b_n = 0.99$. Inequality (3) is not satisfied, and the trajectories, obtained by iteration of equations (A1), (A2), and (A3), converge to one of two alternative locally stable equilibria, $E_p$ or $E_{pn}$, depending on where they start. Note that $E_p$ cannot be reached from small initial values of P.

## APPENDIX 5.1. TECHNICAL DETAILS OF MODELS 1 AND 3

Let $S_p$ and $R_p$ be the frequencies of sibling couples and outbreeding couples that have the meme for punishment of sibling couples of the same generation (P meme). Similarly, let $S_n$ and $R_n$ be the frequencies of sibling couples and outbreeding couples that have the nonpunishment (NP) meme. Punishment causes sterility, or, equivalently, death before reproduction, in a fraction

$\pi(S_p+R_p)$ of the sibling couples. Couples who punish also incur a cost such that a fraction $\gamma(S_n+R_p)$ suffers sterility. Each fertile couple gives birth to one girl and one boy.

The P and NP memes are vertically transmitted from mothers to their daughters. The P meme is transmitted with probability $b_p$, and the NP meme with probability $b_n$. Husbands adopt the meme possessed by their wives (unbiased horizontal transmission). Newborns of the next generation are either inbred (derived from sibling couples) or outbred (derived from outbreeding couples), and belong to a sibship whose female member has either the P meme or the NP meme. The frequencies of inbred sibships with the P meme, outbred sibships with the P meme, inbred sibships with the NP meme, and outbred sibships with the NP meme are denoted $X_p, Y_p, X_n$ and $Y_n$, respectively. Then

$$FX_n = \{b_nS_n+(1-b_p)S_p[1-\gamma(S_n+S_p)]\}[1-\pi(S_p+R_p)], \tag{A1a}$$
$$FY_n = b_nR_n+(1-b_p)R_p[1-\gamma(S_n+S_p)], \tag{A1b}$$
$$FX_p = \{(1-b_n)S_n+b_pS_p[1-\gamma(S_n+S_p)]\}[1-\pi(S_p+R_p)], \tag{A1c}$$
$$FY_p = (1-b_n)R_n+b_pR_p[1-\gamma(S_n+S_p)], \tag{A1d}$$

where

$$F = 1-(S_n+S_p)(S_p+R_p)(\gamma+\pi-\pi\gamma S_p). \tag{A1e}$$

Mating is contingent on survival to reproductive age. Each inbred individual, male or female, has a viability of $1-d$ relative to an outbred individual, and deaths occur independently. Each surviving female wants to mate with her brother with probability $\alpha$, but can do so only if he also survives. If he dies, she mates with an unrelated male with probability $h$. The remaining fraction $1-h$ of incestuously inclined females who have lost their brothers refrains from mating. Then the frequencies of couples in the next generation are

$$TS'_n = [X_n(1-d)^2+Y_n]\alpha, \tag{A2a}$$
$$TR'_n = X_n(1-d)(\alpha dh+1-\alpha)+Y_n(1-\alpha), \tag{A2b}$$
$$TS'_p = [X_p(1-d)^2+Y_p]\alpha, \tag{A2c}$$
$$TR'_p = X_p(1-d)(\alpha dh+1-\alpha)+Y_p(1-\alpha), \tag{A2d}$$

where

$$T = 1 - d[1 + \alpha(1-d)(1-h)](X_n + X_p), \qquad \text{(A2e)}$$

and the primes indicate values in the next generation.
We incorporate the threat of punishment by setting

$$\alpha = \alpha_0(1 - S_p - R_p), \qquad \text{(A3)}$$

where $\alpha_0$ is a constant.

## APPENDIX 5.2. TECHNICAL DETAILS OF MODEL 2

Each daughter inherits her mother's meme, whence we set $b_p = b_n = 1$ in (A1a)–(A1d). Then, using a superscript $v$ to distinguish the frequencies of the four sibships after vertical transmission, we have

$$FX_n^v = S_n[1 - \pi(S_p + R_p)], \qquad \text{(A4a)}$$
$$FY_n^v = R_n, \qquad \text{(A4b)}$$
$$FX_p^v = S_p[1 - \gamma(S_n + S_p)][1 - \pi(S_p + R_p)], \qquad \text{(A4c)}$$
$$FY_p^v = R_p[1 - \gamma(S_n + S_p)]. \qquad \text{(A4d)}$$

The normalization factor F is given by (A1e).
Vertical transmission is followed by horizontal conformist transmission, which we define as

$$X_n = X_n^v + cX_n^v X_p^v[(X_n^v + Y_n^v) - (X_p^v + Y_p^v)], \qquad \text{(A5a)}$$
$$Y_n = Y_n^v + cY_n^v Y_p^v[(X_n^v + Y_n^v) - (X_p^v + Y_p^v)], \qquad \text{(A5b)}$$
$$X_p = X_p^v + cX_n^v X_p^v[(X_p^v + Y_p^v) - (X_n^v + Y_n^v)], \qquad \text{(A5c)}$$
$$Y_p = Y_p^v + cY_n^v Y_p^v[(X_p^v + Y_p^v) - (X_n^v + Y_n^v)]. \qquad \text{(A5d)}$$

Note that the terms in square brackets are differences in the frequencies of sib-ships with the P and NP memes, which explains the pressure to change.
New couples are formed according to (A2a)–(A2e).

## NOTE

We thank Sam Bowles, Melissa Brown, Hill Gates, Joseph Henrich, Hisakazu Hirose, Jamie Jones, Lei Tsuyuki, and Arthur Wolf for suggestions. This research was supported in part by the 21st Century Center of Excellence Program of the Department of Biological Sciences and the Department of Biochemistry and Biophysics, University of Tokyo; and by NIH grant GM28016 to MWF.

# 6

## Sexually Transmitted Infections as Biomarkers of Cultural Behavior

### JAMES HOLLAND JONES

The process of cultural transmission has been likened to the dynamics of an infectious disease. However, cultural transmission models are rarely formulated in terms of epidemic models. In this chapter, I utilize a formal framework for modeling an infectious disease to explore the dynamical implications of population heterogeneity. This approach shows that asymmetric preferences for interaction have a profound effect on the transmission dynamics and behavior of the aggregate population. Empirical research demonstrates that symbolic markers of culture (such as ethnicities) can have a strong effect on the probability of interaction and cultural transmission. The classic anthropological case of such biases is rules for endogamy, and in fact a marriage table can be used to estimate mixing preferences which then become parameters in the system of equations describing transmission dynamics. Conversely, culture models can also inform the study of epidemics. Taking this approach in analyzing recent work on the transmission dynamics of sexually transmitted infections (STIs) reveals that ethnic preferences for sex partners slow epidemic spread, reduce the overall prevalence of an STI, and can produce differential prevalence of these STIs by ethnic group. Culture matters for STI epidemiology, and the formalism of epidemiology provides valuable tools for exploring cultural dynamics.

### CULTURE AND CULTURAL TRANSMISSION

Because the transport and transmission of pathogens occurs on the substrate of human social interactions, the epidemiology of infectious diseases provides a valuable data source for exploring models of human behavior. STIs constitute one particularly striking class of infections. Unlike diseases such as measles or influenza, STIs must be transmitted by intimate contact between two individuals. Patterns in the epidemiology of STIs potentially reveal systematic patterns of human behavior. The trick, of course, is first to understand how these arise.

One clear pattern that has emerged in STI epidemiology in the United States is the differential prevalence of STIs by ethnicity. African Americans, and to a lesser extent, Hispanic Americans, experience rates of STI infection nine to forty times greater than those of non-Hispanic whites. In 2001, 75 percent of reported gonorrhea cases were in African Americans—a prevalence approximately twenty-seven times that of non-Hispanic whites (Centers for Disease Control and Prevention 2001b). Similarly, 62 percent of primary syphilis cases were in African Americans—a prevalence 15.7 times that of non-Hispanic whites. The prevalence of gonorrhea and syphilis in Hispanics, although not as high as among African Americans, was approximately three times as great as the prevalence among non-Hispanic whites.

A variety of hypotheses have been advanced to explain these disparities, including differential access to health care or health care–seeking behavior, differences in sexual behavior between groups, and researcher bias. Recent research has suggested that features of sexual networks—in particular, differential sexual contact by race/sex—can explain these differences (Laumann and Youm 1999). Structured epidemic models in which mixing is posited as preferential can predict epidemic trajectories with large prevalence differentials. Such models can also project much smaller epidemics with slower early growth and slower eventual endemic equilibrium.

Work on STI epidemic models with behavioral heterogeneity has focused primarily on activity level (e.g., number of sexual partners in the last year) as the structuring variable (Hethcote and Yorke 1984; Gupta et al. 1989). However, people do not usually choose their sexual partners by activity level, and although mixing on the basis of activity level clearly has a substantial marginal impact, public-health interventions are directed at people based on their social and demographic attributes, such as sex, age, and race. Overwhelmingly, people choose sexual partners on the basis of a small number of these attributes (Morris 1991; Laumann et al. 1994; McPherson et al. 2001). Morris (1991) and Morris and Dean (1994) used attribute-based mixing models to explore the dynamics of a growing AIDS epidemic among homosexual men. For the most part, however, the dynamics of differential STI prevalence have not been explored using attribute-based models. Accurately ascertaining individuals' sexual-activity level has proven to be a tremendous challenge in STI epidemiology (Stoner et al. 2003). The development of accurate attribute-based mixing models therefore represents a major goal in this field.

Simple epidemic models with structured mixing are, unfortunately, quite limited for realistically modeling STIs in a general population, but they have provided tremendous insights into the qualitative behavior of STI epidemics and have led to the recent development of the next generation of more sophisticated techniques. These models have served as an important starting point for investigating the dynamics of STI epidemics; more germane to the present discussion, they may prove useful for modeling the spread of cultural innovations, which has been likened to infectious-disease epidemics.

## Defining Culture

Following Boyd and Richerson (1985), I define culture as socially transmitted information that can affect the phenotype. According to this definition, the locus of culture is the act of information transmission. Transmission-centered definitions of culture lend themselves to formal treatments and avoid problems with the reification of orthodoxy that potentially beset shared-knowledge definitions (Goodenough 1957).

With the nexus of interest placed in the process of transmission, culture becomes a question simultaneously of individual psychology and of population dynamics (Henrich and McElreath 2003). This is similar in spirit to the "epidemiology of representations" suggested by Sperber (1985), and I will take this metaphor quite literally later in my analysis. One of the great advantages of Boyd and Richerson's approach to culture is that richly developed formal models exist for examining the dynamics of transmission-based cultural information (Cavalli-Sforza and Feldman 1981; Boyd and Richerson 1985).

Heterogeneity is frequently included in models of cultural transmission, and indeed is often fundamental for the establishment of certain equilibria (McElreath et al. 2003). Heterogeneity in cultural-transmission models is typically either spatial or temporal. Here I examine cultural heterogeneity, or, more precisely, heterogeneity in cultural markers such as ethnicity. Biases for (or against) interaction between different subpopulations will affect the dynamics of transmission and, potentially, the ultimate equilibria for the system. This phenomenon is well-known in the game-theoretic literature. For example, cooperative equilibria can evolve in structured systems in which agents can choose their interaction partners (Killingback and Studer 2001; Hauert 2002).

I should note that the replicator dynamics of the models do not describe the cultural markers (in this case, ethnic identities) themselves. Here I take ethnic identity as given and the mixing preferences as fixed. In principle, these preferences can be dynamic (Morris 1991).

## Recursion Models of Cultural Transmission

The dominant framework for modeling cultural transmission is, through analogy to population genetics, a discrete-time recursion equation relating the frequency of a cultural variant in generation $t$ to the frequency in generation $t+1$ (Cavalli-Sforza and Feldman 1981; Boyd and Richerson 1985).

## Epidemic Models of Cultural Transimission

Cavalli-Sforza and Feldman (1981: 32–33) note that "another biological model . . . may offer a more satisfactory interpretation of the diffusion of innovations. The model is that of an epidemic." Cavalli-Sforza and Feldman apparently chose not to pursue epidemic models in detail because at the time of their writing the rigorous mathematical theory of epidemics was still underdeveloped, whereas the theory of population genetics was richly developed and broadly understood. This theory deficit has since been remedied (R. Anderson and May 1991; Diekmann et al. 1991; Dietz et al. 1993; Diekmann and Heesterbeek 2000; Heesterbeek 2002).

The basic models for an infectious disease are the closely related SI (Susceptible-Infected) and SIR ("Susceptible-Infected-Removed") models,, which consist of systems of nonlinear ordinary differential equations (Bailey 1975). By analogy with physiological systems, these models are frequently referred to as "compartmental models" (Jacquez 1996). For the SI model, the compartments are the subsets of susceptible (S), infected (I) in the population; the SIR model adds the category of removed (R) individuals, for which the term "removed" typically refers to recovered individuals who, following infection, have acquired lifelong immunity. In the SI formulation, infected individuals remain infected for life. More complex models that allow for latent or partially immune classes are also easily developed (R. Anderson and May 1991).

Epidemic models have several conceptual advantages over discrete-generation recursion models for modeling the spread of cultural innovations. In a discrete-generation model, the frequency of an allele in generation t is a function of its frequency in generation t-1 and of its relative fitness. We usually expect that the dynamical behavior of a cultural variant will in fact be a function of both the cultural models (i.e., the "teachers") and the culturally naïve segments of a population. In the terminology of compartmental modeling, cultural dynamics are easily conceived as "donor-acceptor controlled systems" (Jacquez 1996). The biased transmission models advocated, for example, by Henrich (2001) are in fact donor-acceptor controlled models of cultural dynamics. In the marketing literature on diffusion of innovations, these are known as "internal influence" models (Mahajan and Peterson 1985).

The process of culture change—particularly with regard to horizontal transmission—is continuous, and the modeling framework should reflect this. A particularly problematic feature of discrete-time recursion models is the fact that they require cultural adoption to occur at some fixed point in the life cycle—typically at birth. Cultural dynamics are characterized by overlapping (cultural) generations, making the interpretations of discrete-time models for innovations (as opposed to intergenerational transfer) difficult.

Early formal treatments of the dynamics of cultural transmission, in fact, used infectious diseases to test models' predictions. For example, kuru in New Guinea and hepatitis B in Taiwan were both used as examples in Cavalli-Sforza and Feldman (1981). However, the epidemiology of both of these diseases suggested vertical transmission, making a discrete-generation model appropriate.

Suppose that a population of N individuals comprises S susceptible, I infected, and R recovered individuals, where $N=S+I+R$. The dynamics of the classical SIR model are given by the following set of three differential equations:

$$\dot{x} = -\beta xy \tag{1}$$
$$\dot{y} = \beta xy - vy \tag{2}$$
$$\dot{z} = vy, \tag{3}$$

where $x = S/N$, $y = I/N$, $z = R/N$, $\beta$ = effective contact rate, and $v$ = removal rate. Because the model assumes a constant population size N, the dynamics are fully described by the first two equations for both the SI and SIR models.

The effective contact rate is a composite parameter that conflates the mean contact rate $c$ of the population (assumed to be the same for all members) and the probability of infection $\tau$ (i.e., the transmissibility of the pathogen). It can be thought of as a rate parameter establishing the scale of contacts that are made at random between susceptibles and infecteds ($\propto xy$).

Epidemic behavior is a function of a threshold parameter known as the basic reproduction number, $R_0$ (R. Anderson and May 1991). The basic reproduction number is the expected number of secondary infections produced by a typical single infection in a completely susceptible population. To determine whether a pathogen introduced into a population modeled by the system of equations (1)–(3), we solve the inequality $dy/dt>0$. The invasibility criterion is satisfied when $R_0>1$. The basic reproduction number for the simple model of (1)–(3) is given by the relation

$$R_0=\frac{\beta x}{v} \approx \frac{\beta}{v}. \tag{4}$$

Because the number of infecteds is assumed to be initially very small ($x(0)\approx1$), the relationship in (4) provides the intuitively appealing result that the spread of a disease is predicated on its infectiousness exceeding its removal rate (i.e., $\beta/v>1$). The definition of $R_0$ has a further intuitively appealing interpretation. The effective contact rate conflates activity and transmissibility: $\beta = c\tau$. Note also that $1/v$ is the average duration of infection before removal. $R_0$ is therefore the product of three terms: the average contact rate, the transmissibility, and the duration of infectiousness.

The system of equations defined by (1)–(3) is highly stylized. Three clear weaknesses include

1  Individuals are homogeneous within compartments. The assumption of mass action is required. This translates into the classic population-genetics assumption of "mating at random." For some infectious diseases spread by casual contact (e.g., influenza, measles), this may be a rea-

sonable assumption. However, for infections transmitted through intimate contact, it is inappropriate.

2  There is no demography. The population is static, with neither births nor deaths. This assumption translates into impossibility of an endemic equilibrium. For SIR models without demography, there are no equilibria with a nonzero number of infecteds.

3  All rates are constant. Both transmission and recovery depend in many circumstances on attributes of the individual (e.g., age, sexual activity, class, sex, duration of infection).

There are numerous refinements that can turn this general system into a practical framework for exploring cultural transmission dynamics. Here I focus on addressing the first two of the weaknesses listed above.

*Incorporating Heterogeneity into the Epidemic Model*

For the SIR model to be a useful tool for analyzing the transmission of culture and of most diseases, it needs to include heterogeneity in behavior as well as the vital dynamics of the population. The standard means of incorporating behavioral heterogeneity into a model is through the use of mixing matrices (Gupta et al. 1989), which are conceptually identical to ethnological marriage tables (Romney 1971). Table 6.1 presents Romney's marriage data gathered in 1964 for three barrios in the village of Aguacatenango in Chiapas, Mexico.

In epidemiology, as in anthropology, it has long been recognized that people have a tendency for endogamous or homophilous sexual and economic unions. Individuals tend to form partnerships with like individuals on any of a number of social-attribute dimensions (McPherson et al. 2001). For example, it is well documented that people in ethnically heterogeneous populations generally form ethnically concordant partnerships (Laumann et al. 1994). Similarly, there are strong patterns of assortative pairing on age and social status (Morris 1995). Note that in table 6.1, the diagonal elements tend to be larger than off-diagonal elements.

To deal with these empirical observations, Gupta et al. (1989) defined a quantity Q to measure the strength of endogamy. For random mixing, $Q = 0$,

**Table 6.1. Aguacatenango marriage data**
(A1, A2, and A3 are different barrios)

|        | Men |     |     |
| Women  | A1  | A2  | A3  |
| ------ | --- | --- | --- |
| A1     | 46  | 8   | 2   |
| A2     | 6   | 24  | 13  |
| A3     | 1   | 5   | 8   |

SOURCE: Romney 1971

whereas in perfectly assortative mixing, $Q = 1$ and in perfectly disassortative mixing, $Q = -1/n$. As a measure of assortativeness, $Q$ has a number of substantial weaknesses, notably, largely unknown statistical properties and an inability to deal with differential homophily.

It is clear that empirical mixing matrices are frequently asymmetric, and that homophily is differential (Morris 1995). That is, the tendency to form ethnically concordant partnerships differs among ethnic groups. A measure of endogamy that cannot account for differential homophily is likely to lead to poor predictions of both the dynamics and the equilibria of a system. All groups display a degree of endogamy, but some groups are more endogamous than others. Furthermore, the likelihood of forming endogamous unions varies by gender and probably by other attributes.

Other measures of selective mixing have been proposed, such as arbitrary mixing functions (Jacquez et al. 1988; Blythe et al. 1991), but even though they allow for the specification of complex mixing, these models still suffer from some of the same problems as $Q$. Specifically, there is no inferential basis for calculating the parameters from data, and the parameters of the models lack ready interpretation. In addition to these difficulties, such solutions tend to be highly over-parameterized, further complicating model identification and interpretation (Altmann and Morris 1994).

Morris (1991) has suggested using loglinear models to estimate mixing parameters for heterogeneous epidemic models. Loglinear models have long been used in anthropological research to describe patterns of endogamy (Romney 1971). Mixing preferences estimated from empirical marriage tables can be distinguished from population constraints. Loglinear models are fully stochastic representations of a variety of discrete phenomena, and the parameters derived from them have well-understood statistical properties (Agresti 1990). Loglinear model estimates of mixing preferences can easily be incorporated into the compartmental modeling framework for epidemic diseases.

Loglinear models have the additional benefit that they apply to multiway tables, allowing interactions across multiple dimensions to be modeled (Agresti 1990). Bayesian extensions are also easily accommodated, allowing information from, for example, previous studies to be combined in a systematic and rigorous way (Fienberg et al. 1999).

A loglinear model is a generalized linear model for counts taken to be Poisson-distributed, and it uses a logarithmic link function. The logarithm of the expected cell count in a (possibly multiway) table is a function of the row and column marginal totals together with all the relevant interactions. For a two-way table this is, in symbols,

$$m_{ij} = \mu + \alpha_i + \beta_j + \upsilon_{ij} \qquad (5)$$

where $m_{ij}$ is the logarithm of the ijth cell, $\mu$ is the overall mean, $\alpha_i$ is the row effect and $\beta_j$ indicates the column effect. For more details, see Morris (1991). A more general discussion of models for categorical data can be found in Agresti (1990), and for generalized linear models can be found in McCullagh and Nelder (1989).

From an observed empirical matrix of partnerships, we can estimate the group-specific effective contact rates, and substitute these into the dynamical system to explore numerically its behavior (see Appendix A). This approach has, in addition to the advantages discussed above, the benefit that $\beta_{ij} \neq \beta_{ji}$, allowing asymmetric transmission. We expect transmission to be more efficient from some people to others. In epidemiology, for example, we know that the probability of HIV transmission is much greater from men to women than it is from women to men through vaginal intercourse (Fischl et al. 1987; Peterman et al. 1988; K. Johnson et al. 2003). Similarly, bacterial STIs appear to be much more efficiently transmitted by men to women than vice-versa in vaginal intercourse.

We can incorporate both transmission heterogeneity and demography into an epidemic model in which infection is not permanent, nor is lifelong immunity acquired through infection (such as for the bacterial STIs). This model is known as a Susceptible-Infected-Susceptible (SIS) system. Denote the fraction susceptible (or naïve) in the ith category by $x_i$ and the fraction infected (or the fraction of models) by $y_i$ so that

$$\sum_i x_i + y_i = 1 \tag{6}$$

$$\dot{x}_i = -x_i \sum_j \alpha_{ij} \tau_{ij} y_j + (\mu_i + v_i) y_i \tag{7}$$

$$\dot{y}_i = x_i \sum_j \alpha_{ij} \tau_{ij} y_j - (\mu_i + v_i) y_i, \tag{8}$$

where $\alpha_{ij}$ is the mixing preference of the ijth partnership class, $\tau_{ij}$ is the probability of i contracting the infection from j, $\mu_i$ is the birth/death rate of i individuals, and $v_i$ is the recovery rate for i individuals.

## STI TRANSMISSION IN HETEROGENEOUS POPULATIONS

### "Race" Is Cultural

Race is not a useful biological concept (R. Brown and Armelagos 2001). Different morphological traits or suites of traits are nonconcordant with respect to racial categories. Starting with Lewontin (1972), a variety of molecular genetic studies have consistently indicated that the great majority of human genetic variation is contained at the level of the individual, with race accounting for less than 10 percent of the measured variation. Despite its status as a biological nonexplanation, however, race continues to occupy an untenably important position in biomedical and public health research (Hammonds 2003).

The argument that the pronounced differences in prevalence of a variety of STIs are due to biological factors attributable to race (e.g., differential transmissibility or susceptibility) is therefore implausible. However, although race may lack biological utility, human behavior is clearly biased with respect to perceptions of racial categories. The social category of "race" may be used as a marker for structuring sexual behavior (Laumann et al. 1994). Furthermore, attitudes toward race are probably learned. These attitudes include both positive/negative feelings of in-group as well as out-group identity (Cameron et al. 2001; Kowalski and Lo 2001).

The aggregate effects of individual preferences for the race of sex partners drive the differential homophily observed in mixing matrices, and help to form the epidemiological patterns observed: differential prevalence, epidemic tempo, and overall population prevalence.

## Data

I demonstrate the differential homophily effects on the transmission dynamics of an STI in a heterogeneous population, using data from the 1992 National Health and Social Life Survey (NHSLS) (Laumann et al. 1994). The NHSLS took an area probability sample of 3,432 respondents. Respondents were asked questions on a range of topics relating to sociological and demographic background, social relationships, sexuality, sexually transmitted diseases, and marriage and residential status.

I calculated a mixing matrix on racial/ethnic identity by cross-tabulating the female respondents' ethnicity (variable "ETHNIC" in the NHSLS data) and the race of their most recent sexual partner (variable "SPRACE1"). Table 6.2 presents the racial/ethnic mixing matrix for three groups: white (non-Hispanic), black (non-Hispanic), and Hispanic. The data that I use from NHSLS differ in one significant respect from those typically employed (Morris 1991, 1995, 1996). Specifically, the numbers of men and women in table 6.2 are equal because of the way in which the matrix was formed. In the work of Morris, mixing matrices are derived from surveys in which focal individuals were asked about how many sexual partners they had in the last year. These matrices, therefore, can potentially contain multiple reported partners for each respondent. In this example, I am modeling the mixing among most recent sexual partners and not among all sexual partners.

Clearly, "white" individuals dominate the table, accounting for approximately 73 percent of the sample, and this makes it difficult to see the full extent of differential homophily present. Using a loglinear model with offset $K_{ij}=N_iN_j/N$ to remove marginal effects (McCullagh and Nelder 1989), we can measure the homophily. Table 6.3 presents the estimated parameters for three different loglinear models: independence, uniform homophily and differential homophily (Morris 1991). For the independence model, the expected cell count is simply the product of the row and column sums divided by the grand sum of the table—no interactions between rows and columns

Table 6.2. Mixing matrix on racial categorization
of last sexual partner from NHSLS

| Women | Men | | |
|---|---|---|---|
| | White | Black | Hispanic |
| White | 1131 | 12 | 16 |
| Black | 5 | 268 | 5 |
| Hispanic | 39 | 1 | 115 |

SOURCE: NHSLS data described in Laumann et al. 1994

are present. A simple way to include interactions is the uniform homophily
(UH) model, in which a single parameter is fit for the diagonal. The differential homophily (DH) model fits a parameter for each cell along the diagonal.
Note that in going from the independence model to the UH model, the
deviance decreases from 1769.2 to 49.9 with only one additional degree of
freedom. Because the matrix is so small, all but one degree of freedom are
used by the DH model. The difference in deviance ($G^2$) between two models
is approximately chi-square distributed with degrees of freedom ($df$) equal to
the difference in parameters between the two models.

The results of the loglinear model indicate clearly that there are interactions
between the rows and columns of the matrix. Interaction terms for both the
UH and DH models are large and highly significant. For example, concordant
black partnerships are 194 times ($=e^{5.269}$) more likely to occur than would be
expected from the marginal totals alone. Note, also, that many of the row and
column effects lose conventional statistical significance once DH interactions
are introduced. Clearly, it is people's interactions that drive, in large measure,
the patterns observed in the mixing matrix.

## Transmission Dynamics Model

Using mixing parameters derived from the loglinear model, we can now
numerically integrate the system of equations (7) and (8) to explore the dynamical properties of a system with such nonhomogeneous mixing.

For illustrative purposes, I will model the dynamics of a hypothetical STI
which use transmissibility and duration-of-infection parameters similar to

**Table 6.3. Fitted parameters for the loglinear models independence, UH, and DH**

| Parameter | | Independence | Uniform homophily | Differential homophily |
|---|---|---|---|---|
| Main effects | | 0.693 | -2.418 | -1.687 |
| Men | | | | |
| | Black | -0.003* | 0.612 | -0.753* |
| | Hispanic | -0.145* | 0.395 | 0.588* |
| Women | | | | |
| | Black | 0.003* | 0.753 | -0.854* |
| | Hispanic | 0.145* | 1.564 | 1.369 |
| Interaction | | | | |
| | Diagonal | — | 3.394 | — |
| | WW | — | — | 2.659 |
| | BB | — | — | 5.269 |
| | HH | — | — | 2.587 |
| $G^2$ | | 1769.2 | 49.9 | 9.7 |
| $df$ | | 4 | 3 | 1 |

* Parameter not statistically distinguishable from 0 at the $p<0.05$ level.

those for *Chlamydia trachomatis*. The biological parameters of chlamydial infection are very poorly understood (Golden et al. 2000), which makes the modeling of chlamydial epidemics with a variety of parameter combinations an important exercise for understanding this endemic infection. I take estimates from Plummer et al. (1996) and Katz (1992).

The numerical results of the model are presented in figure 6.1. The trajectories differ tremendously, though they eventually converge to identical equilibria, and do not reproduce the differential prevalence observed through epidemiological surveillance (Centers for Disease Control and Prevention 2001a). Hispanic women have the highest prevalence in this transient phase because they have the highest degree of differential homophily as estimated from the NHSLS. When we compare the numerical solutions in figure 6.1 to the model with random mixing, presented in figure 6.2, we see that the differential homophily has a very substantial impact on the tempo of the

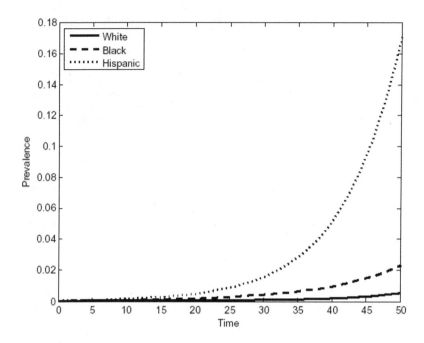

6.1. Dynamics of the compartmental model with mixing preferences estimated from NHSLS data. Plotted here are the prevalence trajectories of a hypothetical chlamydia-like infection among women. The population was initially seeded with three randomly assigned infected individuals.

model epidemic. It also has an eventual impact on the overall prevalence at equilibrium.

## DISCUSSION

Although selective sexual mixing by social attributes does not yield autochthonous prevalence differentials in models of diffusion, it can perpetuate existing differentials (note the slower convergence to the common equilibrium from the random initial prevalence differences in figure 6.1). It also has two other very important effects:

1 It slows down substantially the approach to equilibrium. Differentials in prevalence attributable to differing initial conditions or changing patterns of behavior will therefore persist longer before the system equilibrates.

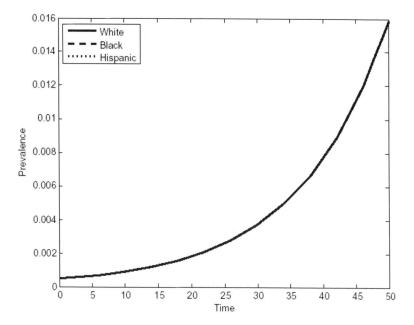

6.2 Dynamics of the compartmental model with random mixing. Plotted here are the prevalence trajectories of a hypothetical chlamydia-like infection among women. The population was initially seeded with three randomly assigned infected individuals. The three trajectories converge almost immediately so that the lines for Black and Hispanic women are not even visible.

2  Selective mixing reduces the overall population prevalence at all points in the epidemic trajectory.

## Mixing and Epidemiology

Models with random mixing will display the most rapid spread of an epidemic and will typically produce the highest prevalence at the endemic equilibrium. Interaction biased by social attributes (e.g., race/ethnicity) will slow down the exponential growth phase of an epidemic and will reduce the overall aggregate prevalence at equilibrium.

However, ceteris paribus, biased interaction will increase the prevalence of highly homophilous subgroups transiently. This amplification due to differential homophily can be confounded by other structural features of the population that change the population constraints within which people form their partnerships. For example, African Americans display a high degree of homophily, and young African American men are characterized by high relative mortality rates, particularly in poor urban areas (Geronimus et al. 1996; R. Cooper et al. 2001). Geronimus et al. (1996) calculated a standardized mortality ratio of 4.11 for black men in Harlem relative to white men for their 1990 sample. An African American boy in Harlem at the age of fifteen had a 37 percent chance of living to his sixty-fifth birthday—at a time when the overall life expectancy for American men exceeded seventy years (Keyfitz and Flieger 1990).

Similarly, for a variety of social and structural reasons, up to one-quarter of young adult African American men will be in prison or jail at any one time—in the National Longitudinal Survey of Youth, for example, one-quarter of black males (not in college) surveyed between 1979 and 1998 were interviewed in prison (Western 2002).

High (differential) mortality and high incarceration rates for African American men lead to a marriage squeeze for African American women (Cready et al. 1997). This squeeze is compounded by the fact that African American men are much more likely to form interracial partnerships than are African American women (Crowder and Tolnay 2000).

Combining a marriage squeeze with an overall high degree of endogamy is likely to amplify the prevalence of STIs within highly homophilous groups—particularly among the more numerous sex (in this case, women). One mechanism by which this can be achieved is through higher rates of concurrent (i.e., temporally overlapping) partnerships. Concurrency can have substantial effects on the dynamics of STI epidemics independent of partner number (Morris and Kretzschmar 1995, 1997). Adimora et al. (2002) note that concurrent partnerships are nearly twice as prevalent among African Americans as among whites, despite the fact that overall partner numbers are statistically indistinguishable between blacks and whites (Adimora et al. 2002).

Unfortunately, classical epidemic models such as the ones that I have discussed are not suited for examining the impacts of concurrency and many other network properties on STI dynamics. However, there are related models with great promise; I will address these in the conclusion of this chapter.

In addition to the negative health consequences for affected groups, amplification of subpopulation prevalence could be bad for the population as a whole. The "core-group" model in STI epidemiology (Hethcote and Yorke 1984) argues that curable bacterial STIs such as gonorrhea are maintained in a population because members of a core group of highly active individuals continually re-infect each other. Thus, even though $R_0 < 1$ for the aggregate population, the fact that $R_0 >> 1$ for the core keeps the pathogen endemic. For noncurable STIs such as HIV/AIDS, the core concept is not strictly applicable because re-infection does not contribute substantially to incidence. This core concept in STI epidemiology is closely related to the so-called "rescue effect" in metapopulation biology. Local populations with net reproductive rates below replacement ("sinks") can continue to exist for the long term because of recolonization from populations, called "sources," with net growth and emigration (Kawecki and Holt 2002).

Biased interaction, by potentially creating pockets of higher prevalence, could contribute to the persistence of endemic infections, and to the development of renewed epidemics when mixing preferences change. Behavior change is known to be important for the dynamics of STIs (Morris 1996).

The aggregate effects of people's preferences for sexual partners have profound implications for the epidemiology of STIs. Clearly, culture matters for STI epidemiology. With the framework for modeling STIs in a heterogeneous population established, it also becomes clear that the epidemiology matters for culture. Sexually Transmitted Infections are a discrete, objective marker for a type of behavior. The patterns that we observe in the aggregate provide information on the cultural norms that underlie transmission.

## Mixing and Cultural Evolution

Compartmental models for infectious diseases have proven to be a powerful tool for examining the dynamics of epidemics and potential public-health interventions for diseases spread by casual contact, such as measles and influenza (Bailey 1975; Longini et al. 1986), and in vector-mediated diseases such as malaria or Lyme disease (MacDonald 1957). Although these models have yielded a great deal of insight into the qualitative behavior of epidemics of infections requiring intimate contact, their usefulness for making quantitative predictions about outcomes and for effectively modeling intervention strategies has been decidedly mixed.

Can these models be of use for studying cultural evolution? The answer to this question is almost certainly "yes," for the reasons mentioned above: donor-acceptor control of dynamics (i.e., biased transmission), and overlapping generations of horizontally-transmitted cultural innovations.

Cultural heterogeneity in the form of discrete, culturally endogamous, ethnically marked subgroups can be expected to have four effects on the spread of cultural innovations:

1    Innovations will spread more slowly than in homogeneous populations. In populations characterized by a high degree of cultural exogamy, cultural innovations will spread rapidly.

2    The equilibrium prevalence of a cultural trait in a heterogeneous population will be lower than in homogeneous populations.

3    Few traits in a heterogeneous population will go to fixation, a criterion that seems to define "successful" cultural traits in some of the culture-evolution literature (e.g., Henrich 2001).

4    Under certain circumstances, the acceptance of a cultural trait will be higher in some subpopulations within the heterogeneous metapopulation.

Specific predictions about equilibrium prevalence depend critically on the details of the cultural transmission process. For example, are adopted traits kept for life, or can adopters revert?

## CODA: MOVING BEYOND COMPARTMENTAL MODELS

With the exception of broadcast transmission, cultural transmission (including disease transmission) is a process that takes place between two actors. Social networks are thus a vital substrate for cultural transmission. The compartmental models I have discussed in this chapter can only deal with very rudimentary types of network structures (e.g., differential association by attribute). Social-network theory is a richly developed area of formal inquiry into the structural properties of human relationships (Wasserman and Faust

1994). The basic SIR framework can be extended to apply explicitly to the graph-theoretic structures that are the bread and butter of social-network analysis (Newman 2002).

Classical network analysis, like most work in compartmental models, is non-inferential. However, there exists a growing body of theory and empirical research into stochastic network models and their estimation (Frank and Strauss 1986; Wasserman and Pattison 1996; Snijders et al. 2004). Using the tools of modern algorithmic statistical computing, network models can be estimated from sexual-survey data. Using the same machinery as in the inferential stage, networks can then be simulated and epidemic processes overlaid. These new methods provide hope for a truly satisfactory formal epidemiology of intimately transmitted infectious diseases.

## APPENDIX 6.1. INCORPORATING HETEROGENEITY INTO THE EFFECTIVE CONTACT RATE

The loglinear modeling approach of Morris (1991) allows for the estimation of mixing preferences from observed mixing matrices, greatly facilitating the development of data-driven epidemic models.

We begin by expanding the effective contact-rate parameter, $\beta$, of equations (1)–(2), which is a composite. This expansion makes explicit the contributions of activity level of the class, as well as the (possibly asymmetric) transmission probabilities between groups. The probability of transmission from infected category j to susceptible category i is the product of the expected number of partners of category i, the probability of contact between i and j people, and the transmissibility from j to i:

$$\beta_{ij} = \frac{c_i \pi_{ij} \tau_{ij}}{N_j} \tag{9}$$

where, $c_i$ is the average number of partners per unit time for members of group i; $\pi_{ij}$ is the conditional probability of a subject from group i having a partner from group j ; $\tau_{ij}$ is the per-partner probability of disease transmission from i $\leftarrow$ j; and $N_j$ is the total number of people in group j.

Note that the total number of contacts between members of the ith and jth categories is simply the product of the first two terms of (9). An estimate of the total number of ij contacts is given by the ijth cell of the mixing matrix. We can therefore rewrite this equation as

$$\beta_{ij}=\frac{\hat{m}_{ij}(0)\,\tau_{ij}}{K_{ij}(0)N(t)}=\frac{\alpha_{ij}\tau_{ij}}{N(t)} \tag{10}$$

where $\hat{m}_{ij}$ is expected number of contacts in cell ij; $\tau_{ij}$ is transmission probability $i \leftarrow j$; $K_{ij}(0)$ is the population weight at $t=0$: $K_{ij}=N_iN_j/N$; $N(t)$ is the total size of the population at time t; and $\alpha_{ij}$ is the assortative preference of i for j.

## NOTE

The work reviewed in this chapter was facilitated by the intellectual and financial support of Martina Morris and King Holmes, the University of Washington Center for AIDS and STI, and the Center for Statistics and the Social Sciences. Mark Handcock, Matt Golden, and Steve Goodreau have also helped me to shape many of these ideas. Libra Hilde helped make the manuscript coherent. I wish to thank Melissa Brown for inviting me to participate in the 2003 conference "Toward a Scientific Concept of Culture," and a number of conference participants for their stimulating comments relating both to this chapter in particular and to the issues raised by cultural transmission dynamics models. Among these are Shripad Tuljapurkar, Marc Feldman, Ken Aoki, Joe Henrich, and Richard McElreath.

# PART III   ETHNOGRAPHIC CASE STUDIES

Where formal models show the possible, these ethnographic case studies, including data from experimental economics and ethnohistory, show the empirical definitive. Documentation of empirical complexity is crucial to understanding actual behavior. For example, ethnographies can help us to understand why behaviors that seem shocking to us as outsiders—such as cannibalism—have been part of the repertoire of human practices. To contribute to a paradigm, however, ethnographic case studies must address theoretically driven research questions. These chapters (as well as Wolf's chapter in Part IV) link ethnographic empirical data to research questions derived from evolutionary theories of culture in order to analyze the relationship between human behavior and culture.

William H. Durham, a biological anthropologist, presents an ethnohistorical analysis of the disease kuru among the Fore peoples of Highland New Guinea, to argue that culture influences behavior through the meanings of cultural ideas themselves, through the power of coercion, and through hegemony (a combination of meaning and power that leads subordinate people to believe cultural ideas supporting the status quo). In considering how ideas cause behavior and how maladaptive practices can spread in a population, Durham suggests that both culture and structure matter. In order to link ethnographic case studies more tightly with formal modeling, Durham advocates "indirect quantification . . . the use of quantifiable indicators or surrogates to allow inferences about the underlying action of structural and semiotic processes."

Melissa J. Brown, a sociocultural anthropologist, presents ethnographic and historical evidence on forming and changing ethnic identities in Taiwan that suggests that structure matters more than culture. She argues that social structure and culture constitute distinct inheritance systems with fundamentally different—but still evolutionary—dynamics. She shows how broad social changes occur at the level of individuals, where cultural ideas appear poor motivators, and the social system constrains which behaviors can be enacted. She suggests, "the influence of social structure on human behavior can be modeled as though individuals are subject to a selection process that leads to differential attainment of social roles or positions."

Biological anthropologist Joseph Henrich uses cross-cultural comparisons of hunting technology, visual perception, and ultimatum game experiments to argue that human psychology and culture so mutually influence each other that neither can be understood independently of the other. Henrich sees this mutualism as an adaptation in humans' evolutionary past that has allowed certain kinds of learning processes: "natural selection will favor cognitive mechanisms that allow individuals to more effectively extract adaptive information, strategies, practices, heuristics, and beliefs from other members of their social group." He suggests that this view yields interesting predictions regarding ethnic-group formation and cooperative social institutions.

Samuel Bowles and Herbert Gintis, both economists, analyze experimental evidence from fifteen foraging and pastoral societies to evaluate a strong hypothesis and a weak hypothesis that culture matters. Their data "for the weak hypothesis seems compelling: faced with a given structure of feasible actions and rewards, individuals in different societies behaved very differently, suggesting that between-group differences in preferences or beliefs are significant." Bowles and Gintis suggest that social systems—via group selection and socialization processes—produce persistent cultural heritages that can potentially influence human behavior more strongly than adaptive principles of self-interest. However, they acknowledge the possibility that "economic structure or other aspects of social relationships" may be a confounding factor in consideration of the strong hypothesis.

# 7

## When Culture Affects Behavior

### A New Look at Kuru

### WILLIAM H. DURHAM

This chapter explores three main pathways through which culture—which I take to be ideas and beliefs in people's minds—influences human behavior. The first pathway is meaning: in this case, the behavior of "A" (an individual or group) complies with culture because of the convincing quality of the ideas and beliefs to A. This pathway marks the "intrinsic power" of culture: believers act. The second pathway is power: here, the behavior of A complies with culture because "B" (another individual or group) uses force or coercion to exact compliance. In this instance, "extrinsic power" produces culture's effect on behavior: subordinates act. The third pathway, hegemony, is a special combination of the other two: the behavior of A complies with cultural expectations because certain ideas are convincing to A even though they run against A's interests. Because of its "hegemonic power," B is able to control the very thoughts and desires of A, the better to serve B's interests: oppressed people act. I explore these three key pathways from culture to behavior in one particularly revealing context: the devastating epidemic of kuru among the Fore peoples of the New Guinea Highlands. Among its other implications, the impact of this deadly disease on the Fore population offers persuasive evidence for each of these ways in which culture affects human behavior.

### THE KURU EPIDEMIC IN HIGHLAND NEW GUINEA

The tragic saga of kuru is already well known in the behavioral and biological sciences, owing to a number of both scholarly and popular accounts of the epidemic (see, e.g., Lindenbaum 1979; Goodfield 1985; Zigas 1990; Rhodes 1997; Klitzman 1998; Ridley and Baker 1998; Schwartz 2003). This lethal neurodegenerative disease spread with devastating impact between 1920 and 1990 through villages of the Fore peoples of Highland New Guinea (of the Eastern Highlands language family) and a few of their immediate neighbors (see fig. 7.1). The disease struck individuals quickly and irreversibly, resulting within weeks in the characteristic trembling and ataxia from which it derives its name (in Fore, kuru means "to shake or tremble"). And in most instances, degenerative decay of the brain and central nervous system brought death to its victims within two years of first symptoms.

The epidemic went largely unnoticed by the outside world until government patrols, missionaries, and anthropologists began to work in the area in the 1950s (see Rhodes 1997 and Poser 2002 for historical details). By the late 1950s, a team of Western physicians and epidemiologists, headed by D. Carleton Gajdusek of the U.S. National Institutes of Health, began a thorough longitudinal study of kuru and investigated many different causal hypotheses—genes, diet, contagious pathogen, autoimmune dysfunction, etc. After numerous false starts and unproductive leads, the team succeeded in transmitting kuru to a chimpanzee. The conclusion was then inescapable: kuru was an infectious disease after all—a disease of long and variable latency. For demonstrating the contagious nature of kuru, Gajdusek received the 1977 Nobel Prize in medicine, even though the pathogen had still not been identified.

Identification of the disease-causing agent came in 1984, when Stanley Prusiner of the University of California in San Francisco presented evidence that kuru was caused by an infectious protein, or "prion," a discovery that later earned him a Nobel Prize as well. Subsequent research (reviewed in Collinge and Palmer 1997; Prusiner 1998, 2001; Collinge 2001) has revealed numerous intriguing properties of the kuru prion and of the prions behind other similar neurodegenerative diseases, known collectively as the TSE's (transmissible spongiform encephalopathies) and including Creutzfeldt-Jacob disease, Gerstmann-Straussler syndrome, and bovine spongiform encephalopathy or "mad cow" disease. One suspects that we are still seeing just the tip of the iceberg of prion-caused disease. However, there are equally fascinating, and more germane, implications of kuru relative to culture.

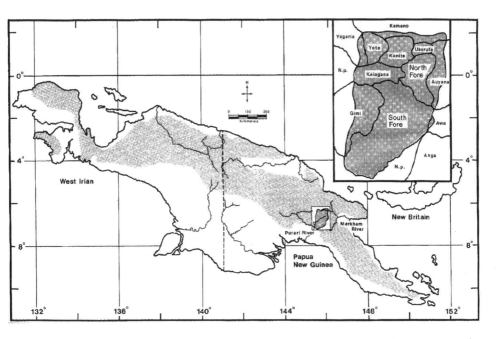

7.1.  The island of New Guinea, showing the Fore region (shading indicates altitudes greater than 200 m). The inset map shows the area affected by the kuru epidemic—namely, the territory of the North and South Fore and neighbors with whom they intermarry. (From Durham 1991: 394)

## CULTURAL IMPLICATIONS

Even before the infectious transmission of kuru had been demonstrated, a number of scholars (among them anthropologists Ann and Jack Fischer, Robert Glasse, and Shirley Lindenbaum; summarized in Rhodes 1997: chap. 6) advanced the hypothesis that kuru was transmitted through the practice of cannibalism, particularly endocannibalism, which had been reported in the Fore region and studied in the 1950s by Ronald and Catherine Berndt (Berndt 1962: chap. 13). ("Endocannibalism" refers to the consumption of the flesh of one's kinsmen, as contrasted with "exocannibalism," the consumption of enemy flesh.) At first, Gajdusek and other medical researchers dismissed the cannibalism hypothesis, both because transmission had not yet been established under controlled conditions, and also because the suggestion was first considered "too exotic" and controversial (Rhodes 1997: 103). However, opinion began to shift in 1966, when contagion was confirmed in the chimpanzee, and by 1977 enough circumstantial evidence had accumulated that Gajdusek (1977) was fully persuaded that cannibalism was key to the epidemic, whether by the "direct route" (the actual consumption of contaminated flesh) or by the "indirect route" (accidental infection from contaminated flesh via mucous membranes or open sores or cuts).

Still, the cannibalism hypothesis remained controversial for several years, in part because of the provocative thesis that William Arens (1979) called the "man-eating myth." Given that few trained Western scholars had actually observed and reported cannibalism first hand (although anthropologist Ronald Berndt was one of them; see Berndt 1962: xii), Arens boldly asserted that cannibalism was a fiction, both generally around the world, and specifically in the Fore region (for a parallel claim, see Steadman and Merbs 1982). In the years since Arens's assertion, however, convincing evidence of Fore cannibalism, and of cannibalism elsewhere, has accumulated (see, e.g., P. Brown and Tuzin 1983, Lindenbaum 2004). Painstaking fieldwork by Robert Klitzman (1984, 1998) in the region affected by kuru, for example, showed that synchronic clusters of kuru deaths were correlated with specific cannibal feasts and even with the distribution and handling of specific body parts. With chilling accuracy, Klitzman's data even predicted the tragic fate, since realized, of then-asymptomatic women who had been exposed at particular feasts in the 1950s. My own comparative study of mortuary customs among speakers of languages closely related to Fore (Durham 1991: appendix A.7.2) confirmed that direct contact with the flesh of the deceased—necessary for kuru transmission—was shunned under former traditional forms of mortuary practice. It was only with the introduction and spread of the relatively new practice of endocannibalism, which diffused into the Fore region from the north in the early 1900s, that locals were brought into deadly contact with the pathogen, whether by the direct route or indirectly through cuts and abrasions (see also Chiou 2006).

## KURU AND CULTURAL EVOLUTION

Looking back over the kuru literature, one finds at least two different kinds of connections between kuru and the scientific study of culture in anthropology. The first connection has to do with evolutionary culture theory and the implications of kuru's rapid and enduring spread, despite its inevitable lethality. As far back as the 1950s, the high incidence of the disease had already made it hard to explain as a product of genetic evolution: How could Darwinian processes ever promote high frequencies of the presumed gene or genes for such a deadly scourge? The enigma then moved over to the cultural side of the ledger when infectious transmission was demonstrated. Cavalli-Sforza and Feldman, for example, in their 1981 *Cultural Transmission and Evolution*, showed that a high rate of horizontal (peer-to-peer) transmission could guarantee the spread of kuru despite its lethal effects. The analysis made the problematic assumption that the disease could be regarded as a "cultural trait." But their effort did open the door to the analysis of actual cultural dynamics behind kuru—that is, to underlying changes in the shared idea system of the Fore people.

From here, kuru's first link to the scientific study of culture is readily told. It was the spread of cultural beliefs supporting and condoning cannibalism, and not the disease itself, that was the real enigma. How could cannibalism have gained its cultural acceptance and prominence within the Fore population when, soon after it diffused in from the north, cannibalism promoted the devastating epidemic of kuru? From the perspective of evolutionary culture theory, endocannibalism among the Fore was a candidate case of opposition between genetic and cultural evolution (Durham 1991: chap. 7): the cultural evolution of beliefs behind human flesh consumption had, with the onset of kuru, sharply impaired the survival and reproduction of many Fore. In some villages in the late 1950s, kuru killed more than 10 percent of the total population in just three years.

Again it was Lindenbaum (1979; see also 2001, 2004) who astutely summarized the enigma, and reviewed key data for its resolution. Lindenbaum showed that cannibalism was enthusiastically received and sanctioned within the Fore region because it provided, in effect, a source of meat at a time when meat was increasingly scarce in the region. As she points out, cannibalism was initially seen by the Fore as a good thing. Cooked human flesh compared favorably with pork, early testers reported, and provided a valuable food supplement at a time when wild game was increasingly scarce. Meanwhile the Fore, who had almost "clinical perception" of more routine disease epidemics (Lindenbaum 1979: 31), were careful not to eat corpses with symptoms of dysentery, leprosy, or yaws. At the height of the epidemic, in fact, the Fore

openly debated whether kuru was indeed an infectious disease, only to dis-
suade themselves of that hypothesis in favor of their preferred explanation:
malicious sorcery. It was the long and variable latency of the disease that hid
from the Fore—and from Western scientists for more than a decade of
research—the causal connection between cannibalism and kuru. In the mean-
time, there was an enduring strong incentive for cannibalism in the desirability
of additional tasty meat. In short, the cultural evolution of endocannibalism
among the Fore qualifies as a case of opposition promoted by cause imper-
ception; it remains one of the most carefully documented cases on record of
opposition sustained by choice. Imposed opposition, where social power is
used to foist a maladaptive tradition onto others, is no doubt more common
(see Durham 1991).

## KURU, CULTURE, AND BEHAVIOR

The second connection between kuru and culture comes from kuru's implications for the scientific study of culture itself. Data collected about the epidemic not only trace the tragic course of virulence in a human population, but
they also plainly demonstrate the enormous power of culture over human
behavior. The kuru data provide remarkably clear illustrations of when culture
affects behavior and when it does not.

Consider, first, the case to be made for the intrinsic power of culture in the
kuru context—that is, the ability of an idea to motivate behavior because of its
inherently convincing quality. Let me simply summarize a few selected examples of intrinsic power: each takes the cannibalism hypothesis as given, and
looks back upon culture and behavior using data originally collected for
another purpose—namely, to understand kuru's etiology. With the etiology
now established, these data allow a number of important inferences about the
way Fore beliefs shaped Fore behavior.

### The Meaning of Cannibalism

One of the initial appeals of cannibalism, quite above and beyond its welcome
reception as a supplemental source of tasty meat, was its symbolic "fertilizing"
or "rejuvenating" effect. Fore believed that eating the flesh of the deceased had
beneficial symbolic effects upon the "strengthening" and "healing" of the

living, and also upon the fertility of their fields: "'Cut my body,' a dying man or woman may say, 'so that the crops may increase.' 'Eat my flesh so that the gardens may grow'" (Berndt 1962: 272). To the Fore, cannibalism was viewed as a source of a generalized vitality or "life force," much as it was among the neighboring and closely related Hua peoples. "In fact," reports Meigs (1984: 110) writing on the Hua, "it is feared that if a person fails to eat the corpse of his or her same-sex parent, that person and his or her children, crops, and animals will become stunted and weak, having forgone the rightful inheritance of vitality."

This vitality or life force, in turn, was a crucial ideological component of belligerence and defense in an area rife with internecine warfare. Local Fore residential groups or hamlets of 70 to 120 people had long been clustered into rivalrous alliances known as parishes or sovereignties, each with its own protective bamboo walls. It was the duty of the able-bodied men from the hamlets to protect the populace, the pigs, and the agricultural land of the parish, and to exact revenge for thefts and deaths of the past. Like many areas of the New Guinea highlands, ongoing warfare was considered a normal, unavoidable part of Fore life, "the main form of intercourse between the [parishes] in this region" and a major source of adult mortality (Berndt 1962: 233). Cannibalism of fallen warriors was a way to reclaim and preserve crucial life force for the parish.

Among the Gimi, other close neighbors and relatives of the Fore, these connections were explicit and powerful, as described by Gillison (1980, 1983). The Gimi, too, were in constant warfare at the parish level, sometimes with Fore parishes. Because the military strength of a Gimi parish depended on the strength of its male lineages (hamlets were virilocal, as among the Fore), the life force of lineages was a central concern. When a warrior was slain in battle, his measure of life force was only available again to his lineage when, following funerary rites, the flesh had decomposed and his bones had been buried in the ground. Cannibalism obviated the wait for decomposition, allowing the deceased warrior's bones a prompt burial, which then sped his life force back to the lineage pool and allowed more new warriors to be born.

Because of universal preoccupation with parish defense in this area, one can infer that the life-force ideology contributed importantly to the rapid and ubiquitous spread of cannibalism throughout the Fore region. But it is the kuru data that tell us just how convincing and powerful this ideology must have been. Fueled by the interest in vitality, life force, and good food, funerary cannibalism became so important to the Fore that, among other things,

- it led to "exponential growth" in kuru deaths from at least 1941 to 1957 (Goldfarb 2002);

- it resulted in a total mortality owing to kuru of about one-third of all adult female Fore by 1977; and

- despite the perception, at the height of the epidemic, that Fore society was coming to an end, cannibalistic practice continued unchanged.

The beliefs behind cannibalism had a strong enough hold on Fore behavior to unleash all of this turmoil.

### The Cause of Kuru

Another revealing example of the power of culture over human behavior comes from the Fores' own theory about the cause of kuru. They view kuru as the result of malevolent sorcery practiced by jealous men, especially those jealous of the wealth of Fore leaders, or "Big Men." Several features of kuru reinforced this belief in the eyes of the Fore: kuru did not follow recognized patterns of infectious disease; its onset came at a time of prosperity and increasing wealth for the Fore, which fostered the jealousy the perpetrators felt; and the families of Big Men, especially those with many wives, were notably among the more heavily affected. Consequently as the epidemic unfolded, the Fore undertook a number of specific actions, some of them quite costly to normal social life, to reduce attacks by sorcerers:

First, the Fore took various steps to avoid ever falling prey to kuru sorcerers, and above all to prevent their getting hold of the personal items—especially clothing, hair, fingernails, and excrement—required to practice the evil magic. Second, "as a measure of their anxiety and frustration," families affected by kuru repeatedly, and often at great expense, sought help from native curers, even while admitting that "no medicine (their own *or* that of Europeans) is strong enough to counteract the work of the sorcerer" (S. Glasse

1964: 43, emphasis added). Third, because the Fore consider sorcerers to be "amenable to public persuasion," they used public gatherings . . . to denounce the continuing acts of malevolence. Indeed, during the crisis years of 1962 and 1963, the South Fore convened special congresses called *kibungs* to formulate and express their local responses. . . .

By the mid-to-late 1960's, however, when all else seemed to have failed, many Fore hamlets instituted a stringent, self-imposed "social quarantine" as yet another attempt to curb the plague. . . . When, largely by coincidence, kuru rapidly declined at the same time, the Fore took this as a sign that their antisorcery efforts had finally succeeded. This, of course, reinforced their belief in the sorcery theory. . . . Not even continued exposure to [Western] medical researchers could shake this conviction. Instead, by 1970 the very ophthalmoscope became a symbol of hope in the campaign against sorcery: if properly used, some Fore suggested, it would surely reveal "the sorcerer's rotting interior." (Durham 1991: 410–13)

Each of these behaviors, especially self-imposed social quarantines, attests to the power of the Fores' own cultural explanation of the disease. When, in the 1960s, the epidemic did begin to wane, The Fore attributed success to these actions, not to the end of cannibalism, which remained, in the Fore view, unrelated. Well into the 1980s, people maintained strict precautionary vigilance over their personal effects, even food wastes, lest sorcerers obtain these materials and use them to create kuru bundles.

But there is another noteworthy feature of this second example: it contains within it a clear example of at least one of the situations in which culture—specifically Western scientific culture, for all its status and prestige—does *not* influence behavior. In this case, Western scientific explanations of kuru completely failed to convince the Fore or to affect their behavior. Even to this day, Fore find their own theory of kuru so convincing that they do not accept the Westerners' theory of infectious prion disease (Lindenbaum, personal communication). Indeed, Western investigators often report feeling exasperated that the Fore stick to their own etiology of the disease and refuse to be persuaded by Western theory and evidence. To the Fore, sorcery is just plain more convincing than the outsiders' story about a hypothetical pathogen that, as late as the 1980s, no one had ever seen and no one could even describe very well (see Klitzman 1998: 166–70). The example is a poignant reminder: for a cultural belief to shape human behavior in the absence of force or manipulation, it has to have convincing meaning.

## The End of Cannibalism

As noted above, endocannibalism was sustained among the Fore even in the face of kuru because it had—or was viewed as having—a number of valued consequences. First, it was seen as a source of fresh, delectable meat at a time when other sources of meat had seriously declined. In other words, it was valued for reasons of taste and nourishment. On eating flesh of a dead kinsman for the first time, one group of Fore women are reported to have said, "'This is sweet. . . . What is the matter with us, are we mad? Here is good food and we have neglected to eat it. In [the] future we shall always eat the dead, men, women and children. Why should we throw away good meat? It is not right!'" (Berndt 1962: 271). Second, it was valued as an accelerator of life-force renewal as discussed above. Cannibalism greatly facilitated defleshing of the bones of deceased warriors, the sooner to bury them and thus, in the local view, to return the fallen warrior's force to the lineage pool, and from there to new warriors in the next generation. In the case of cannibalism of kuru victims, there was also a third valued consequence: the relatively high body fat of kuru victims (who often died quickly, without wasting) made their flesh more similar to pork, long a preferred meat for the Fore. With tragic epidemiologic effects, kuru victims were viewed as especially "good to eat."

With all these internal evaluations promoting endocannibalism, its cultural "fitness" or staying power was high even in the face of kuru, and no doubt the practice would have continued for some time. Instead, cannibalism was brought to an abrupt end in 1957, according to government reports, through the impositions of outsiders, mainly missionaries, New Guinea police, and Australian patrol officers. As Michael Alpers (1992: 318–19), the main epidemiologist working on kuru, summarized:

> Cannibalism was categorically proscribed by the administration and all public consumption of the dead at mortuary feasts quickly ceased. By 1958 the dead were all buried [among] the South Fore, though the practice of cannibalism continued for a little longer among the Gimi. . . . There were some sto-

ries even in the early 1960s of bodies being dug up and partially eaten by older women but certainly by the end of the 1950s no children were engaged in cannibalism. So, in effect, the practice died out very suddenly at this time and has never been resumed. In the North Fore, because of the earlier influence of the administration, it ceased completely at an earlier date [perhaps as early as 1951].

How had the outsiders achieved this elimination in the face of so much favorable cultural valuation by the Fore? The threat of force and physical sanction was very apparent: New Guinea police patrols came through the region village by village saying "if you eat people we will *kalabus* you (jail you)" (Goodfield 1985: 36). Twenty years after the fact, people clearly remembered the use of force to curb the practice: they told Robert Klitzman (during his fieldwork in the 1980s) that armed patrol officers came along and "put people away in *kalabus* for cannibalism" (Klitzman 1998: 160). Other accounts suggest that the outsiders also manipulated local ideology, drawing in clever ways on local respect for deceased ancestors. One Gimi woman explained the local renunciation of cannibalism this way:

> "You see, when a man dies that is not the end (of him). . . . If we do not eat him he will return (to us as a white man). In the past, we did not understand (this) and so we ate the dead. The White man came and explained these things to us . . . (that) our dead go to stay in Australia and come back (here) later as White men. (Knowing this), we do not cut up the dead anymore." (Gillison 1983: 42)

Here, too, cultural beliefs and ideology were clearly influencing behavior, bringing a speedy stop to cannibalism. But they achieved that end by means of a power differential, using threat of force by outside native police mixed with ideological manipulation by whites. In this instance, what I have called "extrinsic power" produces culture's effect on behavior.

If extrinsic power truly had the effect of bringing cannibalism to a halt by the start of 1958, then one would predict that (1) the kuru death toll should begin to taper off in the late 1950s; (2) children born after 1957 would not be exposed, and should not therefore die of kuru; and (3) the age at death of the youngest kuru victim for a given year should generally grow older after 1957 (with perhaps some variation because of variable latency in disease onset). Figure 7.2 outlines data from the kuru epidemic that support the first two of these hypotheses.

The figure shows that deaths from kuru among both adults (top curve, over age ten at death) and children (bottom curve, under age ten) did decline sharply in the early 1960s from a peak in the late 1950s. The data on children under age ten (bottom curve) also show that no child under ten died from kuru after 1968, suggesting that disease transmission had been interrupted by 1958. Figure 7.3 lends support to the third of these hypotheses, as it shows that, as a general trend, the age of the youngest kuru victims did increase after 1957. However, the figure also indicates that transmission occurred as late as 1959, as shown by the several points below the comparison line, rather than ceasing altogether by 1957 as officially reported. One may reasonably conclude that sanctions and threats by armed patrols, together with manipulation of local ideology, were sufficient to cause the cessation of kuru transmission via cannibalism by 1960.

### "Pollution" and the Rights of Adult Males to Pork

To this point I have made the case that in kuru we can see clear signs of the intrinsic and extrinsic power of culture over human behavior. There remains the third pathway through which culture influences human behavior: hegemonic power. In this pathway, the behavior of an actor or group of actors, A, complies with cultural expectations because certain ideas are convincing to A, even though they favor the interests of B (another individual or group) and run against those of A. My focus here is not so much on the ways in which B worked its way into such a position of control over A—a worthy topic to be sure, and a key topic in the realm of social evolution (see, for example, Brown, this volume). Instead, my concern is with the compliance of A with, in effect, the culture of B—the fact that A's behavior is what B expects, even in the face of harm to A's interests.

Again, kuru provides a provocative candidate case. Gender relations among the Fore, and among their relatives in the East New Guinea Highlands language family, are heavily skewed and male biased. Their particular ideology of male supremacy asserts that males, as warrior-protectors of parishes, are in a privileged position within Fore society, exempt from certain kinds of difficult and dirty activity (e.g., the backbreaking work of sweet-potato cultivation) and

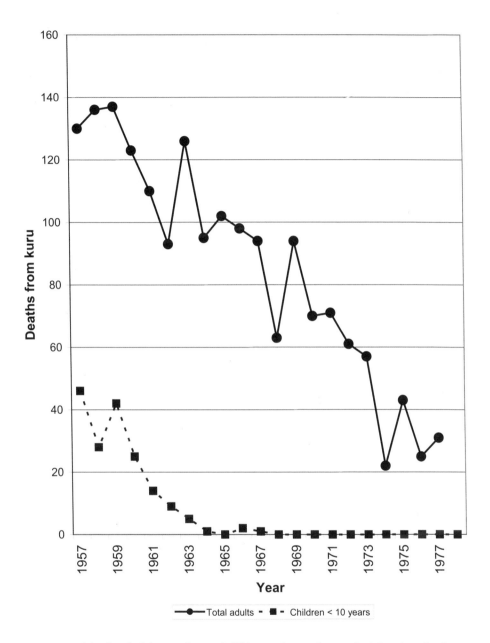

7.2. Annual deaths of adults (top line) and children under ten (bottom line) from kuru in all affected populations (Fore and their immediate neighbors). (Data from Alpers 1965, 1979, 1992; Lindenbaum 1979)

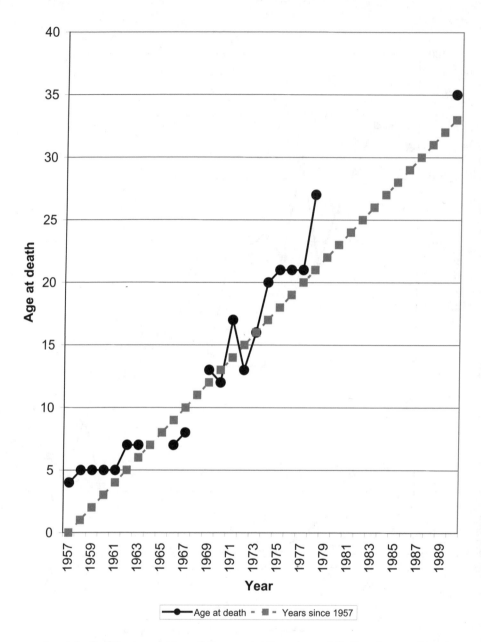

7.3. Age at death of the youngest kuru victim reported for each year (all affected areas, all ages), compared with the age of a hypothetical person born in 1957 (the year, according to government reports, when endocannibalism ended in the Fore region). The increasing age of the youngest victims supports the hypothesis that kuru transmission was interrupted in the 1950s, but suggests that the last kuru-transmitting mortuary feasts took place in 1959, not 1957. (Data from Alpers 1979, 1992)

beneficiaries of certain accepted advantages, such as their claim over pre-ferred foods. At the time of the kuru epidemic, women and children generally accepted this difference, acknowledging that they depended on adult males for protection from hostile neighbors, and accepting a subordinate role and the ideology that came with it. Three aspects of this ideology are especially noteworthy. Under it, (1) men claimed a "prior right" to pig meat, particularly to the pork distributed at mortuary feasts; and did so (2) on the grounds that lesser sources of meat, such as small game, frogs, and insects, were "pollut-ing" and debilitating of their military prowess. The ideology also (3) assigned to woman the task of dealing with corpses, which were traditionally viewed as being "polluted" and dangerous, owing to forms of "supernatural corruption" (for more on "pollution," see Lindenbaum 1979).

Within Fore and other communities of the East Highlands language fam-ily, this ideology appears to extend back in time indefinitely, and it was cer-tainly in place when the idea of endocannibalism diffused into Fore parishes from other peoples to the north. As a result, able-bodied warriors, especially young to middle-aged men, shunned human flesh as it became available, and asserted instead the prior right to pork from domesticated pigs (whose herd-ing was also women's work). Consequently,

> South Fore men rarely ate human flesh, and those who did (usually old men) said they avoided eating the bodies of women. Young children residing apart from the men in small houses with their mothers ate what their mothers gave them [whereas] boys moved at about the age of ten to the communal house shared by the adult men of the hamlet. [In this way, cannibalism was] largely limited to adult women, to children of both sexes, and to a few old men. (Lindenbaum 1979: 20)

To make matters worse, the body parts of deceased kin were also doled out according to specific rights of kinship, as was pork when it was available to women. Although there was some intercommunity variation in these rights, as a general rule specific body parts—including the brain—were reserved for female relatives of the deceased (see Durham 1991: 400). The influence of culture over behavior in this instance was tragically reflected in the high famil-ial prevalence of kuru that outsiders reported, from the earliest visits on.

Given this ideology, some aspects of the kuru plague within Fore parishes would seem readily predictable. As befits their hegemonic power over the food supply, adult men would be expected to be least affected by the outbreak of the epidemic, and adult women and children under ten would be the most immediately and directly affected. The expected "signature" of male-biased hegemony would be a serious sex and age difference during the kuru epidemic. This is, of course, exactly the case. As shown in figure 7.4, adult male deaths averaged fewer than 10 per year before 1965, during which period the tally for adult females averaged in excess of 100 deaths per year. During the peak years of the kuru epidemic, the impact of male hegemony resulted in a more-than-tenfold difference between female and male mortality rates. These data also show that during peak epidemic years (e.g., before 1963), there were fewer deaths among adult males than even among their children under age ten, whose numbers are shown in figure 7.2.

A study by anthropologist Gillian Gillison (1983) of cannibalism among the Gimi, neighbors to the west of the Fore, documents one of the institutional mechanisms that ensured low rates of kuru transmission to adult males. (The kuru epidemic affected some Gimi villages, but only in eastern valleys of their territory, where certain Gimi men had taken Fore wives.) Gimi men explained to Gillison that local mortuary ritual had changed from the very first episode of cannibalism, when women seized a dead man's body from the normal mortuary platform in a garden and took it with them to cut up and distribute the meat in seclusion within the men's house (a communal meeting and sleeping structure for the men, normally off-limits to women). Interpreting the occupation as a deliberate, polluting provocation, the men as a group remained outside the structure for the duration, but "kept track of the women's meal by sending into the men's house as observers several men of low status and several boys who also ate the human meat" (Gillison 1983: 36). In this way, not only did high-status men shun human flesh, but they also remained physically separated from what might otherwise have been an episode of kuru transmission.

### Summary: From Culture to Behavior

Some of anthropology's most important variables, especially those having to do with cultural meanings and social structure, are notoriously difficult to quantify.

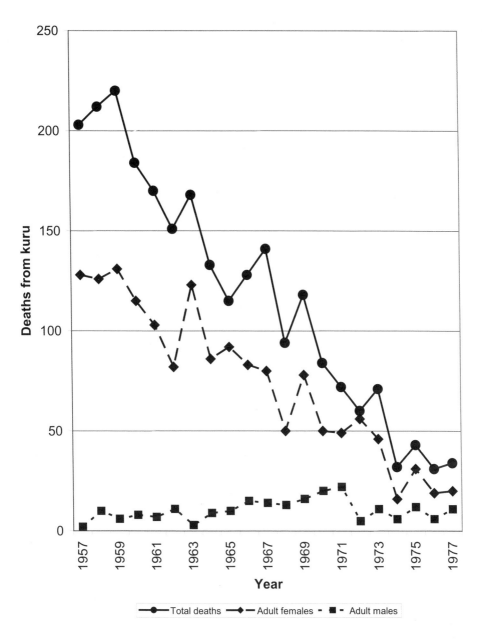

7.4. Annual kuru deaths in all affected populations (Fore and their immediate neighbors: total victims (top line), adult females (middle line), and adult males (bottom
line). (Data from Alpers 1965, 1979, 1992; Lindenbaum 1979)

In these cases, one good way forward is to try "indirect quantification," that is, the use of quantifiable indicators or surrogates that allow inferences about the underlying action of structural and semiotic processes. The preceding analysis uses indirect quantification to document the importance of meaning and social structure in the epidemic of kuru in Highland New Guinea.

The kuru example shows in a clear and compelling way how culture influences behavior through meaning, power, and hegemony. The influence of "meaning," of course, permeates all three pathways, because in each case culture shapes human behavior because of its convincing quality to someone. The difference between the three pathways hinges on just who that "someone" is. In the cases of power and hegemony, the pertinent ideas are not among those intrinsically convincing on their own to all individuals or to a broad spectrum of society. Rather, they are ideas of significance to those specific individuals and/or groups in positions of social power. In this manner, the study of culture's effects on behavior always brings up key questions about the society in which the behavior is observed—namely, who is it that finds this particular aspect of culture compelling? And whose interest does this part of the greater cultural system serve? Time and time again in the study of culture and cultural evolution one is thus brought back to the importance of social structure (which I take to mean the existing set of relations between individuals and groups in society) and the social distribution of power (Durham 1991; also Brown, this volume; A. Wolf, this volume). In other words, close study of cultural dynamics commonly leads to the study of social dynamics, the analysis of changing social relations within a human population. If the kuru case demonstrates, as I have argued, that culture matters, it also shows that social structure matters.

## COEVOLUTION OF GENES, CULTURE, ENVIRONMENT, AND SOCIAL SYSTEM

It is beyond the scope of this chapter to attempt a general treatment of the interplay between social system and cultural system, but a simple schematic rendering of the two—plus genes and environment for the sake of completeness—will help to clarify my arguments here about "meaning" and "structure." For this purpose, I have drawn upon a schematic framework representing "niche construction" in organic evolution (Laland, Odling-Smee, and Feldman

1999), which was based upon my own coevolutionary schema (Durham 1991: 186), itself derived from Lewontin's depiction of organic evolution (1974: 14). The framework, sketched in simplest form in figure 7.5, proposes that there are four different lines or "tracks" of inheritance affecting any given human population. First, there is the familiar track of *genetic inheritance*, consisting of the transmission through reproduction of the genetic information of the population. Second, there is the parallel track of *cultural inheritance*, distinct from but interacting with the genetic track of inheritance, consisting of the socially transmitted ideas in people's minds. These first two tracks of inheritance are featured prominently in coevolutionary theory as it has developed over the last couple of decades (Durham 1991). The third track, *ecological inheritance*—consisting of the historically derived ecological conditions of the population at any given time—has been emphasized by the proponents of niche-construction theory (see Feldman, this volume), where it is portrayed as interacting with the track of genetic inheritance. In niche-construction schemas, the ecological-inheritance track is often depicted as affecting genetic inheritance by way of natural selection, meaning that existing environmental conditions affect the differential reproduction of genes in the population and thus shape genetic inheritance across generations. In like manner, the genetic inheritance track is depicted as having a reciprocal effect on ecological inheritance through niche construction, that is, through human-induced ecological change in the given habitat. As certain genotypes become more prevalent in a given environment through natural selection (for example, genotypes conferring resistance to malaria), so the theory goes, the needs and wants of their carriers will influence surrounding environmental conditions (e.g., clearing more rainforest for agricultural fields), which may then feed back via natural selection (clearing the forest promotes increased malaria) to further influence genetic inheritance, and so on (on the malaria example, see Durham 1991, chap. 3).

The fourth track shown in figure 7.5, "social inheritance," represents existing social relations within the human population at a given time, particularly the various forms of inequality and power asymmetry as may characterize its social structure. These inequalities and power differences within the social system may then influence the cultural track of inheritance through the "imposition" of ideational phenomena that serve the interests of powerful subgroups. In the kuru example, I have reviewed evidence for two distinct but related

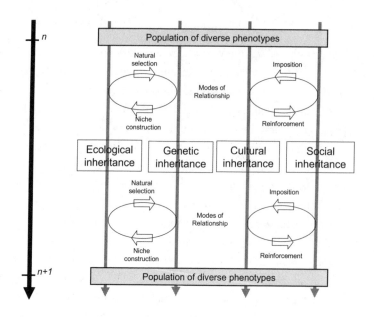

Figure 7.5. Four lines of inheritance influence the distribution of observable properties, or phenotypes, in a given human population. In this highly simplified schema, existing environmental conditions, genes, culture, and social system are all depicted as lines or "tracks" of inheritance that contribute to the distribution of phenotypes in a hypothetical population. Time runs vertically, with two reproductive generations ($n$ and $n+1$) shown for reference. In this rendering, each line of inheritance can influence phenotypes both directly and indirectly, through interactions over time with other lines of inheritance. Thus ecological inheritance is shown as influencing phenotypes directly or through the action of natural selection over time. "Modes of relationship" refers to the complex set of relations between genetic and cultural inheritance, as described in coevolutionary theory (see Durham 1991).

forms of imposition: the use of direct force, or threat of direct force, to exact compliance (e.g., the end of cannibalism under police and missionary threats), and the use of power and social position by one segment of society to influence the values of another segment, producing preferences that are not in the latter's best interests (e.g., Fore men claiming "prior rights" to pork, resulting in women's "best choice" option of endocannibalism, with the devastation that followed). The reciprocal influence of cultural inheritance on social inheritance can be called "reinforcement," a shorthand label for all the ways in which the shaping of cultural values and beliefs can feed back into the consolidation or expansion of power by the powerful.

In terms of figure 7.5, then, I have explored in one particularly revealing microcosm the kinds of direct and indirect influences that cultural inheritance may exert over human phenotypes. We have seen both culture's direct effects on phenotypes—in promoting cannibalism even as it caused the spread of a deadly epidemic—and its indirect effect on phenotypes, working through the power differentials of social inheritance in this context, as in the imposition of values and beliefs by police, missions, and the Fore men with their claim to pork. Thus, I have focused largely on the right-hand half of figure 7.5. I have suggested here, and more explicitly in Durham (1991, 1992, and 2002), that the cultural-inheritance track exhibits its own special form of Darwinian or variational evolution, based on the value-guided preferences of individuals with the power to choose. The kuru example has also opened the door to questions about the dynamics of change in the social-inheritance track, or social evolution. However, the data and time frame of the kuru example do not take us very far through that door. All I have suggested in this regard is that cultural reinforcement, as schematized in figure 7.5, is one important player in the processes of social evolution. How social evolution works, what its guiding forces are, and whether it is also typically a matter of Darwinian variational evolution, are all topics that remain largely unresolved at present (see also Brown, this volume).

I leave for another context analysis of the left-hand side of the diagram, except for a few concluding points. As Goldfarb (2002) has succinctly summarized, there is today clear and striking evidence that kuru has acted as a powerful force of natural selection in the Fore population. The PRNP gene (the gene for the prion protein whose abnormal configuration causes kuru) existed in diverse forms among the Fore in the 1950s, with one form more susceptible to triggering the disease and another form less susceptible or "resistant" to causing kuru. As Goldfarb notes, kuru-resistant genes and gene combinations (genotypes) in the Fore population increased dramatically between 1959 and 1979—that is to say, within the one-generation time frame depicted in figure 7.5. One resistant genotype nearly doubled in frequency during that interval, making kuru surely one of the most, if not the most, intense of selection pressures ever documented in human populations. There is little doubt that natural selection is another process amply illustrated by the kuru case.

Natural selection via kuru is closely related to the ecological inheritance of the Fore, the far-left "track" of the four in figure 7.5. Judging from the arguments and evidence of Lindenbaum (1979: chap. 2), the linkage is clear: cannibalism (and the kuru that ensued) is concurrent with change in forest cover and game supply in the Fore region. Citing evidence for changes in Fore hunting and agricultural practices over the decades of the 1900s, Lindenbaum argues that

> Population increases in the region and the conversion to the sweet potato as a dietary staple thus appear to have led to the progressive removal of forest and animal life . . . and to the keeping of domestic pig herds which compensate for the loss of wild protein. As forest protein sources became depleted, Fore men met their needs by claiming prior right in pork, while women adopted human flesh as their supplemental *habus*, a Melanesian pidgin term meaning "meat" or "small game." (1979: 24)

In effect, niche construction by the Fore, in the form of forest clearing and agricultural expansion, with the help of existing cultural beliefs and social structure, triggered dietary changes that were later to bring on the plague of kuru and the natural selection that it generated. One can thus see all four of figure 7.5's lines of inheritance at play in the kuru epidemic.

Another recent finding suggests that lessons from kuru may have even wider significance for understanding human history. Simone Mead et al. (2003; see also Stoneking 2003) suggest that there is a widespread genetic signature in human populations suggestive of pervasive natural selection by "prehistoric kurulike epidemics." Although one should be cautious about claims that cannibalism "may have been rampant" in ancient human populations (as proposed by Pennisi 2003), we may at least view this as evidence for meat consumption in human prehistory and associated changes by natural selection in the genetic-inheritance track. The finding adds a genetic exclamation point to a conclusion that seems inescapable: the kuru case, as remote and exotic as it may at first appear, has proven rich with lessons for humanity that reach far beyond the upland valleys of Highland New Guinea.

## NOTE

I thank Shirley Lindenbaum and Ron Barrett for discussion and feedback on the arguments of this paper. Thanks also to Richard Pocklington and the students in our autumn 2002 course, "Evolutionary Theory for the Anthropological Sciences," for fruitful discussion of the themes of this paper.

# 8

## When Culture Does Not Affect Behavior

*The Structural Basis of Ethnic Identity*

MELISSA J. BROWN

The end of the twentieth century saw myriad examples of ethnic identity motivating human behavior in the form of ethnic violence in Rwanda, the Balkans, the Middle East, India, and the United States. Why does ethnic identity motivate action? Many commentators accept the political rhetoric that culture and/or ancestry determine ethnic identity, and they assume that culture serves as the impetus for ethnic action.[1] Culture cannot serve as such a motivational force, however, because, contrary to ideological claims, it does not determine ethnic identity. Social structure—or rather, people's experience of it—forms ethnic identity (M. Brown 2004). The motivational force behind ethnic action, then, must be a social one.

By "social structure," I mean the social relations of individual members of a society, which are hierarchical, networked, and dynamic.[2] These relations aggregate to a population-level social order (which may be characterized as "capitalist," for example). To say that a motivational force is "social" or "structural," then, suggests that individuals are motivated by their hierarchical social position and/or network connections.

An analysis of ethnographic and historical data from Taiwan shows that structure, rather than culture, is the basis of ethnic identity. Two examples indicate the disjuncture between ethnic identity and culture. In one, a multitude of changes in culture did not lead to a change in identity; in the other example, identity changed regardless of little change in culture. The history of Taiwan also offers more generalizable insights on the close relationship between social structure and ethnic identity. Through five regime changes on the island from the seventeenth century through the present, major changes in social structure led to changes in ethnic identity for large numbers of people; this consistent correlation suggests that social structure is the basis of ethnic identity. Moreover, because the correlation spans almost 400 years, distinct political systems, and vast cultural differences in the ruling elite, it suggests there may be a general structural basis to ethnic identity.

The historical evidence of structure as the ultimate cause of identity change limits our understanding of the social dynamics, merely sketching a rough picture with broad strokes. At such a level, social change may appear transformational, the result of periodic external shocks that shift the system from one state to another. But how does that shift occur at the level of individuals? Detailed ethnographic evidence demonstrates how broader changes in ethnic identity, and subsequently in cultural practices, resulted from variation in actions. It shows that, because individuals must always socially negotiate the behaviors they enact, variation in actions derives from social negotiation. The empirical data on social dynamics also support an evolutionary explanation of human behavior that uses multiple inheritance systems.

## ETHNIC GROUPS IN TAIWAN

"Ethnic identity" can mean either a person's individual sense of self or the socially negotiated label that classifies people into one ethnic category or another. These different definitions are related, not because a sense of self leads one to socially negotiate for a particular label, but because a person's living under a socially defined ethnic label develops a sense of self linked to that label (M. Brown 2001a, 2004). Taiwan's history shows that ethnic identity—in the latter sense of how people are classified in their society—has important ramifications. These labels are associated with cultural ideas, including stigma or stereotypic attributes, but they are also linked to social experiences. Classification as Han or Aborigine, Taiwanese or Mainlander has influenced how individuals treat each other in social interactions, including whom they seek as sexual or marriage partners (see Jones, this volume). Labels have also influenced institutional treatment of individuals, in areas such as taxation, employment, and education. Ethnic identities, then, are important because they affect people's options and limitations in the specific structural relations of their society.

Table 8.1. Ethnic groups in Taiwan

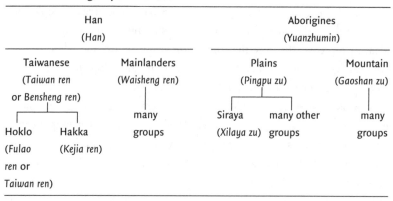

| Han (Han) | | Aborigines (Yuanzhumin) | |
|---|---|---|---|
| Taiwanese (Taiwan ren or Bensheng ren) | Mainlanders (Waisheng ren) | Plains (Pingpu zu) | Mountain (Gaoshan zu) |
| Hoklo (Fulao ren or Taiwan ren) · Hakka (Kejia ren) | many groups | Siraya (Xilaya zu) · many other groups | many groups |

There have long been several levels of ethnic distinction among the peoples of Taiwan (see table 8.1). "Han" is the Mandarin Chinese term for the people most Westerners think of as ethnic Chinese. Hoklo and Hakka are two regional varieties of Han found in Taiwan, with differing customs (especially in earlier centuries) and mutually unintelligible spoken dialects. Although Han demonstrate enormous social, cultural, and linguistic variation—as great as or greater than different nations in Europe—Han themselves and others classify Han as a single ethnic group. Hoklo have always constituted the ethnic majority in Taiwan, since at least the end of the seventeenth century.

Mainlanders came to Taiwan after 1945, with Chiang Kai-shek and his Nationalist government (see table 8.2 for the order of political regimes in Taiwan). Most Mainlanders are Han, but because they came from all regions of China, they include a range of native languages and customs. Under Nationalist martial-law rule, the government classified Hoklo and Hakka together as Taiwanese, distinct from Mainlanders. The discrimination and political repression faced by most Taiwanese during this period fostered the construction of Taiwanese identity as an ethnic identity (see, e.g., Gates 1981, 1987; Chun 1994; Chang 2000; M. Brown 2004). Not all Mainlanders were wealthy and powerful but Mainlanders controlled the government well into the 1980s, when the Nationalists finally made efforts to "Taiwanize" their party (see, e.g., Chang 1994).

"Aborigines" is the preferred English appellation of the indigenous peoples

Table 8.2. Regimes in Taiwan and China

| Regimes in Taiwan | | Regimes in China | |
|---|---|---|---|
| Dutch | 1624–1661 | Ming | 1368–1644 |
| Zheng | 1662–1683 | Qing | 1644–1911 |
| Qing | 1683–1895 | | |
| Japanese | 1895–1945 | Republican | 1911–1949 |
| Nationalist, martial law | 1945–1987 | Communist | 1949–present |
| Nationalist, post–martial law (transitional period) | 1987–1996 | | |
| Full electoral democracy | 1996–present | | |

of Taiwan, who are related to Polynesians. The broad category of Taiwan Aborigines encompasses many historic Austronesian languages and cultures recorded by the Dutch, as well as those identifiable today. Zheng, Qing, and Japanese rulers classified Aborigines into two categories of "barbarians": (1) "raw," later politely called "mountain," Aborigines who adopted few Han customs, and (2) "cooked," later called "plains," Aborigines who adopted much of Han culture, including language. By the martial-law period, plains Aborigines had disappeared as a category because the people had assimilated into the Han category (Brown 2004). Democratization fostered the nationalization of Taiwanese identity to include Aborigines as well as Mainlanders, valorizing Aborigines as worthy ancestors and leading many to reassert their Aborigine heritage.

## ETHNIC IDENTITY NOT BASED ON CULTURE

If ethnic identity were based on culture, we would expect that major cultural changes would lead to identity change, and that, conversely, ethnic change would not occur in the absence of much cultural change. These expectations would be particularly strong for Han identity, because Confucian ideology emphasized the importance of culture (M. Brown 2004: 24, 29, 31). Practicing Han customs was supposed to be enough to make one Han; beliefs did not matter (e.g., Watson 1988, 1993). The two examples that follow contradict both of these expectations, belying the assumption that ethnic identity is based on culture.

At the beginning of the twentieth century, residents and neighboring Hoklo classified Toushe, Jibeishua, and Longtian in southwestern Taiwan as plains Aborigine villages, even though the villagers and the nearby Hoklo shared many customs, some Hoklo ancestry, and some cultural ideas (see M. Brown 2004: chap. 3). People in these villages spoke Minnan—the same Chinese dialect spoken by their Hoklo neighbors—as their native language, and had done so for decades. Villagers grew the same cash crops—rice and sugarcane—under the same rental terms as their Hoklo neighbors. They dressed like neighboring Hoklo of the same economic class. They used the same forms of marriage (though in different proportions, a difference unlikely to be visible to their neighbors). Except for a Christian minority, villagers buried their dead in the same way as did the Hoklo, with a Daoist priest presiding. Differences in grave-tending practices would likely have gone unnoticed, in part due to Hoklo religious taboos regarding other people's dead. Absence of a Han-style temple would not have been remarkable for such poor communities, and worship of an Aborigine deity by most villagers, and Christian worship by a minority, appear to have gone largely unnoticed outside these villages at that time. Although most villagers linked surnames with property inheritance, as Hoklo elsewhere did, some expressed ambivalence about it. Also, marriage forms did not always carry the same meanings across the two groups. In Hoklo villages, poor men who married uxorilocally (into their wife's home), would have been scorned, but they did not face such stigma in Toushe, Jibeishua, and Longtian. The local ideational system of culture had already begun to change toward a Hoklo model (and completed the change later in the century), but ideas changed after practices were introduced, and some ideas spread more quickly than others.

Given all these shared cultural practices and ideas, and even shared Hoklo ancestry (a common way of claiming Han identity), why were these villages still considered plains Aborigine? Absence of a single cultural practice marked these communities as non-Han: they did not bind women's feet. In interviews, people spoke repeatedly of the importance of this marker in the most pejorative terms: "We savages didn't bind feet." Ironically, this comment shows the degree to which local people were culturally Hoklo, because they had internalized enough Hoklo cultural ideas to believe that lack of foot-binding proved them less than civilized, less than Han. Being culturally Hoklo, however, was not sufficient to win them Hoklo ethnic identity.

Lest anyone think these villagers were not considered Hoklo because they had not undergone sufficient cultural change—after all, there were differences in custom and ideas—let us consider plains Aborigines who succeeded in taking on Hoklo identity in the seventeenth century (see M. Brown 2004: chap. 4). Evidence of identity change here comes from discrepancies between Dutch-period census records of 1650 and Zheng-period records circa 1680 (which involve between 16,000 and 22,000 individuals). It appears that 40 to 50 percent of the total population in one area—probably a "mixed" population—changed identity from Aborigine to Han under Zheng rule.

There are two kinds of evidence here for a disjuncture between culture and identity change, both of which are circumstantial. First, practices of Aborigines—as a mixed population would have been categorized under Dutch rule—indicate little Han influence in Dutch-period records (see M. Brown 2004: 36–48, 74–94). Aborigines did not participate in cash-cropping, despite encouragement. The title of "interpreter" for tax collectors in Aborigine villages, combined with the fact that Dutch missionaries used Austronesian languages in Aborigine communities, suggests that, regardless of whether some Aborigines learned Chinese, villages maintained primary use of Austronesian languages. Although missionaries succeeded in dismantling the indigenous women-led religious system and changing marriage practices (see, e.g., Shepherd 1993, 1995), evidence suggests that cultural change among Aborigines was largely toward a Dutch model, and it did not affect their classification as Aborigines.

Further evidence of the disjuncture between culture and identity comes from differential treatment by the Dutch of Han and Aborigines. The Dutch viewed Aborigines as allies and Han Chinese as threats. Although a mixed population would have had the opportunity to be bilingual—and bilingualism would have offered some economic opportunities, especially if most Aborigines were still not native speakers of Minnan Chinese—the mixed population could not simply adopt Han customs, even if they were willing to do so. The Dutch missionaries who lived in these communities had the power to fine people whose behavior they did not approve, and they required Han men who wanted to marry Aborigine wives to convert to Christianity. "Mixed" people in the seventeenth century, then, appear to have taken on few Han customs before they acquired Han identity. At the very least, it seems certain that they had not taken on as many Han customs as the villagers of Toushe, Jibeishua, and Longtian had by the early twentieth century, when the latter villagers were still labeled Aborigines.

These examples contradict the assumption that ethnic identity, especially Han identity, tracks cultural practices or customs. Instead, in the early twentieth century, practicing many Han customs was not enough to make people Han, and in the seventeenth century, absence of most Han customs did not prevent people from becoming Han.

## ETHNIC IDENTITY BASED ON STRUCTURE

If every major structural change in Taiwan's documented political history led to a subsequent change in ethnic identity, this consistency would strongly suggest that ethnic identity is based on structure. Of five structural changes so major as to constitute regime change, there is evidence of ethnic identity changes for four. Absence of documentary material for the fifth prevents any conclusion, although conditions make an ethnic change plausible.

The first regime in Taiwan began in 1624, when the Dutch established a fort at a trading site popular with plains Aborigines, Han Chinese, Japanese, Dutch, Portuguese, and Spanish, and set about consolidating control of the plains (see table 8.2). The primary trade between Aborigines and outsiders throughout the Dutch period consisted of the hides, dried meat, and other parts of an indigenous form of deer (hunted into extinction). By 1650, the Dutch colony also exported rice and sugar, cultivated largely by the Han population who had immigrated under Dutch encouragement. Dutch revenues came from taxation of products, land rents, hunting licenses, interpreter fees, mill fees, and a head tax for Han. Although the entire Dutch population never numbered more than 1,200, they controlled the much larger Han population (estimated at 25,000 men in 1661), by providing muskets to the Dutch militia, using Aborigine militia to support Dutch troops, and even benefiting from information from Dutch missionaries and Han merchants, as when the latter warned of an impending rebellion of Han farmers in 1652.

The first regime change occurred in 1661, when Zheng forces took Taiwan from Dutch control. They were losing ground in their fight against the new Qing dynasty in China, so an estimated 25,000 Zheng troops invaded Taiwan to establish a fallback position for their continued war with the Qing. Lacking the military reinforcements they had requested from the larger Dutch colony in Indonesia, the Dutch surrendered Taiwan, turning over documents of all outstanding rents, licenses, and taxes. Although there were Portuguese mercenaries and possibly freed Africans among the invading troops, the Zheng leadership and its forces were primarily Han from southeast China (like most of the earlier Han immigrants to Taiwan). The Zheng regime kept in place the structure of the Dutch taxation system, but raised tax rates and added labor-service requirements for all males. Han men had to provide military service, and Aborigines had to provide corvée labor (e.g., in construction, driving carts, and running messages).

Identity change following the invasion was both structurally derived and structurally spread. The virtually overnight change in Aborigine social position from government allies to potential rebels against a new, more powerful government motivated identity change. The ability of individuals to negotiate a Han identity probably came from the structural relations of intermarriage, because records suggest high intermarriage rates between Han immigrant men and Aborigine women, and patrilineal ancestry was a commonly accepted way to claim Han identity (see, e.g., Ebrey 1996). Many identity changes probably occurred quickly, as the Zheng began collecting taxes and fees, but ongoing changes throughout the period would have been possible as Zheng soldier-settlers married local Aborigine or mixed-ancestry women (Han women were rare in Taiwan at that time).

The second regime change occurred in 1683, when the Zheng surrendered to Qing forces. The rulers of the Qing dynasty were Manchu, a non-Han ethnic group. Nevertheless, their regime can be considered largely Han because they continued the bureaucratic system of the earlier Han-ruled Ming dynasty. In Taiwan, the Qing maintained the ethnically biased taxation system used by the Zheng, although they lowered rates across the board. Nevertheless, structural shifts did accompany regime change. Although most Qing bureaucratic and military officials were Han, they viewed the Han in Taiwan with some suspicion, because many were the remnants of Zheng troops they had been fighting for decades. Aborigine militia once again supplemented regime forces in controlling the Han population in Taiwan.

There is no known documentary evidence regarding identity change during the period of this regime change. Such a shift is plausible, given that structural conditions were very similar to those of the early Zheng regime: a "mixed" population, most with Han patrilineal ancestry, facing a Han-style regime with an ethnically biased taxation system. Such people could easily argue that the Zheng had gotten their identity wrong. Such identity change would have been limited to the early years of Qing rule, as the Qing banned Han-Aborigine intermarriage in the aftermath of the 1736 Aborigine uprising.

The Qing regime ruled Taiwan until 1895 (although it never controlled areas held by mountain Aborigines). The western plains became lush agricultural fields, producing sufficient rice for export. By the late nineteenth century, the structural position of plains Aborigines shifted significantly. Plains Aborigine militia were no longer needed to supplement Qing troops. Deer had been driven to extinction, so Aborigines no longer provided sought-after trade goods. Plains Aborigines were probably viewed with some suspicion because many converted en masse to Christianity in the 1860s and 1870s. In the late 1880s, the Qing governor of Taiwan restructured the taxation system, greatly reducing a special form of income due to Aborigine landlords, and indicating that any claim that plains Aborigines might have had to Qing favor had long since ended.

The third regime change occurred in 1895, after the Qing ceded Taiwan to Japan upon losing the Sino-Japanese War. The Japanese colonial government restructured taxation and land tenure yet again. Han and plains Aborigines were taxed at the same rates, but most remaining Aborigine land rights were wiped out. The Japanese government gathered income from its monopolies on camphor, salt, opium, sugar, and sulphur. There were other major structural changes, as, through 1930, Japan invested heavily in developing Taiwan's infrastructure (e.g., major dams, irrigation systems, and railroads), its population (e.g., medical services and schools), and its production (e.g., industrialization). As one way to increase production, Japan banned footbinding in

1915. Japanese police responsible for enforcement of this law told rural Hoklo that the ban was so women could work in rice paddy fields.

Identity change in the villages of Toushe, Jibeishua, and Longtian was structurally derived from the 1915 footbinding ban. Before then, largely because women's feet weren't bound, village women did not marry out to Hoklo men as brides, and Hoklo women did not marry into the villages as first-time brides. Around 1930, however—when first-time Hoklo brides would no longer have had bound feet—marriage patterns changed. Suddenly, women from Toushe, Jibeishua, and Longtian were marrying out to neighboring Hoklo communities and Hoklo women were marrying into these communities as first-time brides. At the same time, people reported that use of the ethnic slur of "savages" referring to these villagers disappeared from its previously common usage. When I interviewed elderly people there in 1991 and 1992, grandchildren who were present for the interviews often expressed shock that their community and families had ever been considered anything but Hoklo—the people of Toushe, Jibeishua, and Longtian had spent the remaining fifteen years of Japanese colonial rule socially accepted as Hoklo.

The fourth regime change occurred in 1945, when, as part of the terms of Japanese surrender to the U.S. at the end of World War II, Taiwan was returned to Chinese rule. During Taiwan's time as a Japanese colony, the Qing dynasty had ended, and China, which had nominally become a republic, was torn apart by feuding warlords, Japanese invasion, and civil war between the Nationalist and Communist parties. The Nationalists took control in Taiwan after the Japanese withdrawal and, like Zheng forces several hundred years earlier, used the island as a fallback position, removing to Taiwan one to two million of their people and as many of China's moveable resources as they could, before their complete defeat on the Chinese mainland in 1949.

Hoklo and Hakka in Taiwan, who had been treated as second-class citizens and viewed themselves as (Han) Chinese under Japanese rule, initially welcomed Nationalist rule. However, the extensive structural changes that ensued changed those sentiments. Nationalists continued the corrupt practices that had lost them popular support in China, such as confiscating property, seizing people for ransom, and selling pieces of portable infrastructure for personal profit. They changed the official language overnight (from Japanese to Mandarin, a language not previously spoken in Taiwan), and generally treated

Han in Taiwan as part of the hated Japanese empire. When Taiwanese rebelled in 1947, the Nationalist government suppressed the uprising brutally with executions (an estimated at 8,000 to 23,000 of them), prison sentences, and the beginning of forty years of martial law. Anyone with relatives or close friends involved in the "incident" was subsequently blackballed from all government positions, including primary and secondary school teaching.

This regime change led to the formation of two new ethnic identities: Taiwanese and Mainlanders. Previously, Hoklo and Hakka had viewed themselves as Han and distinct from each other, but the Nationalist government lumped them (and people descended from plains Aborigines) together as Taiwanese (*bensheng ren*, literally, people from the province of Taiwan). This group was defined structurally by their historical presence under Japanese rule, not on the basis of culture. What Taiwanese continued to have in common was structural: their shared social experience of discrimination in education, employment, and political participation under martial law solidified ethnic Taiwanese identity.

People accompanying the Nationalists' retreat came to Taiwan from all over China—speaking all the different, mutually unintelligible Han dialects of China (and some of the minority languages) as well as the Mandarin they had learned for the military, industrial, or educational work that earned them removal to Taiwan. Although they likely thought of themselves as sharing only the category of Chinese, they too were combined into a new single category that stood in opposition to Taiwanese: Mainlanders (*waisheng ren*, literally, people from outside Taiwan). Culture was not a basis for this classification either. Mainlanders from Fujian, who shared the Minnan language and many customs with Taiwanese Hoklo, were still considered Mainlanders, not Taiwanese. What Mainlanders shared was structural: the social experience of fleeing the Communist unification of China.

These distinctions were ubiquitous in Taiwan throughout the martial-law years. For example, at the beginning of each school year, every child was required to state for the class which province his or her father came from, marking Taiwanese from Mainlanders. Because most Taiwanese had a recognizable accent in speaking Mandarin, in part from having had to learn it so quickly (see, e.g., Kubler 1985), language skills were used to exclude many Taiwanese from government positions. Individuals caught speaking the native dialects of the Hoklo or Hakka in public places, such as university campuses, were subject to fines. Forty years of such government-sanctioned distinctions between Taiwanese and Mainlanders led to self-identification along these lines. Taiwan's fifth regime change had a ten-year transition period. In 1986, then-president Chiang Ching-kuo (Chiang Kai-shek's son) responded uncharacteristically to a group of Taiwanese activists who had acted illegally by founding a political organization outside the Nationalist Party. Knowing he was dying of cancer, Chiang legalized the formation of political parties, lifted martial law, and set in place the transition to full electoral democracy. Even though Chiang's vice-president Lee Teng-hui—who is Taiwanese Hoklo and a member of the Nationalist Party—won the presidential election in 1996, I consider this transformation of the political system a regime change.

Throughout this period, the political transition had dramatic consequences in the social sphere. For example, not only was Minnan—still the native language of most Hoklo Taiwanese—used in many radio and television programs, but politicians, including Lee and other Nationalists, began giving speeches in the language. These speeches moved ordinary people who remembered being fined for speaking Minnan in public. The numeric majority in an electoral democracy, Taiwanese were no longer oppressed; instead, they were the largest voting constituency, to be courted by politicians.

Such structural changes contributed to the formation of a new Taiwanese identity, bringing together Hoklo, Hakka, and Mainlanders (all Han categories) with both plains and mountain Aborigines, and standing in contradistinction to Chinese in the People's Republic of China (PRC). Many Taiwan Mainlanders visited the PRC after martial law was lifted, and subsequently identified themselves more strongly with Taiwan than China. Aborigines also came to be lauded as indicative of the cultural and ancestral differences between Chinese (in the PRC) and Taiwanese. Many people claimed plains Aborigine identity in the mid-1990s as the "true" Taiwanese—a mixture of Han and Aborigine. This inclusive Taiwanese identity, however, has been challenged since 2000, primarily by increasing social exclusion of Mainlanders (see Corcuff 2000). This new Taiwanese identity, like the democracy that gave rise to it, is still a work in progress.

At least four, and possibly all five, of the regime changes in Taiwan's history led to changes in ethnic identity for large numbers of people. Political shifts major enough to be considered regime changes affect social structure by definition. For example, each regime change affected which group(s) controlled the government. Shifts in access to political power had cascading social effects, from taxation to employment to marriage. They also led to shifts in the treatment of individuals seen as belonging to or excluded from the group of those in power. Structural aspects of regime changes consistently influenced identities, despite the fact that these regimes spanned almost 400 years, distinct political-economic systems, and huge differences in the culture of the ruling elites.

## THE PROCESS OF STRUCTURAL CHANGE AT THE LEVEL OF INDIVIDUALS

Regime changes may appear as periodic shocks to the system—often externally derived, via invasion or colonial annexation, but sometimes internally derived, as with Chiang Ching-kuo's decision to democratize. Such population-scale transformations occur because individuals act, and their actions have consequences that aggregate to constitute the structure of society. The key here, widely recognized by historians and social anthropologists, is that the transformation is not a foregone conclusion—it is not simply a state-dependent transformative process.[3] Consequently, the social interactions of individuals do matter. Detailed ethnographic evidence at the level of individuals indicates how variation in actions that individuals socially negotiated resulted in broader changes in ethnic identity and subsequently in cultural practices.

As already discussed, when the Japanese colonial government banned footbinding among Hoklo in 1915, it removed the last major, visible marker of plains Aborigine identity for Toushe, Jibeishua, and Longtian villagers. An external observer can see removal of the boundary between Hoklo and plains Aborigines as the logical implication of the footbinding ban, but this logic is not sufficient explanation for the process of change. Hoklo did not collectively discuss this logic and so agree to grant Hoklo status to these communities. How, then, was the boundary actually removed? The suddenly elevated rate of bride exchange around 1930 shows that the removal did occur. Prior intermarriage involved primarily poor Hoklo men in a form of marriage (uxorilocal) stigmatized by Hoklo. The increased numbers of Hoklo women marrying (virilocally) into Toushe, Jibeishua, and Longtian in the standard Hoklo marriage form, and the increased numbers of women from these villages marrying (virilocally) out to neighboring Hoklo villages constitute a shift in marriage practices. This shift is the aggregated result of many individual-level decisions about marriage partners.

In Taiwan at that time, parents generally arranged marriages with little or no consultation of the bride and groom, so this shift in marriage practices represents the individual decisions by many parents to bring in daughters-in-law from villages once considered Aborigine, and decisions by many parents to marry daughters off to these same villages. Individuals making these decisions had multiple potential partners, with Hoklo parents having a wider range to choose from than did the villagers of Toushe, Jibeishua, and Longtian before intermarriage took off. Once such marriages occurred, Hoklo would not have wanted to admit or emphasize the Aborigine heritage of their in-laws, and eventually the reputations of entire villages changed.

This new marriage pattern did not emerge without effort and social negotiation. One woman, who told me she was the first Hoklo woman to marry into Toushe as a first-time bride, reported that it took three years for her husband's father to negotiate the marriage agreement with her parents, even though he was an uxorilocally married Hoklo man. We cannot know whether the change to Hoklo identity would have occurred if this particular Toushe man had not been so persistent, but we have to consider that possibility. The household he headed was one of the wealthiest in Toushe at the time (though Toushe was impoverished) and, as a Hoklo man himself, he likely understood the perspective and goals of the Hoklo family with whom he was negotiating. Nevertheless, it required the intervention of his friend—the bride's brother—to finally win the agreement of the bride's parents to the marriage. If he had not been able to accomplish the marital negotiations, I think it possible that most others, in Toushe at least, would have been reluctant to try. In pursuing the marriage, this Toushe man contributed significantly to the ethnic change that occurred, by setting a precedent and creating virilocal, affinal social ties between Toushe and the natal village of his Hoklo daughter-in-law. He took advantage of a social opportunity to solidify Hoklo identity for his son, an opportunity created unintentionally by the footbinding ban, but it was not a foregone conclusion that his negotiations would be successful.

The shift in marriage patterns also instigated a fundamental structural change in the daily lives of individuals in Toushe, Jibeishua, and Longtian. Households there, most of which had previously brought in daughters- or sons-in-law from within their own community or at least one of these three communities, began bringing in Hoklo daughters-in-law from nearby Hoklo villages. These women knew Hoklo culture and would talk about the practices of their husbands' families to their natal family on home visits. At the same time, women from Toushe, Jibeishua, and Longtian could no longer expect to remain in their natal homes, or even their natal villages, after marriage. Most would marry into Hoklo families and villages, where they would observe Hoklo practices and could report them to their natal families on home visits, but where they would also be strangers, without the local social network of their natal home or community. Stranger status affected one's voice in family decisions as well as the degree of cooperation one could expect from other members of the household and neighbors (see M. Wolf 1992). (Women are well aware of the pervasive influence of postmarital residence arrangements.)

As these structural changes moved across the population, they affected cultural practices. For example, the shift away from uxorilocal marriages in Toushe led to men taking over a locally important religious role—spirit medium to the local Aborigine deity (M. Brown 2003, 2004: 108–18). One woman whom the last woman spirit medium wanted to train explained her refusal of the role based on her expectation to marry out of the village. Her marrying out was not a foregone conclusion—she could have asked the spirit medium to intercede with her parents and arrange a local marriage that would have accommodated her accepting the role. In fact, she did marry within the community, but she still did not accept the position because she married virilocally, so she could not necessarily expect successful negotiations with her sisters-in-law to care for her children while she fulfilled the duties of spirit medium. As a result of her refusal, men took over this spirit-medium role, and they instituted changes in the content of the religious practices for which that spirit medium was responsible.

Another example of cultural impact can be seen in the adoption of the common Hoklo practice of tomb-sweeping—visiting ancestral graves annually to clean them and present offerings. Tomb-sweeping was introduced to Toushe, Jibeishua, and Longtian after the shift to virilocal bride exchange with neighboring Hoklo (M. Brown 2004: 90, 104). Initially, villagers only tomb-swept after events that would bring them into close contact with their Hoklo in-laws, primarily a death, virilocal marriage, or birth of a son, and then they only tomb-swept for three years after one of these events. These rites spread because they helped maintain the newly achieved Hoklo identity by meeting Hoklo cultural expectations.

I was not able to locate the pivotal innovator(s) in this case, but I want to emphasize that adoption of tomb-sweeping rites was not a foregone conclusion that followed necessarily from the structural change in marital practices. Although Toushe, Jibeishua, and Longtian residents may well have desired to impress their Hoklo in-laws with their "normalcy" by following important Hoklo customs such as tomb-sweeping, enacting these rites involved more than a desire to do so. It required much more than having sufficient money to provide appropriate offerings (of spirit money and food) and more than coordinating the extended family to spend hours on the appropriate day of the lunar calendar to haul the offerings around to the various grave sites, although designating funds for this purpose and gaining family members' participation would have required social negotiation, too. More dauntingly, initiating tomb-sweeping required years of forethought, because, prior to 1930, few people in Toushe, Jibeishua, and Longtian marked the graves of their dead in any way. A family cannot make graveside offerings if no one can locate the grave sites. Many people I interviewed reported that, into the early 1990s, they still did not tomb-sweep older graves because they had no idea where they were located. Moreover, Taiwan's lush vegetation makes it quite a challenge to mark graves sufficiently to be able to locate the grave a year later (with no tending before then). It is not surprising, then, that the many people who told me that they first marked graves with an uncarved rock also reported that they were often unable to locate the marker (and the grave) when they initially went to perform tomb-sweeping rites some years later. Once those desiring to tomb-sweep realized the need for more substantial grave markers in order to practice this rite, there would still have been a need to negotiate setting aside family funds for such a marker (at the time of the funeral, which would already have been quite expensive by local standards), and the subsequent need to coordinate family offerings and participation on the appropriate festival day each year. The forethought, persistence, and social negotiation skills that innovators must have had to enact tomb-sweeping rites in Toushe, Jibeishua, and Longtian should not be underestimated.

Population-level social change means that many individuals across the population take up the new practice and that some individuals initiated the change by adopting the new practice before it was common. Sociological studies discuss the importance of these initiators (and identify an S-shaped adoption curve across a population, e.g., Rogers 2003), but ethnographic analysis also demonstrates that the spread of practices requires much more than a decision on the part of a single individual to adopt the practice. A process of social negotiation—for example, negotiations about brideprice and dowry and other preexisting, locally relevant marital contract issues—is often required for the potential initiator to be able to enact the desired practice. We have to consider not only who wants to innovate but which of those people are socially able to accomplish the innovation and under what circumstances they are able to do so. I suggest that the process by which a social change moves across a population—that is, which people in the society are successful in enacting the new practice at different points in its trajectory—follows an evolutionary dynamic.

## EVOLUTIONARY DYNAMICS OF SOCIAL STRUCTURE

The empirical data on ethnic changes in Taiwan have shown us interactive influences between social structure and individual actions related to ethnic identity, which have often been misconstrued as derived from the meaningful ideas and beliefs of culture. These data suggest a theoretical revision (or reversion—see Parsons 1951) worth further consideration: social structure and culture are distinct systems, each with its own dynamics, that can interact to influence human behavior. Recent evolutionary perspectives have devoted much attention to the transmission of behaviors and beliefs (see Laland and Brown 2002), but have generally given limited, if any, consideration to how the structure of a human society interacts with such transmission of cultural practices and ideas (see Wolf, this volume; Boyd and Richerson 1985; Durham 1991, this volume; Feldman and Laland 1996; Ihara and Feldman 2004; Aoki, Ihara, and Feldman, this volume; Bowles and Gintis, this volume; Feldman, this volume; Jones, this volume; Starrett, this volume). The empirical data on ethnic changes in Taiwan suggest that consideration of structural influences on actions and, thereby, on cultural practices and ideas, is important to understanding the connection between micro- and macro-level processes of change. Indeed, for ethnic changes, structural influences appear to be more important than cultural ones in motivating individual actions that aggregate to population-level social and cultural changes.

What gives social structure its motivational force? In hierarchical social relations, dominant individuals can impose their will on subordinate individuals (see Durham, this volume). However, the motivational force of social structure is not limited to what one individual may be able to impose on others. It is ubiquitous, and it derives not solely from the imposition of some individuals over others, but in the power dynamics of the social system more broadly (e.g., Bourdieu 1977, 1990; Foucault, 1977, 1980, 1983; Durkheim 1912; Marx 2000; see also Roughgarden 2004).

Social structure is a major part of the environment for human beings, one that we "inherit" from previous generations. Odling-Smee, Laland and Feldman (2003) argue that the ecological environment has an evolutionary dynamic in how it influences populations via niche construction, even suggesting that niche construction offers the potential to link microevolutionary theory with macro-level ecosystems analysis. I suggest that social structure is analogous to the ecological environment in this way, with the same potential for linking individual actions to broad-scale societal change.[4] Social structure exerts a selection influence on culture, because structure influences which cultural ideas can be enacted behaviorally by which individuals, and it is potentially subject to a feedback effect from culture, because cultural ideas influence which practices individuals attempt to enact (M. Brown 1995, 1997b, 2004, in progress). The social system, moreover, in its structure-based dynamic, can influence access to social as well as environmental resources (see also Roughgarden 2004).

Social hierarchy is fluid. Every individual's social position constantly changes—usually subtly, sometimes dramatically, up or down or laterally—at all times, based on his or her own actions or the actions of others. Individuals constantly jockey for position, sort of a Red Queen phenomenon, with people—and perhaps primates and other species more generally (see Byrne and Whiten 1988, Worden 1996, Roughgarden 2004)—acting to maintain or improve their social positions insofar as they can, and not necessarily consciously. Actions are always socially negotiated. Individuals cannot just perform any action they want: one individual tries to carry out an action and other individuals in the social system tolerate, facilitate, or hinder it. Where an individual stands in the social structure—her hierarchical position and networked contacts—at any given point influences which actions are possible for her.

I argue that the influence of social structure on human behavior can be modeled as though individuals are subject to a selection process that leads to differential attainment of social roles or positions (M. Brown 1995, 1997, in progress; derived from Boyd and Richerson 1985: 187; Richerson and Boyd 1992: 76; Durham 1991: 181, 372–73). In this process, the existing social relations cull individuals' attempts to achieve and/or maintain a dominant social position. That is, the ways that individuals seek and/or hold power are based on the specific existing social structure in which they live. Individuals' projections of future social structure are also salient, and these projections are predicated upon their understanding of historical changes in social structure (see also Strauss 1992b).

For a human empirical example of social selection, consider the village monopoly system and historic Aborigine-Han relations in Taiwan (see also M. Brown 2004: 222–23). A Han Chinese middleman paid the Dutch East India Company (VOC) a fee for exclusive rights to trade and collect taxes in a particular Aborigine village. Initially, many Aborigine villages had ties with Dutch missionaries who could mediate leaseholders' exploitation. (Missionaries frequently disagreed with VOC governing policies and greatly distrusted Chinese.) The monopoly system continued under Zheng and Qing rule, but under these regimes there was no mediating group Aborigines could use to reduce exploitation, and abuses became both rampant and extreme. Some Aborigine villages seeking to reduce the extortion nominated their own people to be middlemen, but nominations had to be confirmed by officials. Most Qing officials viewed Aborigines as barbarians but saw identity as dependent on one's father, so Aborigine villages were more successful in getting the son of a Han father and Aborigine mother (who was often the daughter of the Aborigine leader) into the role of middleman.

This example illustrates two important contingencies. Social selection is historically contingent—what happens at one point in time affects the range of possibilities available at another point in time. The monopoly system set up by the Dutch was continued by the later regimes. Social selection is culturally contingent. In other words, there is overlap between the influences of social and cultural selection. The patrilineal ideas of those with social power, the Qing officials, constrained the range of maneuvers Aborigines could make to manipulate social organization in their favor. At the same time, intermarriage between Han and Aborigines expanded the range of possible maneuvers by allowing Aborigines to identify their nominees as Han according to the cultural beliefs of the Qing officials themselves.

Such a form of social selection may require a universal goal of seeking to improve one's social rank as an evolved cognitive mechanism (M. Brown 1995, 2004: 228–31).[5] This goal could be possible because the culling mechanism of social relations—the selection process that determines who within a population actually achieves a particular role—is located in the external social structure, not within individuals' minds. The cognitive ability to perceive what actions are required to achieve a particular role is not the same thing as the ability to socially carry out those actions successfully. The latter ability depends both upon the social roles themselves—which roles exist and what their relative degree of influence is—and upon where the individual in question starts out within the existing social structure of the specific population.

This view of social selection provides connections between micro- and macro-level social changes by explaining that the existing social structure affects who achieves influential social roles, and that role-holders affect what others have to do or be in order to achieve those roles in the future. A holder of an influential role may change the future criteria, thereby changing the dynamics of social selection. For example, Taiwan's late president inherited his position from his father, but instituted reforms leading to electoral democracy. In changing the rules about how to gain office, he also changed expectations about who was likely to hold office, and aspirations about holding office.

Such analysis also provides a new perspective on people who do not act so as to improve their rank. Such people are sometimes treated as unable to accurately judge how to achieve that goal, problematically implying that cognitive abilities differ between those who do and those who do not seek to achieve influential social roles (see also Durham 1991: 463–64). However, some individuals, especially those low in a social hierarchy, may not aspire to influential social roles because the criteria are, for them, impossible to meet (many ordinary roles, such as parents or teachers, are socially influential). Gender, ethnic identity, or school ties may affect the possibility of achieving specific roles, regardless of whether these criteria are explicitly stated in the society (Durham 1991: 427). Aspiration to social roles may be affected by cultural ideas about which categories of people should hold particular roles and by rule changes.[6] As a direct result of Chiang Ching-kuo's rule changes, Chen Shui-bian, who had gone to prison as a political dissident during the martial-law period, was elected president in 2000.

This view of the evolutionary dynamics of social structure is a work in progress. It could potentially contribute to understanding when culture does not affect behavior, and to linking micro- and macro-level changes, but it needs, among other things, independent testing and mathematical expressions that capture the features of the model. Because this view was developed to explain Taiwan's ethnic changes (M. Brown 1995), these data cannot test the selection model. However, two other empirical studies support this view. In the first, research on ethnic change in central China (M. Brown 2001b, 2002, 2004) found that the same processes occurred in an area independent of Taiwan. Nelson Polsby (2004) also arrived at similar conclusions about the importance of social structure in his analysis of the U.S. House of Representatives over six decades. These cases suggest that viewing social structure as a distinct inheritance track that interacts with other inheritance tracks deserves further consideration.

## NOTES

I want to thank Kate Barrett, Ron Barrett, Bill Durham, Marc Feldman, Joanna Mountain, Joan Roughgarden, Jim Truncer, and Arthur Wolf for productive comments and/or discussion related to this chapter.

1. I focus on culture in this essay, rather than ancestry, because when people assume that ancestry motivates ethnic action, it is really the idea of having a common ancestor—thus a cultural idea and not the genetics of ancestry—that is usually assumed to cause behavior.

2. I view "social structure" and "hierarchical social relations" as synonymous terms for the networked and dynamic social system.

3. An example of a state-dependent transformative process is songbirds who will sing in the morning if they are in a state with sufficient energy (from having enough forage).

4. Odling-Smee et al. (2003: chaps. 6, 9) focus on agricultural effects on the ecological environment as "human cultural niche construction." They do not explicitly discuss social structure as distinct from cultural ideas, even in examples that rely fundamentally on social structure (2003: 351, 367; for similar examples, see Boyd and Richerson 1985; Lansing et al. 1998).

5. Worden (1996: 604) suggests the existence of such a goal in some nonhuman primates (see also Boyd and Richerson 1985: 187; Barkow 1992: 632; Cronk 1995: 188; Flinn 1997: 47–48; Bliege Bird et al. 2001; Bliege Bird and Smith 2005). Such an innate goal might derive from an earlier evolutionary need for a social means of gaining access to resources necessary for reproductive fitness (e.g., Roughgarden 2004: 69–74, 146, 175–81).

6. Individual variation in temperament can affect aspiration to social roles. However, individuals themselves may not recognize cultural influence on their interest (or lack thereof) in influential roles. Cultural ideas about gender roles, for example, can inculcate lack of interest in power and label it feminine modesty.

# 9

## A Cultural Species

JOSEPH HENRICH

The effects of culture—understood as learned information stored in brains—
run deep for humans and influence an enormous range of behavior and psychology. I address the importance of human culture in two steps. First, using
examples from foraging, visual perception, and social behavior, I show how
culture is essential for understanding fundamental aspects of human action
and cognition, and suggest that our brains are shaped by learning and developing in culturally evolved environments. In describing this, and setting up the
next step, I argue that much of this enculturation occurs through an ongoing
process of imitation and practice that, while it continues throughout the
human lifespan, has its most substantial influence during our extended ontogeny (approximately the first two decades of life). The second step sketches a
way to approach the cultural nature of humans by examining our evolved psychological mechanisms for cultural learning that allow individuals to effectively extract adaptive behavior from the wash of information available in the
social world. In closing, I highlight how formal theoretical models that integrate these learning processes with social interaction can illuminate a variety
of sociological processes, including the formation of ethnic groups and the
emergence of large-scale cooperation.

### CULTURAL LEARNING IS OUR PRIMARY MODE OF ADAPTATION

In contrast to other primates, humans have successfully spread to nearly every
corner of the globe in a relatively short period of time, from the dry savannahs

and tropical forests of equatorial Africa to the frozen tundra of the Arctic and the humid swamps of New Guinea. Humans are unique in their range of environments and the nature and diversity of their behavioral adaptations. Although our species displays many local genetic adaptations, it seems certain that the same basic genetic endowment produces arctic foraging, tropical horticulture and desert pastoralism—a constellation of adaptive patterns that represents a greater range of subsistence behavior than the rest of the Primate order combined. When it comes to hunting tools, for example, some social groups use blowguns, others use bows, and still others rely principally on boomerangs, clubs, or atlatls. As for social organization, different human groups arrange themselves by clans, matrilines, moieties, phraties, and age-sets (to name just a few). Some social groups are segmentary, linking groups into larger and larger networks of relations; others recognize few relations beyond the immediate family. Similar lists marking the extraordinary range of human behavior—compared to all other species—could be constructed for an immense range of cultural domains.

If it were plausible that these variations were due to genetic differences, and human evolutionary history was tens of millions of years deep, the story would perhaps not be so complicated. Yet humans have less genetic diversity than most other primates (e.g., chimpanzees) and substantially more behavioral variation. Furthermore, the natural experiments resulting from migrations and contact periods demonstrate that many if not most of these differences are maintained by social learning, not simply by exposure to different physical environments (Boesch, this volume). This observation suggests that the immense success and diversity of the human species is rooted in capacities for social learning—our cultural capacities.

The essentially cultural nature of human behavior can be demonstrated with a comparison of bow-and-arrow hunting technology from two groups of nomadic foragers—the Hadza of Tanzania and the San of the Kalahari. Because both groups rely primarily on bows and arrows for bringing down some of the same large game, these groups provide an interesting comparison. Table 9.1 lays out some of the details of Hadza and San hunting technologies, highlighting their differences. To begin, note the substantial difference in the sizes of the bows made by the two groups. Hadza bows are at least twice as long as the bows of San hunters, and it takes at least six times as much

**Table 9.1. Comparison of Hadza and San bow-and-arrow technologies**

| Item/Process | San | Hadza |
|---|---|---|
| Bow length | 1 m or less | 2.0–2.25 m |
| Bow pull | 8–10 kg | 60 kg |
| Bow shaft material | *Grewia flava* | *Dombeya kirkii* |
| Bow string material | Tendon of gemsbok, kudu, or eland | Nuchal ligament of zebra, eland, or buffalo, or the sinew of a giraffe |
| Bow string processing | Soaked and separated into fibers that are twined into a string 4 m long, which is shortened via "gravity twisting" | Fibers are chewed until soft and then rolled on the thigh |
| Securing and protective materials | Sinew bindings attached near end of bow to secure bowstring | Fresh skin from the tail or metapodials of impala, eland, or giraffe |
| Arrow shaft material | Reeds, grasses, or *Grewia flava* | Light woods with a pith core |
| Tuning bow string to correct tension | Based on musical pitch | Unknown |
| Fletching | None | For poisoned arrows: feathers from Guinea fowl or vulture |
| Fletching attachment | NA | Mastic and helically wound single fibers of sinew plus glue from plant bulb |
| Arrow head | Fence wire (formerly ostrich bone) | Metal with one or two barbs (two barbs for female game, one for males) |
| Arrow poison source | *Diamphidia* sp. (beetle larvae and protective | Two types: shanjo seeds (*Strophanthus eminii*) and |

Table 9.1. (continued)

| Item/Process | San | Hadza |
|---|---|---|
| | casings), often mixed with *Acacia* gum | panjube sap (*Adenium* sp.) |
| Location of poison source | 20 cm to 1 m below ground near rare *Commiphora* bushes; typically harvested in late summer | NA |
| Processing of poison | Larval casing are rolled to homogenize them, mixed with saliva, then baked to a crust on the arrow shaft | For *Panjube* sap: chopped-up branches of *Adenium* are squeezed and slowly cooked into a tarry black substance |
| Poison application and protection | Poison applied to upper shaft | Poison applied to arrowhead, which is then wrapped in impala hides |
| Poison use | Almost always | Primarily for big game |
| Quiver size | 75 cm | No quivers |
| Quiver materials | Outer bark of lateral roots of *Acaci luederitii* tree | NA |
| Quiver materials acquisition | Dug up and cut out of ground, avoiding sections with emerging rootlets | NA |
| Quiver materials processing | Any rootlets must be drilled out; root length is roasted; steam allows outer covering to be separated with pounding and twisting action | NA |
| Quiver assembly | Root sheath is bound with wet sinew; one end is sealed with moist hide | NA |

SOURCES: Compiled from Woodburn 1970, Lee 1979, Silberbauer 1981, Liebenberg 1990, Bartram 1997

force to draw them. The Hadza fletch their poisoned arrows with feathers from either guinea fowl or vultures, whereas San do not fletch at all. Both groups use poison for big game, but use very different poisons. The San manufacture their poison from chrysomelid beetle larva and casings found in the ground around corkwood bushes. The Hadza, despite living around numerous cork-wood bushes, predominantly use two kinds of vegetable poisons, one made from shanjo seeds (*Strophanthus eminii*, a tree-bush), and the other (their pre-ferred poison) from sap that is extracted by boiling panjube (*Adenium* sp.) branches until a tarry black substance emerges. San go through an extensive process to manufacture a quiver from the large lateral roots of an Acacia tree, whereas among the Hadza, only the western populations use quivers, which they craft from hide (Woodburn 1970: 31).

Clearly, the skills, detailed knowledge, and procedures that go into manu-facturing this equipment, which is essential to bringing down big game, is acquired predominately through some kind of imitative learning process. There is no way an individual can figure out all the details (such as where to find the beetle larva, or which branches to boil) that go into making a success-ful hunting kit, without learning extensively from others. In all the ethnogra-phy on Kalahari San foragers, we see no evidence that Kalahari hunters have experimented with longer (more than two meters) bows and fletched arrows, only to later reject these alterations. Woodburn (1970: 14) reports that enter-prising Hadza have occasionally manufactured their bow staves from woods other than *mutateko* (*Dombeya kirkii*), but have always returned to *mutateko*—that is, most Hadza have never experimented with alternative woods. Kalahari foragers have not been observed to routinely test a range of beetles, seeds and branches for their poison-making possibilities—they just learn to gather and process chrysomelid beetles from other group members. If the acquisition of the adaptive repertoire were principally a product of individual learning, every person would have to go through a trial and error process in which fletching, various potential poisons, and different bow sizes were tested. This stochas-tic process, based on a small number of trials and riddled with errors, would generate massive within-group inter-individual variation, as different hunters would find different poisons and different-sized bows most effective. Instead, although individual variation certainly exists, much of the variation is between groups. More importantly, detailed ethnographic observations corroborate this inference by showing that such manufacturing skills are acquired through a process of imitation and practice, not by free-ranging individual experimentation (Fiske 1998). In our species, cultural learning is essential to even the most basic elements of foraging.

One obvious question arises from the observation that individuals acquire much of their behavioral repertoire (such as making a hunting kit) from other individuals in their social group. If most people imitate, how did these intricately integrated cultural adaptations arise in the first place? This problem can be solved by understanding the psychology of human cultural learning and attention. If individuals, in the course of learning, pay particular attention to the most skillful or successful individuals in their groups (e.g., to the best hunters or arrow makers), and people make some errors in imitation—or even occasionally innovate—culturally learned repertoires can become increasingly adapted to local environments. Furthermore, this can occur without cost-benefit analysis, rationality, genetic change, or individual learning (Henrich 2002). More generally, the cumulative nature of human knowledge, practices, and technology cannot be understood without examining cultural learning.

A critic might be quick to suggest that the technologies discussed above could be explained as "optimal" given the details of the local ecology. Although this may be true (it's impossible to say), the claim is irrelevant for deeming such practices "cultural," or for studying them as cultural evolutionary features produced by the psychological processes that allow humans to acquire behavioral information by observation and imitation.

## CULTURALLY CONSTRUCTED COGNITIVE ARCHITECTURE

The danger in the above examples is that, although they illustrate both the culturally learned and adaptive nature of human practices and behavior, they run the risk of suggesting that the mind can be effectively divided into structure (the acquisition machinery) and contents (the knowledge and practices). However, a wide variety of evidence suggests that learning involves the construction of brain structures (the "wiring"), which occurs throughout the human life course but particularly during ontogeny (Quartz and Sejnowski 1997; Quartz 1999, 2002). Cortical gray matter continues to increase in the frontal and parietal lobes up to age twelve, in the temporal lobe until age seventeen, and in the occipital lobe through the first two decades of life. White matter increases in all areas until around age twenty-two (Giedd et al. 1999). This extended ontogeny, which allows human brains to construct and adapt themselves to their local social and physical environments (the EOA—environment of ontogenetic adaptiveness), may lead to a wide array of cross-cultural variation in people's susceptibility to visual illusions, hunting skills, memory, notions of fairness, and tastes for punishing. In short, people from different cultures—having experienced different social and physical environments while their brains were developing—likely possess different cognitive architectures. The criticality of ontogeny in adult performance is consistent with field data such as those on the acquisition of foraging skills among the foraging Aché— foraging skills take a long time to perfect, and depend critically on childhood environments.

Experience in a particular environment shapes brain circuitry. Feldman and Knudsen (1997) compared the brain topography and circuitry of two sets of barn owls. The first group had prism goggles (which inverted the visual field) cemented to their heads soon after birth, while the second (control) group was left unadorned. After a period of weeks, the goggle-wearing owls had adapted their ability to visually locate the source of sounds in space to equal the sound-locating performance of their non-goggle-wearing brethren. After this period, the researchers compared the optic tectums—the portion of owl midbrain involved in visual processing—of the two groups, and demonstrated a topographical reorganization and the formation of new anatomical axonal projections (DeBello et al. 2001; Hyde and Knudsen 2002). That is, the goggles caused the juvenile owls' brains to adaptively rewire themselves (Quartz 1999).

## CULTURALLY CONSTRUCTED ENVIRONMENTS ALTER VISUAL PERCEPTION IN HUMANS

As in owls, human visual perception seems to respond to experience, and by conjectural inference from the owl data, human brains may rewire themselves via experience. Consistent with this hypothesis, cross-cultural experimental data demonstrates substantial differences in visual perception across populations. Building on W. H. R. Rivers's pioneering work, Segall et al. (1966) performed one of the few rigorously controlled cross-cultural experimental

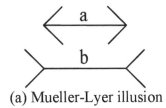

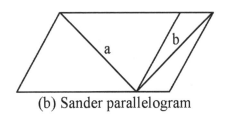

(a) Mueller-Lyer illusion          (b) Sander parallelogram

9.1. Two of the illusions used by Segall et al. (1966) in their cross-cultural study of illusion susceptibility. The lines labeled *a* and *b* in each figure are the same length, but some subjects perceive line *b* as longer than line *a* in the Mueller-Lyer illusion, and *a* as longer than *b* in the Sander parallelogram.

projects in the history of anthropology and psychology. This interdisciplinary project gathered data from both children and adults in a wide range of human societies about their susceptibility to five "standard illusions." Their results are numerous, so I will summarize only their findings for two of these visual stimuli, the Mueller-Lyer and Sander parallelogram illusions.

In the Mueller-Lyer illusion (figure 9.1a), subjects from industrialized societies typically perceive that the horizontal line segment marked *b* is longer than the horizontal line segment marked *a*, when in fact both are the same length. By varying the lengths of lines *a* and *b* and asking subjects which of the two is longer, researchers can estimate the magnitude of the illusion for each subject—by determining the approximate point at which an individual perceives the two lines as being the same length. For the Sander parallelogram (figure 9.1b) subjects must again determine which line segment is longer, *a* or *b*, and subjects from industrialized societies typically perceive *a* to be longer than *b* even when they are the same length. Again, using a series of different figures that vary the relative lengths of *a* and *b*, researchers can assess the illusion's strength by finding the point at which subjects perceive the line segments as equal.

Figures 9.2 and 9.3 summarize the results for the Mueller-Lyer and Sander parallelogram illusions, respectively, for the societies studied by Segall et al. The seventeen societies include eleven groups of African agriculturalists (some of whom also rely on foraging and pastoralism), one group of African foragers (San), one group of South African Europeans (in Johannesburg),

one group of Australian Aboriginal foragers (Yuendumu), one group of
Filipino horticulturalists (Hanunóo), one mixed group of South African gold-
mine-laborers ("SA Miners" in the figures) and two groups of "Westerners"
(Europeans and Americans). In twelve of these seventeen societies, data were
gathered from both adults (equal proportions of males and females, ages
18–45) and children (ages 5–11). In figures 9.2 and 9.3, the left-hand vertical
axis gives the "point of subjective equality" (PSE) and corresponds to the vertical bars for adults (white) and children (black). PSE is a measure of the
strength of the illusion for each group. It represents how much longer segment $a$ must be than segment $b$ before people perceive them as equal (until
there is a fifty-fifty chance that people from that group will choose either $a$ or
$b$). The right-hand vertical axis gives the difference between the PSE of the
adults and the PSE of children for each group, and refers to the scatter of tri-
angular data points above the vertical bars.

The results for the Mueller-Lyer stimuli show substantial differences
among these social groups in their susceptibility to the illusion. American
adults in Evanston (Illinois) are the most susceptible. On average, these adults
require that segment $a$ be about a fifth longer than $b$ before they perceive
them as equal (PSE = 19 percent). At the other end of the susceptibility spectrum, San are virtually unaffected by this so-called illusion, requiring that segment $a$ be only slightly longer than segment $b$ before seeing them as equal
(PSE = 1 percent). Figure 9.2 shows that although there is significant variation
across the range of social groups, there is a notable jump between Evanston
and the rest of the world.

Comparing across societies, children show a pattern similar to that of adults
on the Mueller-Lyer illusion. PSE scores range from over 20 percent among
Evanston children to 0 percent among Suku children. The PSE scores for
children correlate with those for their adult counterparts: $r = 0.81$, indicating
that most of the cross-cultural effect is in place by age eleven. Moreover, the
amount of intergroup variation drops from a standard deviation of 5.5 among
children to 4.5 for adults—that is, there is more cross-cultural variation in
children than adults, so adolescence must act to reduce this variation. Developmentally, the scores show a fairly robust pattern: adults are consistently less
susceptible to the illusion than children. This is illustrated by the scatter of triangles on the upper portion of figure 9.2. The triangles (which refer to the

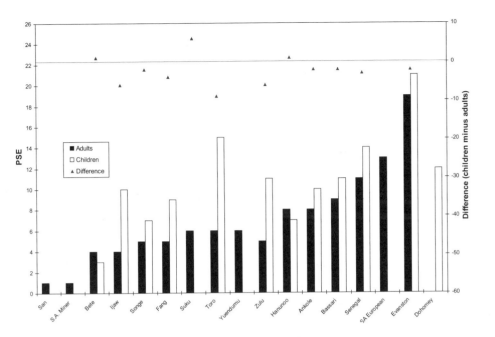

9.2. Mueller-Lyer results for the Segall et al. cross-cultural project. PSE is the percentage by which segment *a* must be longer than segment *b* for individuals to perceive them as equal. There is no data for children among the San, S.A. miners, Yuendumu, and S.A. Europeans. There is no adult data for the Dohomey. The Suku children's data has a mean of zero, as shown.

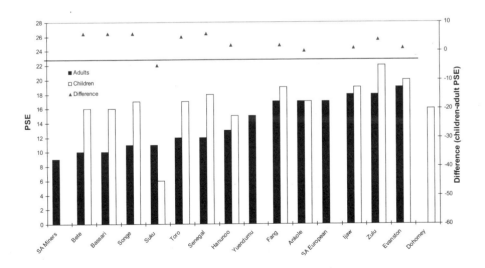

9.3. Sander parallelogram results from the Segall et al. cross-cultural project.

right-hand vertical axis) plot the difference between the PSE scores of the adults and children in each society. With three exceptions, the adults' scores are lower than those of the children in their society. Of the three exceptions (the Bete, Suku, and Hanunoo), only Suku children were, on average, not susceptible to the illusion (and provided the lowest score of all the groups—PSE = 0). These findings are consistent with more detailed developmental data from U.S. populations showing that adults are less susceptible to the illusion than children (Walters 1942; Wohlwill 1960). I return to this pattern below. Note also that although children were generally equal to or greater than adults *from their social group* in susceptibility to the illusion, this pattern does not hold if we compare children and adults from different societies. Many child samples are less susceptible to the illusion than adult samples from other societies.

Figure 9.3 shows similar findings for the Sander parallelogram. As with the Mueller-Lyer illusion, susceptibility to this illusion varies from a high of 19 percent in Evanston to a low of around 10 percent for the SA Miners and Bete (agriculturalists/foragers from the Guinea coastal area). The PSE scores for children and adults show the same patterns discussed above. Children's scores correlate with adults at 0.60, and if we remove the outlying Suku, the correlation jumps to 0.76. The variation among groups in PSE scores drops from children (std. dev. = 5.8) to adults (std. dev. = 3.47). Thus, children's susceptibility to this illusion also is greater than or equal to that of the adult members of their social group for all groups (except the Suku). Interestingly, there is a gradual decrease in the size of the adult adjustment with increasing children's PSE scores, leading to the same general point: most of the adult-level variation is in place by age eleven, and adolescence (ages twelve to twenty) acts only to reduce the size of the variation.

Detailed developmental data from several studies in the United States on the Mueller-Lyer illusion show that susceptibility generally decreases from ages five to twelve, reaching its lifetime low at the onset of adolescence, and then increases from ages twelve to twenty. The decrease in susceptibility from age five to age twelve is larger than the subsequent increase, leaving adults less susceptible to the illusion than five-year-olds, but only because of the preadolescent decrease (Wapner and Werner 1957). After age twenty, susceptibility to this illusion does not change again until old age (Wapner et al. 1960; Porac and Coren 1981).

Together these data suggest several important inferences. First, carpentered environments (built environments with lots of right angles, in construction, furniture, etc.) and perspectival art (artistic techniques to convey depth), or whatever it is that accounts for the cross-group differences in illusion susceptibility, likely have their effects up until age twenty, but not afterwards. Second, whatever causes the variation has much of its effect before age eleven, otherwise the pattern for children in the cross-cultural sample would not mirror the adult pattern. Third, these data suggest that variables such as "experience in a carpentered environment" can be misleading. What matters is not "experience in carpentered environments" (or whatever the relevant variable), but rather "experience in a carpentered environment *before* age twenty." This is relevant for critics of the "carpentered environment" hypothesis, who have suggested that it fails because males and females in many of these societies have experienced substantially different amounts of contact with carpentered environments (for example, males spend much more time seeking wage labor in cities), yet males and females consistently show little or no difference in illusion susceptibility. However, from the cultural developmental perspective, the observation confirms rather than contradicts the hypothesis, as females and males live in nearly identical visual environments from infancy to age twelve, when much of the effect seems to occur.

While different visual experiences during ontogeny affect both owls (as seen in the Feldman and Knudsen 1997 experiment discussed above) and humans, there is an important difference: the human effect is the product of growing up in a culturally evolved environment. As with foragers' bow-and-arrow technology, rectangular houses, "carpentered corners," angular furniture, and perspectival drawings are all products of particular cultural evolutionary trajectories. These environmental elements evolved non-genetically, through social learning, over centuries. There was surely a time in human history when—assuming that Segall et al.'s hypotheses are correct—the Mueller-Lyer illusion was not an illusion at all. Illusions such as this probably came into existence through an interaction between a particular line of cultural-technological evolution and the ontogenetic processes of brain development. Cultural evolution has likely generated changes in brains without any changes in genetics.

## CULTURALLY EVOLVED SOCIAL BEHAVIOR

As with their visual environments, human groups also experience different culturally evolved social environments. It is within these social environments that human brains continue their ontogenetic development. Through the ontogenetic processes, which deliver most of their effects before the age of twenty, individuals acquire cultural rules, preferences, and beliefs, and the mental models of the social and physical world that propel their behavior. This learning process is fundamentally social: A combination of cross-cultural anthropological work (Lancy 1996; Fiske 1998), developmental psychology (Karmiloff-Smith 1994; Meltzoff and Prinz 2002), and neuroscience (Quartz and Sejnowski 1997, 2000; Quartz 1999, 2002) suggest that children acquire their cultural understandings of the world though a process of imitation and practice that builds an increasingly hierarchical, integrated, and abstracted understanding of the world. First, learners acquire the rules of behavior by observation and direct imitation of the simplest components (a pattern that parallels language learning). Then they practice these bits of behavior in an ongoing process of rehearsal. As they master the bits, they gradually connect them through both higher-level forms of imitation (e.g., imitating strategies) and direct experience. Some developmental psychologists have characterized infants and children as "imitation machines" (Tomasello 1999a). These developmental patterns, which strongly differentiate humans and nonhumans, suggest that imitation is one of the essential developmental tools that natural selection has deployed to ratchet up human cognitive abilities and adapt us to the vast range of local social and physical environments (e.g., New Guinea swamps and dual moieties systems) mentioned at the outset of this chapter.

A combination of recent experiments deployed both cross-culturally and developmentally suggests that growing up in a particular place has a sub- stantial impact on social behavior. More specifically, experimental techniques designed to measure an individual's social preferences—e.g., their altruism, sense of fairness, and taste for punishing unfairness in anonymous others— yield evidence of important cultural variation. The experiments have suggested, among other things, that these social preferences are principally acquired during the first twenty years of life, although relatively smaller modifications may occur later. It has also become clear that growing up in different places results in quite different patterns of adult behavior, and these cultural patterns are likely largely in place before adolescence. The learning sequence for social behavior follows a pattern of learning the rules (or cultural models) first, and later integrating those rules with strategic considerations that operate within the context of the rules and associated expectations. As a corollary to these learning processes, having grown up in a particular place has a substantially larger impact on adult behavioral variation than do individual-level variables such as sex, age, income, wealth, education, and wage-labor participation. Thus, the substantial variation in social preferences that we observe across human social groups likely results from experience in different EOAs (environments of ontogenetic adaptiveness) and from the human propensity for cultural learning, not principally from differences in adult experiences (in parallel with illusion susceptibility). Of the various experiments that have yielded these observations, the ultimatum game is a particularly good example.

The ultimatum game (UG) is a two-person bargaining experiment that has been extensively tested on college undergraduate populations by experimental economists, psychologists, and economic sociologists (for reviews, see Roth 1995; Camerer 2003). In the baseline experimental setup, two players are anonymously paired to divide a sum of real money (the games are played with cash, which they players actually receive, and there is no deception). The first player, often called the proposer, must decide how much of the total sum (the pot) to offer to a second player, who is called the responder. Both players know how much money there is in the pot. Upon receiving an offer from the proposer, the responder must decide whether to accept the offer or reject it. If the responder accepts the offer, he/she receives the amount offered, and the proposer gets the remainder. If the responder rejects the offer, both players get zero (the pot vanishes). Both players are fully informed of the situation: they know the game is one shot (it will not be repeated) and that they will never know the identity of the other player. From the perspective of self-interested rational actors, this experiment leads to a straightforward prediction: responders should accept any nonzero offer, and proposers, realizing this, should offer the lowest nonzero amount possible.

Using adult participants from industrialized societies, the UG show robust results. The strong modal offer is a fifty-fifty division. Offers above 50 percent of the total pot and below 30 percent are rare, and low offers are often rejected. Among undergraduates, for whom we have by far the largest database, the overall patterns are similar, although offers tend to be slightly lower. This research also shows that stake size (the amount of money in the pot), sex and adult age (for those over twenty-two) do not significantly influence game behavior (Camerer and Hogarth 1999; Camerer 2003).

Recently, Harbaugh et al. (2002) have begun administering these experiments to children in rural Oregon in order to explore the developmental trajectories of social preferences. As an additional point of comparison, I have included data from a sample of graduate students from UCLA that provide a point of reference for adults (over the age of twenty-two). Interestingly, the youngest children (mean age seven, in second grade) conformed most closely to the economists' model of rational self-interest—they made smaller UG offers than did older children (and adults), and were more likely to accept lower offers. Figure 9.4 provides a comparative plot for these data. Each age cohort is labeled along the vertical axis such that the distributions of offers for each age cohort can be examined by reading horizontally across the possible offer amounts, which are marked on the x-axis. The relative sizes of the bubbles graphically show the proportion of the samples for each cohort that made the corresponding offer. In the 50 percent offer bubbles (0.5 on the horizontal axis) I have included the actual percentage of the cohort who made that offer. The plot illustrates a gradual movement from higher variance and a greater proportion of low ("selfish") offers among seven-year-olds to lower variance and a higher proportion of fifty-fifty offers among older individuals. The trajectory is not linear, however. By age nine (fourth and fifth grades), 68 percent offer 50 percent, and only 18 percent are making offers of 20 percent and below. However, by age eighteen (twelfth grade), the lower offers have entirely disappeared, yet the fraction of fifty-fifty offers has also dropped to 43 percent, reflecting a shift from 50–percent to 40–percent offers—as we'll see, this may reflect an increase in self-interested strategic thinking. In total, 88 percent of the eighteen-year-old cohort makes either fifty-fifty or sixty-forty offers. These data, in light of previous work on moral development (e.g., Kohlberg 1976), suggest that by age nine (fifth grade) most children in our society have learned the cultural rules or cultural models that govern behavior in the UG situation, and, lacking a higher-level integration, they stick close to these rules. After age twelve, they display an increasing amount of strategic reasoning within the context of these cultural rules.

Figure 9.5 examines both the learning trajectory of the taste for punishment (as measured by responder's behavior) and the amount of strategic behavior on the part of proposers in different age cohorts. This plot includes data from adults in rural Missouri and from Chaldeans (Catholic immigrants from Iraq; see Henrich and Henrich, 2007) in Detroit to provide adult points of reference (the UCLA data lacked any rejections). The black portions of the bars in figure 9.5 give the income-maximizing offer (IMO) for each age cohort. The IMO is the amount a proposer would offer if he wanted to maximize his income from the game, and he had full knowledge of the probability of rejection at each possible offer amount. Thus, a group's IMO captures the strength or willingness of its responders to punish (by rejecting) offers they deem "unfair." The black bars in figure 9.5 show the development of an increasing taste for punishment with increasing age. The black and white bars together (stacked) reach to the mean offer for each group; thus, the white portions of the bars express the difference between what the proposers give and what the responders demand. In the Missouri sample, the IMO (50 percent) is higher than the mean offer (48 percent)—hence the top-down black-to-white shading.

It is worth noting that the sense of fairness (as represented by the UG offer) reaches near its adult value (the modal offer) by age nine, and the acquisition of a taste for punishment lags behind it. This sequence is interesting, as some economists have supposed that it is an awareness of the taste for punishment that drives the apparent fair-mindedness of proposers among university populations. Developmentally, this is not the case, as a sense of fairness precedes the corresponding taste for punishment: children first learn the normative behavior (give half), and only later develop the motivation for punishing norm violators at a cost to themselves.

Some researchers might seek to view the above trajectories as a more or less fixed pattern of moral development that is part of a reliably developing aspect of species-specific cognitive architecture. For example, the robust results of the UG among university students were initially interpreted by some as the operation of evolved cognitive modules for social interaction (Hoffman et al. 1998; Nowak et al. 2000) that arose in the human lineage via natural selection

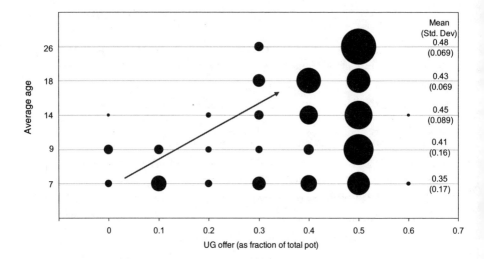

9.4. The distribution of ultimatum-game offers for five different age groups. The distribution reads horizontally, with the size of the bubble illustrating the proportion of the sample making the corresponding offer.

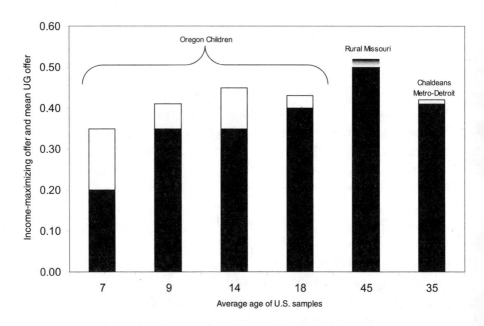

9.5. Income-maximizing offer (IMO) means UG offers compared for a range of age cohorts in the United States. IMO provides a measure of each group's taste for punishment.

operating through the logic of repeated interaction (Trivers 1971; Axelrod 1984). This evolved piece of cognitive machinery, which would have allowed humans to take advantage of cooperation among small groups of repeated interactants, is assumed to "misfire" in the context of the one-shot UG, thereby causing people to behave fairly and punish unfairness (this argument has severe empirical and theoretical problems, however; see Fehr and Henrich 2003 and Hruschka and Henrich 2006). However, UG data gleaned from fifteen small-scale societies as part of a unified inquiry into human social behavior makes this position increasingly difficult to maintain (Henrich et al. 2005; Henrich et al. 2006). The results show substantially more variation across the human spectrum than is observed in comparing U.S. samples ranging in age from seven to seventy.

Figure 9.6 presents the UG offer distributions in the same format as used for the developmental data in figure 9.4. The social groups represented include three groups of hunter-gatherers (Lamalera, Aché, and Hadza), four of pastoralists (Orma, Sangu herders, Kazakhs, and Torguuds), six of horticulturalists (Quichua, Machiguenga, Achuar, and Tsimane) including two (Gnau and Au) who incorporate significant foraging, and three of small-scale agriculturists (Shona, Mapuche, Sangu farmers). Focusing on proposers' offers, the results show substantial variation, both within and across groups. The mean offers range from 25 percent among the Quichua of the Ecuadorian Amazon to 58 percent among the whale hunters of Lamalera (Indonesia). The modes span the range from 15 percent among the Machiguenga of the Peruvian Amazon to 50 percent among a wide range of groups. However, although the range of mean offers is large, they span less than 50 percent of the possible spectrum. More than 80 percent of all offers fall between 10 percent and 50 percent (inclusive). Analysis of individual offers using multivariate linear regressions that include age, sex, income, education, wealth, and social group (group-level dummies), shows that only the variable "social group" captures any significant portion of the variance.

Comparing adult behavior from these societies to both U.S. adults and children reveals substantial differences, with some societies favoring lower offers than the U.S. seven-year-olds, and showing less preference for the fifty-fifty offers. Thus, the learning trajectories that produce adult social behavior among Machiguenga, Quichua, Hadza, Tsimane, and Mapuche likely fol-

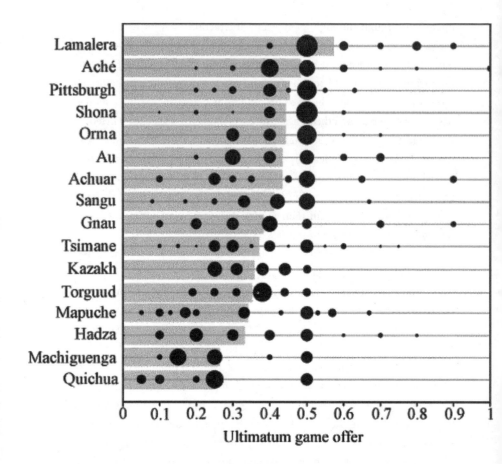

9.6. Ultimatum-game offers among adults in a variety of social groups from around the world. Bubble size represents the proportion of offers made at each amount for each group. The lightly shaded horizontal gray bar represents the mean offer for that group.

low quite different paths than those observed in Oregon, with substantial divergence in place by age nine.

As among the proposers, the behavior found among responders shows substantial cross-group variation. Some groups, such as the Tsimane and Machiguenga in the Amazon, have few or no rejections at all, despite a large number of low offers. Thus, the hypothesis that humans will reliably develop a taste for punishing unfairness (or inequity), and that this taste will be extended to include individuals in one-shot, anonymous transactions, is not supported. Yet, by age seven, children in Oregon have already acquired some taste for punishing anonymous others. The combination of these results suggests that the cultural differences in tastes for punishment are emerging by age seven and substantial by age nine. At the other end of the spectrum, groups such as the Hadza, Gnau, and Au show rejection rates as high as, or higher than, those found for university students. Interestingly, the two New Guinea groups (the Au and the Gnau) both show a willingness to reject offers of more than 50 percent. Subsequent comparative experimental work has both replicated these general findings, and revealed a widespread willingness to reject high offers (i.e., offers greater than 50 percent; Henrich et. al. 2006).

For groups with more than two rejections, we estimated their income-maximizing offers and plotted them in figure 9.7, using the same approach used for U.S. age cohorts in figure 9.5. Note that IMOs could not be estimated for several groups because of a lack of rejections (despite low offers); thus, the societies represented in figure 9.7 likely represent a high end of the IMO spectrum. As in figure 9.5, the black-to-white shading used on the Sangu farmers' bar indicates that their IMO was higher than their mean UG offer. These cross-cultural results show that IMOs in adult populations vary from below that found among seven-year-olds in the United States to roughly the same as that found among U.S. adults. The difference between UG means and IMOs (the white section) also varies substantially, from near zero for Hadza foragers and Sangu farmers to large values among the Achuar and the whale hunters of Lamalera.

The responder data also show, consistent with the proposer findings, that age, sex, income, education, and wealth do not significantly predict the likelihood of a rejection. Examining the data across all groups, only social group and amount offered by proposer are important predictors of responders' decisions to accept or reject. The likelihood of a rejection increases as the proposer's offer decreases, but even controlling for this, the likelihood of rejection still varies significantly across our fifteen groups. The combination of these develop-

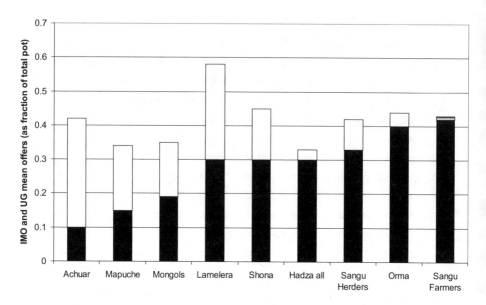

9.7.   IMO and UG mean offers for several societies. The black bars give the IMO and the stacked black-white combination reaches the mean UG for each group.

mental and cross-cultural results suggests that, like the sense of fairness, a taste for punishment is—at least partially—an acquired taste, whose flavor and intensity strongly depend on the cultural environment one grows up in.

I have highlighted here two age trends in UG behavior. Among Oregon children, we observe that both a sense of fairness (i.e., a preference for equitable distributions) and a taste for punishment increase (non-linearly) with age, between the ages of seven and twenty-six. However, among adults from both industrialized societies and a wide range of small-scale societies, no age effects have been observed. In fact, individual-level variables such as age, sex, income, education, and wealth are generally nonpredictive. In light of the above data, I argue that the first process (cultural learning during ontogeny) explains the second observation (little importance of adult individual-level variables). If individuals acquire their "taste for punishment" and "sense of fairness" principally during ontogeny, as Harbaugh et al. (2002) suggest, then adult variables such as age, income, and wage-labor participation which may vary over an individual's adult life course, will show relatively small effects vis-à-vis the cultural environment of ontogeny—because social preferences are mostly constructed during ontogeny.

This explains another pattern in the cross-cultural UG data: the presence of group-level effects of market integration in the absence of any significant individual-level effects of market participation (for further discussion, see Bowles and Gintis, this volume, and Henrich et al. 2004). The analysis shows that a *group's* market integration is a significant predictor of its mean UG offer: higher levels of market integration predict higher UG offers. However, looking at the within-group analyses across the field sites, we find that individual-level measures of market exposure (such as wage-labor participation or cash cropping) consistently fail to reveal predictive effects. Typically, the argument about the effects of markets runs like this: the experience of working or operating in markets and market institutions alters an individual's beliefs or preferences vis-à-vis the kinds of anonymous transactions that characterize the UG, thereby leading to higher offers (i.e., more fifty-fifty offers). Thus, market exposure affects individuals directly as adults, and group-level differences merely represent the aggregate effects of the adults' market exposure. From such theoretical premises, our group-level effects of market integration stand as a partial confirmation of the hypothesis, but the failure of individual-level measures presents a prickly puzzle.

However, now consider this theoretical proposal from the perspective of the nine-year-olds in rural Oregon, 77 percent of whom offered 40 percent or 50 percent of the total pot in the UG, and who had a mean offer of 41 percent (equal to or greater than most of the adult samples in small-scale societies). It seems likely that if we were to measure the market integration of these children using variables such as wage labor or commercial selling, we would find them to be one of the least market-integrated groups in our fifteen-society sample, yet they have one of the highest mean UG offers and a substantial IMO compared to many of our social groups. Although they do not actually participate in any wage labor themselves (or do so very little), they are growing up in one of the most market-integrated societies in the world. They are indirectly acquiring the cultural models, preferences, habits, and beliefs that will allow them to function effectively as adults in this society. Thus, this combination of cross-cultural and developmental data suggests that growing up in a more market integrated society may make one more likely to offer half in the ultimatum game as an adolescent or adult, but adult market experience may not substantially alter behavioral patterns acquired during ontogeny. This approach predicts that group-level measures, to the degree that they reflect and correlate with the environments of onto-genetic adaptiveness of the experimental participants, will predict the mean behavior of the group. At the same time, this cultural learning approach predicts that post-adolescent adult variations in income, age, and wealth will not substantially predict game behavior, although small effects should be anticipated.

## PROGRAMMED FOR ADAPTIVE CULTURAL LEARNING

Unfortunately, these observations of cultural differences alone do not get us very far: People learn their culture by growing up in a particular place, and this learned culture affects their decisions and behavior. These observations, which have been essentially ignored in the fields of economics, biology, political sci-ence, and most of psychology, have been at the foundation of anthropology since early in the twentieth century, and have provided the unquestioned point of departure for most anthropological inquiry. Yet, although they are likely true in the broadest sense, they have not proved particularly useful, in my view. To remedy this problem, evolutionary theory can be applied to the ontogenetic dilemma faced by individuals growing up in a particular place, in order to understand the psychological learning processes that individuals use to adap-tively acquire their behaviors, skills, preferences, tastes, knowledge, and men-tal models of how the world works.

The claim here is that an understanding of culture and cultural learning requires an understanding of the psychological processes that construct our minds, making them "self-programmable" (Pulliam and Dunford 1980). From this point of departure an immense variety of questions can be posed. One question that allows theorists to address a variety of anthropological issues related to cultural adaptation, diversity, and history focuses on how individu-als figure out who, what, and when to imitate. An evolutionary approach suggests that natural selection will favor cognitive mechanisms that allow individuals to more effectively extract adaptive information, strategies, prac-tices, heuristics, and beliefs from other members of their social group at a lower cost than that demanded by vertical transmission or individual learning (e.g., trial and error learning). One such mechanism is known as prestige-biased transmission.

## PRESTIGE-BIASED TRANSMISSION

If individuals vary in skills (e.g., tool making), strategies (e.g., game-tracking techniques), and/or preferences (e.g., for foods) in ways that affect fitness, and at least some components of these differences can be acquired via cultural learning, then natural selection may favor cognitive capacities (or biases) that cause individuals to preferentially learn from more skilled or knowledgeable individuals. The greater the variation in acquirable skills among individuals, and the more difficult those skills are to learn via individual learning, the greater the pressure to preferentially focus one's attention on, and imitate, the most skilled individuals. In laying out this evolutionary process, Francisco Gil-White and I (2001) have called this learning capacity "rank-biased transmission": individuals rank potential cultural models (individuals they may learn from) along dimensions associated with underlying skills (e.g., hunting returns), and focus their social learning attention on the most highly ranked models (those most likely to possess acquirable skills, practices, etc.).

This theory suggests that because underlying traits such as skill and knowledge are often difficult to observe directly, success and achievement measures are used as proxies. This explains the widespread observation that people copy successful individuals. Further, because the world is a noisy, uncertain place, and it's often not clear which of an individual's many traits have led to his or her success, this approach suggests that humans have evolved the propensity to copy a wide range of cultural traits from successful individuals, only some of which may actually relate to the individuals' success. If information is costly, this strategy will be favored by natural selection, even though it may allow neutral and somewhat maladaptive traits to hitchhike along with adaptive ones. In a world of costly information, cognitive adaptations don't always produce adaptive behavior (even in ancestral environments), but this approach allows one to systematically predict the circumstances of maladaptation (Boyd and Richerson 1985).

In addition to using cues of skill and success, natural selection should favor any other reliable cues that allow imitators to more effectively focus their attention on models likely to possess behaviors and strategies that are both readily learnable and adaptive for the imitator. For example, assuming that the sexual division of labor is relatively old in the human lineage, imitators should preferentially focus their attention on successful members of their own sex, because these individuals are most likely to have skills and knowledge that will be useful to learners in their roles in later life. Furthermore, children should not only preferentially copy members of their own sex, but should pay particular attention to children who are somewhat older than themselves. By preferentially imitating individuals who are skilled, older, and of their same sex, imitators increase their likelihood of acquiring adaptive behaviors and strategies, and can effectively build themselves up to increasingly more complex skills: If you are an eight-year-old, you would be wise to first master the skills of the most successful eleven-year-old you know before aspiring to imitate the most successful sixteen-year-old. Once you've mastered the skills of the sixteen-year-old, perhaps then you will focus your attention on learning from the most skilled adults in your village. Other criteria, such as self-similarity and shared ethnic markings can also facilitate more adaptive cultural learning (McElreath et al. 2003).

This line of evolutionary thinking suggests that once rank-biased transmission (as an innate cognitive ability) has spread through a population, highly skilled individuals will be at a premium, and social learners will need to compete for access to the most skilled models. This creates a new selection pressure on rank-biased learners to pay deference to those they assess as highly skilled (those judged most likely to possess useful information), in exchange for preferential access (and perhaps assistance, hints, etc.). Deference may take many forms, including coalitional support, general assistance (e.g., chores), caring for the offspring of the skilled, gifts, and so on. Such deference patterns provide a costly cue of which individuals are generally considered highly successful or skilled: deference is paid to such individuals in exchange for copying opportunities—faking the cue would require paying deference to the unskilled, which would carry little or no payoff.

With the spread of deference for highly skilled individuals, yet another opportunity is presented for natural selection to save on information costs.

Naïve entrants (say, immigrants or children), who lack detailed information about the relative skills and successes for potential cultural models, may take advantage of the existing pattern of deference, and use received deference as an indicator of underlying skill. Assessing differences in deference patterns provides a best guess of the skill ranking until more information can be accumulated over time. This also means that skilled individuals will prefer deference displays that are public, and thus easily recognized by others. Along with the ethological patterns dictated by the requirements for high-fidelity social learning (proximity and attention), deference displays also include diminutive body postures and sociolinguistic cues. The end point of this process gives us the psychology, sociology, and ethology of prestige, which must be distinguished from those of the phylogenetically older dominance processes (see Henrich and Gil-White 2001 for details).

From this theory, we (Henrich and Gil-White 2001) derived and tested twelve predictions about the interrelationships of preferential imitation and influence with prestige deference, age, sex, memory, and ethological patterns (e.g., gaze and skill). With regard specifically to imitation and influence, these findings show that (1) both adults and children preferentially imitate more skilled and prestigious individuals, usually unconsciously; (2) this imitation occurs even when the thing being imitated is not clearly connected to the imitatee's domain of skill or prestige; (3) children preferentially imitate older, same-sex models across a wide range of behavioral domains; and (4) people remember what prestigious individual say more than what same-status individuals say, and unconsciously align their opinions with prestigious individuals, even when the individuals' opinions are not related to their domains of prestige (e.g., people care about Tiger Woods's car preferences). Combined with numerous related findings, this work indicates that prestige transmission is part of the cultural learning processes that build our brains and adapt them to local social and environmental circumstances.


## FROM CULTURAL LEARNING TO SOCIOLOGICAL PHENOMENA

Not only do cultural learning mechanisms provide cognitively realistic and developmentally plausible mechanisms for acquiring adaptive behaviors and strategies in complex, variable environments, they also provide a foundation for building a higher-level set of theories about population-level phenomena. Using formal modeling techniques, ideas about cultural learning mechanisms that are both theoretically and empirically grounded—such as prestige-biased transmission—can be combined with population structure, environmental and ecological factors, and social interaction to rigorously study a wide range of population-level phenomena (Richerson and Boyd, this volume). One example of this can be found in the study of ethnicity. Research by McElreath et al. (2003) that examines social interactions in coordination dilemmas (such as whether to adopt beliefs and practices for bride price or dowry) within a framework of prestige-biased transmission has shown that symbolically marked social groups (ethnic groups) will spontaneously arise under a wide range of conditions. This study also shows that an in-group learning bias—a bias to learn from and interact with people who share your symbolic markings—can genetically coevolve through this process, as natural selection acting on genes takes advantage of the changes in social environments wrought by cultural evolution. Cooperative institutions offer another example: a combination of conformist transmission and prestige-bias transmission can stabilize cooperative strategies in large social groups in a manner not possible in an acultural species (Henrich and Boyd 2001). This can lead to a process of cultural group selection and culture–gene coevolution that will generate increasingly complex social, political, and economic institutions along timescales of hundreds and thousands of years (Richerson and Boyd 1998, 2000; Henrich 2004a).

# 10

## Culture Matters

*Inferences from Comparative Behavioral Experiments and Evolutionary Models*

### SAMUEL BOWLES AND HERBERT GINTIS

Gary Becker and George Stigler's classic paper "De Gustibus Non Est Disputandum" expresses what was until recently the unquestioned perspective on the nature of preferences among economists: "one does not argue about tastes for the same reason that one does not argue about the Rocky Mountains—both are there, and will be there next year, too, and are the same to all men." (Becker and Stigler 1977: 76) By contrast, other students of human behavior have doubted the universality of preferences, and instead have embraced the view that people differ in their motivations and that preferences are situation-dependent and evolve under the influence of socialization and other life experiences (Ross and Nisbett 1991; Bowles 1998). Becker's later book, *Accounting for Tastes*, suggests a reconsideration of his earlier work (Becker 1996). Recent experiments have challenged the standard economic assumption of self-interest (Camerer 2003; Camerer and Fehr 2004; Gintis et al. 2004; Falk et al. 2005), but none of these studies has addressed the extent to which preferences differ across societies.

Here we address one of the issues raised by Richerson and Boyd (this volume)—the influence of culture on behavior. We ask what light may be shed on this question by recent behavioral experiments reported by Henrich (this volume) and elsewhere, and described briefly below. Our experiments were designed to measure the extent of sharing behavior and the propensity to punish those who did not share, in a common experimental setting implemented among subjects of vastly differing cultures. In virtually all of the fifteen small-scale societies in which our behavioral experiments were carried out, a large fraction of the subjects acted in ways that cannot be explained by self-interest defined over the outcomes of the experimental game. Although individual differences (e.g. age, sex, wealth) explain little of the differences in experimental behavior, the experiments revealed large differences in group-average behaviors among the populations studied. These group effects seem to be closely associated with the typical social interactions involved in the production and distribution of the goods constituting the livelihood of each population. But it is also true that groups with similar livelihoods in similar ecologies also exhibit large differences in the distribution of behaviors.

What does this evidence say about the influence of culture on behavior? Why do we observe group differences in behavior even between groups occupying similar environments? Our response is that the experimental behaviors reflect adherence to general-purpose social norms that have proliferated in human societies at least in part because of their contribution to the survival and expansion of those groups that have adopted them. This may lead to group differences in behavior for two reasons. First, the norms that are internalized by group members may be the result of a process of socialization characterized by conformism, winner-take-all political processes controlling the institutions of socialization, and other strong group effects (to put it formally, positive feedbacks). These group effects allow the persistence of distinct norms among different groups in the same environment (representing multiple equilibria in the system of cultural evolution). Second, the economic and social institutions that humans create are to some extent culturally transmitted, and these affect the rate and direction of evolutionary processes, whether genetic or cultural. Because these institutions influence who one meets, to do what, and with what rewards, distinct institutions will support different patterns of behavior.

## THE WEAK AND STRONG HYPOTHESES THAT CULTURE MATTERS

Richerson and Boyd (this volume) wonder "if any contemporary scholars actually doubt that culture is important in humans." Perhaps they are excepting economists, for Becker and Stigler (above) appear to be saying that culture—in the sense used by most anthropologists, meaning distinct beliefs and values among groups—is unimportant for the explanation of behavior. Traditional

Marxist and other materialist approaches have relegated "culture" to the "superstructure" of society representing a consequence of the organization of the mode of production. What Boyd and Richerson (1985: 157–58) term the "pure genes" version of sociobiology affirms that "we would be able to predict cultural variation by asking what increases genetic fitness." Thus, whereas many economists see cultural differences as being of minor importance (but see Katzner in process), adherents to other approaches (e.g., Marxism, sociobiology) see cultural differences as the epiphenomenal expression of underlying determinants.

What we call the "weak hypothesis that culture matters" holds that along with genes and the natural environment, culture differs across societies and is among the proximate causes of behavior, but it remains silent on the possibility that culture itself may be entirely explained by the interaction of genes and environments. The strong hypothesis, by contrast, holds that culture is necessary to the explanation of behavior because culture affects behavior directly or indirectly and is not simply an expression of the interaction of genes and natural environments.

To clarify the distinction, suppose that we have continuous scalar measures of an individual behavior ($b$), and the individual's culture ($c$), genes ($g$), and past and present natural environments ($e$). We might predict the behavior of members of a group of individuals by using a statistically estimated structural equation: $b = f(c,g,e)$. Suppose we could predict culture by another equation: $c = k(g,e) + \lambda$, where $\lambda$ represents influences on $c$ other than $g$ and $e$. Because $c$ appears in the function $f(c,g,e)$, we say that culture is a proximate cause of behavior, confirming the weak hypothesis. Knowledge of culture helps to predict behavior.

To test the strong hypothesis, we need to compare the information provided by the structural function $f(c,g,e)$ with that provided by a reduced-form equation in which culture does not appear directly but rather is represented by its genetic and environmental determinants. Thus we substitute $k(g,e)$ for $c$, which yields the reduced-form equation $b = \phi(k(g,e),g,e)$. If this equation is as informative about the behavior as is the structural equation, then the strong hypothesis is false. In this case, $\lambda = 0$, and as a result, knowledge of the culture provides no additional information for the explanation of behavior beyond that provided by knowledge of genes and the natural environment. If the structural equation provides a significantly better prediction of behavior than the reduced-form equation, then the strong hypothesis is sustained.

It is clear from the above analysis that the strong hypothesis could be true and the weak hypothesis false. This would be the case if the behavior in question were entirely the expression of gene distributions that evolved under the influence of culture, as posited by gene culture coevolutionary theory (Cavalli-Sforza and Feldman 1973a; Boyd and Richerson 1985; Durham 1991). In this case, genes would provide the proximate cause of the behavior (knowledge of culture would be redundant if one had information on genes and environment). But earlier values of genes, culture, and environments would explain the evolution of genes.

The above illustrative model is intended to clarify what we mean by the strong and weak hypothesis, not to serve as a strategy for testing either. Available evidence does not allow such clear distinctions. However, the implementation of a given experimental game with the same structure of feasible actions and associated rewards across a large number of cultures provides a novel and quite clean test of the weak hypothesis that culture matters. The substantial differences between groups in experimental play described below are consistent with the weak hypothesis. And given that genetic differences between groups are likely to be small (Rosenberg et al. 2002), group differences among peoples occupying similar natural environments and engaging in similar livelihoods may provide some support for the strong hypothesis.

## PREFERENCES, BELIEFS, AND CONSTRAINTS

When we explain individual behavior, we use the terms *preferences*, *beliefs*, and *constraints* (Bowles 2004). We think that this standard decision-theory terminology is adequate for most purposes, but given that other chapters in this volume refer to *cultures*, we should attempt a translation. Here is our lexicon.

Preferences are an individual's evaluations of states that may result from her actions. A breakfast of two fried eggs and bacon in the company of one's cat and a bagel breakfast with a friend are states. Preferences include norms, behavioral rules of thumb, and any other facts about individuals, other than their beliefs and capacities, that contribute to explaining behavior. Constraints define the set of actions that are feasible; constraints include factors such as the individual's wealth and capacities, the prices relevant to the actions under consideration, and other limitations. Beliefs are the individual's understandings of the causal relationship between her actions and the states that the actions will bring about. Beliefs are important to the explanation of behavior when the best action to take depends on the actions taken by others, such as driving on the right side of the road in the United States and the left in the United Kingdom. In the "preferences, beliefs, and constraints" approach, behavior is explained as the attempt by an individual to bring about a preferred state, given her beliefs and feasible set of actions

Preferences may depend on the current state in which the individual finds herself, or on past experiences of the individual or others. Preferences need not be self-regarding, but may concern the states experienced by others. Likewise, beliefs may be endogenous (through a process of learning) and they may be false. All three—preferences, constraints, and beliefs—may be influenced by custom, law, the exercise of power, and other aspects of social structure. Neither the self-regarding *Homo economicus* nor a reductionist dictum that scientific explanation must begin with the individual is implied by this approach. We make these disclaimers because in classroom practice most economists subscribe both to a form of reductionism and to the axiom of calculative self-interest, and many anthropologists and others therefore have come to see these as integral to the approach that we wish to pursue. They are not.

Where does culture come in? We do not attempt to define culture, but note that it must be the component of preferences, beliefs, and constraints that are learned from others, rather than being an aspect of the individual's genetic, economic, or other endowments, or the physical environment in which the individual lives. We do not think that the term *culture* supplies anything necessary to the explanation of behavior that is missing from the preferences, constraints, and beliefs approach, though the term is certainly useful in making some distinctions that would be cumbersome within the conventional decision-theory framework. The common usage, in which *culture* refers to ideas that are shared among members of a group, does not strike us as fruitful, at least not for the purposes of the investigation at hand, because preferences and beliefs are often not uniform in a population. The figures presented by Henrich in this volume make it clear that the variance of experimental play within groups in the cross-cultural experiments was substantial in most societies, and in some the distribution of play was bimodal.

Using this framework, we ask if culture matters for behavior, and if so, how.

## THE CROSS-CULTURAL EXPERIMENTS PROJECT

Our primary data are from the cross-cultural behavioral experiments project reported in the previous chapter, in Henrich et al. (2004), and elsewhere. The main game implemented at each of fifteen field sites in Africa, Asia, Latin America, and New Guinea was the ultimatum game, a simple bargaining interaction that has been extensively studied by experimental economists. In this game subjects are paired anonymously, and the first player, often called the "proposer," is provisionally allotted a divisible "pot" (usually money). The proposer offers a portion of the pot to a second person, often called the "responder." The responder, knowing both the offer and the total amount in the pot, then has the opportunity to either accept or reject the proposer's offer. If the responder accepts, he or she receives the amount offered, and the proposer receives the remainder. If the responder rejects the offer, neither player receives anything. In either case, the game ends there; the two subjects receive their winnings (if any) and depart.

Players typically receive their portions of the pot in cash, and remain anonymous to other players but not to the experimenters (although experimental economists have manipulated both of these variables). In the experiments described here, players were anonymous, and the games involved substantial sums of the local currency or, in one case, goods. For this game, the canonical assumptions—that all participants maximize their income and that this is known by all of them—would predict that responders, faced with a choice between zero and a positive payoff should accept any positive offer. Knowing this, proposers should offer the smallest nonzero amount possible.

In seven field sites we also implemented a public-goods game, which is simply an n-person prisoner's dilemma: total payoffs are maximized when each player contributes to the public good, but irrespective of what the others do, individual payoffs are maximized by contributing nothing.

The societies in which we conducted our experiments (indicated in figure

10.1) exhibited a wide variety of economic and cultural conditions, consisted of five groups that engage in a significant amount of foraging (the Hadza of Tanzania, the Au and Gnau of Papua New Guinea, the Aché of Paraguay, and the Lamalera of Indonesia), four horticulturists who engage in very little foraging (Machiguenga, Quichua, Tsimane, and Achuar of South America), four pastoralist groups (the Turguud and Kazakhs of Central Asia, and the Orma and Sangu herders of East Africa), and three sedentary, small-scale agricultural societies (the Mapuche of Chile, Sangu farmers and the Shona in Zimbabwe). The experimental games were played anonymously, without misleading or deceiving subjects, and for real stakes—the local equivalent of one day's wages or more.

## RESULTS

The results of the experiments, some of which are presented by Henrich in the preceding chapter, can be summarized in four points.

There was no society in which experimental behavior in the ultimatum game was consistent with the self-regarding-actor model. In some societies there were virtually no rejections, even of very low offers, but proposers still made high offers to respondents. In others, responders were quick to reject low offers, and proposers correspondingly made relatively high offers. In still other societies, neither proposers nor responders conformed to the self-regarding-actor model.

There was markedly more variation among groups than had been reported in previous similar experiments. The mean ultimatum game offers seen in experiments with university-student subjects are typically between 43 percent and 48 percent, but the mean offers from proposers in our sample range from 25 percent to 57 percent. And whereas modal ultimatum game offers among university students are consistently 50 percent, modes for the samples in our experiments range from 15 percent to 50 percent. In some groups, rejections were extremely rare, even in the presence of very low offers, whereas in others, rejection rates were substantial, including rejections of offers exceeding half of the pot. The distributions of play among the Aché and Tsimane resemble those of American students, but with very low rejection rates.

In previous studies, typical distributions of contributions in a one-round

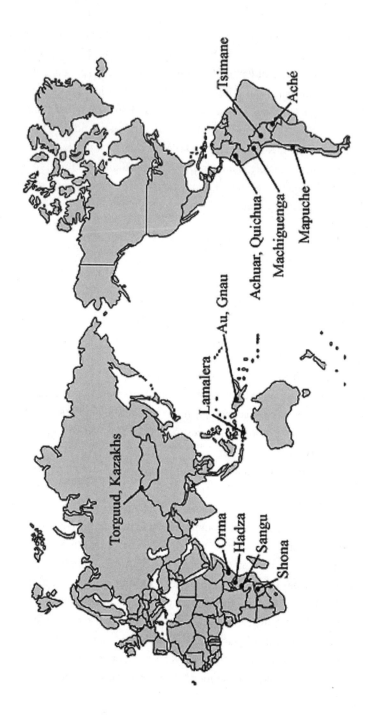

10.1.   Locations of the fifteen small-scale societies included in the ultimate-game and public-goods-game experiments.

public-goods game have a U-shape, with one mode at no contribution and a secondary mode at universal contribution, with an average contribution between 40 percent and 60 percent of the maximum possible. In our experiments, average contributions ranged from 22 percent among the Machiguenga to 65 percent among the Aché. In contrast to typical student play, the Machiguenga have a single mode of no contribution, with no subject contributing fully. At the other end of the distribution, among the Orma, the Mapuche, and the Huinca (non-Mapuche Chileans living among the Mapuche) all subjects made contributions to the public good.

Differences among societies in their market integration and in the payoffs to cooperation explain a substantial portion of the behavioral variation between groups. The societies were rank-ordered in five categories: market integration (how often do people buy and sell, or work for a wage), payoffs to cooperation (is cooperation in the pursuit of one's livelihood beneficial), anonymity (how prevalent are anonymous roles and transactions), privacy (how easily can people keep their activities secret), and complexity (how much centralized decision making occurs above the level of the household). Using statistical regression analysis, we find that only the first two characteristics, market integration and payoffs to cooperation, are significant, and together they account for about two-thirds of the variance among societies in mean ultimatum-game offers (see table 10.1).

Behavior in experimental games tended to mirror patterns of interaction found in everyday life. In a number of cases the parallels between experimental game play and the structure of daily life are quite striking. Nor was this relationship lost on the subjects themselves.

The Orma immediately recognized that the public-goods game was similar to the *harambee*, a locally initiated contribution that households make when a community decides to construct a road or school. They dubbed the experiment "the *harambee* game," and gave generously. Among the Au and Gnau, many proposers offered more than half the of the pot in the ultimatum game, and many of these "hyper-fair" offers were rejected! This would appear to reflect the Melanesian culture of competitive status seeking through gift giving. Making a large gift is a bid for social dominance in everyday life in these societies, and rejecting the gift is a rejection of being subordinate.

Among the whale-hunting Lamalera, 63 percent of the proposers in the

Table 10.1. How differences among groups in market integration and in payoffs to cooperation correlate with mean offers in the ultimatum game

|  | Coefficients | | Standardized coefficients | | |
|---|---|---|---|---|---|
|  | B | Standard error | Beta | t | Significance |
| Constant | 0.261 | 0.036 | — | 7.323 | 0 |
| PC[a] | 0.021 | 0.007 | 0.528 | 2.922 | 0.011 |
| AMI[b] |  | 0.005 | 0.448 | 2.479 | 0.027 |

[a] Payoffs to cooperation

[b] Aggregate market integration

ultimatum game divided the pot equally, and most of those who did not offered more than 50 percent (the mean offer was 57 percent). In real life, a large catch, always the product of cooperation among many individual whalers, is meticulously divided into predesignated portions and carefully distributed among the members of the community. Among the Aché, 79 percent of proposers offered either 40 percent or 50 percent, and 16 percent offered more than 50 percent, with no rejected offers. In daily life, the Aché regularly share meat and honey, which is distributed equally among all households, irrespective of which hunter made the catch. The low offers and high rejection rates among the Hadza in the ultimatum game seems to reflect the tendency of these foragers, like the Aché, to share meat and honey, except that they also include a high level of contentious bargaining in the process.

Both the Machiguenga and Tsimane made low ultimatum-game offers, and there were virtually no rejections. These groups exhibit little cooperation, exchange, or sharing beyond the family unit. Ethnographically, both show little fear of social sanctions and care little about public opinion. The Mapuches' social relations are characterized by mutual suspicion, envy, and fear of being envied. This pattern is consistent with their post-game interviews after they played the ultimatum game. Mapuche proposers rarely claimed that their offers were influenced by fairness, and rather cited a fear of rejection. Even proposers who made hyper-fair offers claimed that they feared spiteful responders (who were rare), who would be willing to reject even fifty-fifty offers.

## ECONOMIC INFLUENCES ON PREFERENCES AND BELIEFS

Do cultural differences explain why members of different groups behaved so differently in these experiments? Addressing this question requires an explanation of why and how preferences and beliefs spread in different groups, and how they are maintained. An important clue is the fact that behavioral patterns co-vary with economic and social structures in the ways we have observed.

We suspect that both individual-level and group-level selection pressures, operating on both genetic and cultural variation, are at work. Groups of whale hunters (such as the Lamalera) who did not embrace a sharing norm would waste resources in conflicts over prey. And individual whale hunters who resist the sharing norm are likely to incur shunning, ostracism, and other forms of punishment from their fellow group members.

There is substantial evidence that behavioral predispositions are often closely aligned with the demands of the tasks entailed by making a living (Bowles 1998). Studies of the relationship between economic roles and child-rearing practices are among the most telling. Over a period of three decades, Melvin Kohn and his collaborators have studied the relationship between a person's position in the authority structure of her workplace—that is, whether she gives or takes orders—and the individual's valuation of self-direction and independence in her children, as well as the individual's own intellectual flexibility and personal self-directedness. They concluded that "the experience of occupational self-direction has a profound effect on people's values, orientation, and cognitive functioning" (Kohn et al. 1990: 967).

Kohn's collaborative study of Japan, the United States, and Poland yielded cross-culturally consistent findings: people who exercise self-direction on the job also value self-direction more in other realms of their life (including child-rearing and leisure activities) and are less likely to exhibit fatalism, distrust, and self-deprecation. Kohn and his coauthors reason that "social structure affects individual psychological functioning mainly by affecting the conditions of people's own lives." They conclude, "The simple explanation that accounts for virtually all that is known about the effects of job on personality is that the processes are direct: learning from the job and extending those lessons to off-the-job realities" (Kohn et al. 1990: 59).

The personality dimensions described by Kohn et al. are part of individuals' preferences, which influence how they raise their children, what kind of leisure activities they engage in, and the like; this is therefore strong evidence that behavioral predispositions learned in the workplace are generalized to other arenas of life.

The evidence is not confined to industrial societies. Edgerton (1971) found a large and statistically significant relationship between pastoral as opposed to farming livelihoods and the valuation of independence, as well as a number of other behavioral predispositions. These data are consistent with causation running in the other direction: independent-minded people may become herders rather than farmers. But Edgerton's results are robust even when the pastoralism/farming measure is not the actual means of livelihood (e.g., whether an individual is a herder) but rather the geographical suitability of the relevant locale for each of these two pursuits (whether the land could be used for herding). The relevant correlations are only slightly diminished when the underlying, and presumably exogenous, geographical measure is used. McElreath (2004) has reproduced Edgerton's results and was able to identify the mechanisms that preserve cultural differences among groups.

Barry, Child, and Bacon (1959) categorized seventy-nine mostly nonliterate societies according to their prevalent form of livelihood (animal husbandry, agriculture, hunting, or fishing) and the related ease of food storage or other forms of wealth accumulation, the latter being a major correlate of dimensions of social structure such as stratification; food storage is common in agricultural societies but not among foragers. The researchers also collected evidence on aspects of child rearing, including obedience training and the inculcation of self-reliance, independence, and responsibility. They found large differences in the examined child-rearing practices. These co-varied significantly with economic structure, controlling for other measures of social structure such as unilinearity of descent, extent of polygyny, levels of participation of women in the predominant subsistence activity, and size of population units. Barry et al. (1959:59) conclude, "knowledge of the economy alone would enable one to predict with considerable accuracy whether a society's socialization pressures were primarily toward compliance or assertion." The causal relationship is unlikely to run from child rearing to economic structure, as the latter is dictated primarily by geography in the sample of small-scale societies under study.

## EXPERIMENTAL EVIDENCE FOR THE STRONG HYPOTHESIS

It is entirely possible, then, that the group differences in behavior observed in our experiments are not evidence for the independent causal force of cultural differences, but rather are a reflection of differences in economic structure or other aspects of social relationships. The fact that group membership predicts behavior independently of individual-level characteristics is consistent with the weak hypothesis that culture matters. But these data do not exclude the possibility that although they are part of the proximate causes of behavior, cultural differences are simply the epiphenomenal expression of differences in the livelihood or other aspects of a group's ways of living. By contrast, the strong hypothesis holds that culture not only is a proximate cause, but also is among the ultimate causes of behavior.

A thought experiment shows why this is a very difficult hypothesis to test. In the cross-cultural experiments, we observed in the ultimatum game a very high level of sharing among the Lamalera whale hunters in Indonesia, whereas the Amazonian Machiguenga horticulturalists shared much less. Suppose a group of Lamalera were to be transported to the Amazon, taking up a family-based social structure and slash-and-burn agricultural economy as their livelihood. Would their sharing norms come to approximate those of the Machiguenga? Or, were the Machiguenga to take up whale hunting as their livelihood, would they come to approximate Lamalera sharing norms?

History does not perform experiments from which unambiguous conclusions might be derived. When previously isolated peoples come into contact, it is generally the case the cultures, livelihoods, and other aspects of a society change simultaneously, making the identification of causal relationships difficult, if not impossible, and permitting competing explanations to flourish, if not to convince.

Our experiments provide some suggestive evidence. In Ecuador, the Achuar and Quichua of Conambo, who live side by side and make their livings in similar ways, played the ultimatum game quite differently—Achuar proposers offered a mean of 43 percent, whereas Quichua proposers offered only 25 percent. This difference is especially notable because Quichua and Achuar subjects were randomly paired, so proposers from the two groups faced similar probabilities of rejection. Experimental play among the two near-archetypal hunter-gatherer groups in our sample—the Hadza in Tanzania and the Aché in Paraguay—deviated significantly and in opposite directions from the familiar university-student subject pools. Among the Hadza, offers averaged 27 percent and there were a significant number of rejections even of offers close to 40 percent or more. Aché proposers offered on average about 50 percent, and no offers were rejected (though there were a significant number of offers of 40 percent).

## MODELING THE STRONG HYPOTHESIS

Group differences in experimental play are the proximate result of different preferences and beliefs that the subjects have learned. But are these differences in preferences and beliefs also part of the ultimate explanation of group differences in behavior? In the absence of group differences in environments (and hence in livelihoods), why would these culturally transmitted influences on behavior persist over time? Two processes seem most likely.

For one thing, in a standard model of cultural transmission, a necessary condition for group differences such as this to be large and persistent in the absence of differing environments is that the behaviors in question are strategic complements. This means that the behaviors generate positive feedbacks, so the material or other rewards to an individual who adopts a given behavior increase in the fraction of a population adopting the behavior. In this case, then, a very large number of equilibria are possible, each characterized by different behaviors even in a population that is otherwise identical. Many norms governing behavior are of this type and hence are conventions, that is, mutual best responses.

A prime example of such positive feedback is the socialization processes by which behaviors are internalized. These processes generally privilege some behaviors over others, inducing people who have internalized these norms to penalize deviant behaviors. The cost for deviating from any particular norm thus increases along with the number of individuals who have internalized it. Unless the personal benefits accruing to those who fail to internalize the dominant norms are substantial, populations with otherwise identical environments but distinct systems of socialization may exhibit substantially different behavioral norms. The distinct systems of socialization may themselves persist because individuals produced by the system tend to value its results.

The punishment of deviants is not necessary to generate positive feedbacks. Similar results arise from what Boyd and Richerson (1985) term "conformist cultural transmission," whereby behaviors are more likely to be copied if they are more common, independently of the material payoffs associated with them. Furthermore, if the system of socialization reflects the preferences and beliefs of the dominant group within a society (winner-take-all socialization), many distinct cultural equilibria are possible. In all of these cases, which norm becomes common may be decided by historical accident: a given norm may persist over long periods even if a different norm would have been a better payoff-enhancing adaptation to the environmental conditions of the group. Group selection pressures may militate against the long-term persistence of norms imposing inferior payoffs on members of a group, but the ethnographic evidence of maladaptive norms suggests that these pressures may be weak or absent in many cases (Edgerton 1992).

The second process whereby cultural differences may be ultimate rather than merely proximate causes of social behavior exists as a result of group selection, not despite it. Consider an individual predisposed to a social behavior that is the expression of a hypothetical allele that, because the altruistic nature of the behavior, suffers a selective disadvantage compared to the alleles of non-altruistic individuals. As is well known, if the adverse selection differential operating against the bearer of the altruistic allele is sufficiently small relative to the benefits conferred on other group members by the altruistic behavior, and differences in group composition are sufficiently large, the altruistic allele may proliferate (G. Price 1970).

Although these conditions for group selection are considered unusual for most animals, culturally transmitted human behaviors may reduce within-group selection differentials and enhance group differences in composition. Examples are the egalitarian sharing of resources and information within a group, small group size, consensus decision making, between-group migration limited by ethnocentric or other parochial behaviors, mating systems that reduce reproductive skew, and lethal intergroup competition. These are all examples of what Feldman (this volume) and Odling-Smee et al. (2003) call "the neglected process in evolution"—namely, niche construction. Culture matters in this case because it creates a selective environment that alters the dynamics of selection processes acting on alleles. An example of such a process involving a culturally transmitted rule concerning food sharing and a hypothetical genetically transmitted altruistic predisposition is given in Bowles et al. (2003). Culturally transmitted niche construction may thus constitute an ultimate cause of an altruistic behavior without being a proximate cause of that behavior.

## CONCLUSION

Prior to the cross-cultural behavioral experiments project, evidence pointed to strong similarities in behavior in the ultimatum game across two dozen or more countries, including the United States, Switzerland, Japan, the former Yugoslavia, Israel, Indonesia, and Germany. Modal offers were typically half of the pot, mean offers were somewhat less, and offers below a quarter of the pot were routinely rejected (Roth et al. 1991; Camerer 2003). To some, these results suggested either that culture did not matter, or that the cultures of the entirely student subject pools were similar in Djakarta, Jerusalem, Pittsburgh, Ljubljana, and the other experimental sites.

We selected our cross-cultural experimental sites deliberately to include peoples whose livelihoods and everyday lives differed across groups, unlike university students worldwide. The much greater between-group behavioral differences that we found have been interpreted as testimony to the importance of culture. We think this view is correct. The evidence for the weak hypothesis seems compelling: faced with a given structure of feasible actions and rewards, individuals in different societies behaved very differently, suggesting that between-group differences in preferences or beliefs are significant.

The evidence for the strong hypothesis is less clear. As we have seen, between-group social and economic differences are associated with large and statistically significant behavioral differences in the experiments. Thus, group behavioral differences in the experiments could reflect the influence of the economic livelihoods in which the groups engaged rather than the independent causal importance of culture.

Our experiments do provide evidence, consistent with that of Kohn and his collaborators, that behaviors learned from the task structures in one sphere of life are generalized to other spheres. We conclude this because, unless the behavioral rules derived from the job of making a living are internalized and generalized across different spheres of life, it is puzzling why they should affect play in our experiments. That is why dividing up the pot in the ultimatum game apparently seemed to the Lamalera equivalent to dividing a whale, and contributing in the public-goods game reminded the Orma of joining a *harambee* to build a primary school.

But this in no way supports the strong hypothesis. Here the available evidence is at best suggestive, as only a few paired comparisons of experimental play by peoples in similar natural environments are possible. Even though these do favor the strong hypothesis, the statistical power of such comparisons must be regarded as minimal, in light of the small sample sizes and inadequate controls of environmental differences.

## NOTE

This essay draws on experiments and models presented in Henrich et al. (2004). We are grateful to all of our collaborators. We would also like to thank Melissa Brown, Marcus Feldman, Toshio Yamagishi, and Elisabeth Wood for insightful comments, the University of Siena, and the MacArthur Foundation and the Behavioral Sciences Program of the Santa Fe Institute for support of this project.

# PART IV  CHALLENGES TO A SCIENCE

# OF CULTURE


The chapters in this part raise challenges to the development of a scientific approach to culture on the basis of ethnographic case studies that show contradictory empirical results and methodological biases. These challenges bring us back to many of the issues examined in the first section—causation, cognition, structure, and units—but with additional concerns regarding variation, diffusion, power, and verification. Skepticism and critical evaluation produce good science by inspiring new studies, analyses, and conclusions. These chapters, then, should be seen as a contribution to building a scientific paradigm, because they spur us to reformulate our approaches to account for these empirical data and resolve these methodological cautions.

Arthur P. Wolf, a social anthropologist, presents an empirical challenge to the causal power of culture in a specific case. He raises the question of whether cultural transmission models get at cultural ideas by comparing field data on uxorilocal marriage in two different Han Chinese contexts: data collected by Li, Feldman, and Li from late twentieth-century north-central China, and data analyzed by Wolf and his colleagues from early twentieth-century northern Taiwan. Wolf suggests that the Taiwan data support a more structural causal interpretation of transmission, in contrast to the cultural interpretation offered of the China data. This comparison explicitly raises questions about cultural transmission modeling (especially Cavalli-Sforza and Feldman 1981, but see also Boyd and Richerson 1985; Durham 1991, this volume;

Odling-Smee et al. 2003; Feldman, this volume; Richerson and Boyd, this volume) and about structural effects (see the chapters in this volume by D'Andrade; Durham; Brown; Bowles and Gintis). Wolf asks whether Li, Feldman, and Li have found not cultural transmission but "a social transformation the source of which was not visible to most people and not controllable by them [that] changed irrevocably the conditions that forced some people into uxorilocal marriages"—that is, transmission of the social conditions for uxorilocal marriage instead of transmission of a cultural idea or preference for uxorilocal marriage. For example, if the absence of a man's father increases his likelihood of marrying uxorilocally, and uxorilocally married men are more likely to engage in labor with high rates of mortality, then the sons of uxorilocally married men are more likely to marry uxorilocally even in the absence of a preference, because they are more likely to lack a father when they reach marriageable age. Wolf raises additional issues relevant to the work of other authors. How do we identify culture (see the chapters by D'Andrade; Boesch; Feldman)? Is it a collection of traits (in either ideational or behavioral form), or is it an integrated system? Even if the ability to imitate is innate (see the chapters by Feldman; Richerson and Boyd; Henrich; Aoki et al.), under what limited conditions do individuals practice imitation or not? Wolf implies a methodological caution here as well. The first correlation between a model and empirical data does not necessarily explain causation; further assessment using case studies is also necessary.

Gregory Starrett, a cultural anthropologist, challenges our uncritical usage of the concept of culture itself, by exploring how the content of cultural ideas affects their own transmission and the perception of their transmission by social scientists and others. After reviewing current anthropological rejections of culture, Starrett provides evidence of how one cultural idea—the Western social sciences' concept of functionalism—has influenced Egyptian understandings of the purposes and effects of standard Islamic beliefs and rituals as well as local explanations for the growth of Islamic fundamentalism. He asks, "What does it mean for anthropology when our analytical vocabulary is fastened onto by people who deploy it to build their own institutions, goals, understandings, and experiences of the world? Can we use these concepts to understand institutions organized around these concepts themselves?" Although Starrett rejects abandoning the concept of culture, he nevertheless warns, "where our analytical vocabulary has been appropriated as bricks and mortar for the construction of a way of life, we encounter the difficulty that our analyses may be self-confirming." This methodological warning applies to us all: can we ensure that interview subjects are not reporting the importance of cultural ideas because the Western idea of culture's importance diffused to them, and, if not, then how should we understand our data?

Robert Borofsky, a cultural anthropologist, raises questions about the scientific process, the concept of culture, and the influence of social structure. In reviewing the Ninov case of scientific fraud, he considers the failure of scientists to verify and challenge results, instead acquiescing to results by scholars with impressive credentials. Borofsky argues that anthropologists presenting culture as a shared attribute constitutes such an affirmation of the intellectual status quo. He not only documents intracultural diversity but also quotes some authors whose own early work demonstrated that diversity, referring to cultural groups as though they were homogeneous. The masking of intracultural diversity derives from the hegemonic efforts of nation-states; the term "culture" itself originated in nineteenth-century German nationalism. Borofsky challenges us to return to the intellectual promise of science—to insist on verification and critical evaluation of results: "The reality of challenges—not simply the possibility of challenges—is central to science. It is what gives data their scientific validity." He urges us to "speak truth to power"—to demonstrate the cultural variability within peoples and nation-states, to "illuminate . . . the organization of diversity—how people of different perspectives effectively deal with one another without killing them"—and thereby lay bare nation-states as "politically, not culturally, driven."

# 11

## Cultural Evolution and Uxorilocal Marriage in China

### A Second Opinion

### ARTHUR P. WOLF

After a hiatus of nearly a century, social and biological scientists are again showing an interest in the possibility of an evolutionary account of culture. This is not, however, a case of "descent with modification." The ideas being discussed today bear no family resemblance to their nineteenth-century predecessors. The nineteenth-century theories were what I call "transformation theories." The authors were primarily interested in when, how, and why one social order replaces another. Their units of analysis were what the Marxists among them called "social formations." The current theories are what I term "transmission theories." Their authors are primarily concerned with how small bits of culture spread in a population and are then passed from generation to generation. They evince little interest in how these units relate to one another, and none at all in how they combine to create such distinctive social types as feudalism and capitalism.

The real intellectual forebears of modern transmission theories are the early twentieth-century antievolutionists—most obviously Franz Boas and his many students and disciples, but also Fritz Graebner and the Viennese *kulturkreis* school (Lowie 1937: 177–95; Schmidt 1939). The Boasians argued, contra the nineteenth-century transformationalists, that similarities between cultures are due not to their developing along similar paths, but to transmission between neighbors or from a common ancestor. Like the contemporary transmission theorists, they took as their units small bits and pieces of culture— the designs on Alaskan needle cases, rules that require a man to avoid his mother-in-law and a women her father-in-law, ritualistic complexes such as the sun dance. Their view of the world is brilliantly summed up in Edward Sapir's (1916b) "Time perspective in aboriginal American culture."

The source of modern transmission theories is Luigi Cavalli-Sforza and Marcus W. Feldman's 1981 *Cultural Transmission and Evolution*. Variant versions of the theory have since been offered by William H. Durham (1991) and by Robert Boyd and Peter Richerson (1985), but I will confine my attention to Cavalli-Sforza and Feldman's original formulation. My reason for doing so is that Feldman and two of his Chinese colleagues have recently applied their version of transmission theory to marriage customs in China (Li S. et al. 2000, 2003). This makes the theory relevant to my own research and allows me to assess how well it accounts for familiar phenomena (A. Wolf and Huang, 1980).

Cavalli-Sforza and Feldman (1981: 7) use the term "cultural" to describe "traits that are learned by any process of nongenetic transmission, whether by imprinting, conditioning, observation, imitation, or as a result of direct teaching." Cultural evolution is simply "changes within a population of the relative frequencies of the forms of a cultural trait" (5). Three of the examples given are the innovations in food production that spread from the Middle East to Europe in prehistoric times; the difference in the pronunciation of the vowels A, E, I, O, and U in English and French; and the Japanese style of ancestral tablets that gained wide acceptance in Taiwan during the Japanese occupation (15, 21, 42).

Cavalli-Sforza and Feldman (1981: 7) argue that "the evolution of traits that are cultural depends ultimately on the way in which traits are transmitted among individuals within a generation and between generations." Inspired by epidemiologists' work on the spread of disease, they distinguish two modes of transmission—vertical and horizontal. For epidemiologists, "*vertical* transmission is used to denote transmission from parent to offspring and *horizontal* transmission denotes transmission between any two (usually unrelated) individuals." Cavalli-Sforza and Feldman restrict horizontal transmission to "members (related or not) of the *same* generation," and introduce "the word *oblique* to describe transmission from a member of a given generation to a member of the next (or later) generation who is not his or her child or direct descendant" (54).

What we have, then, is an argument that will be familiar in many ways to social scientists interested in such topics as imitation, social facilitation, and diffusion. It assumes that for some reason or other people modify their behavior or invent new forms of behavior. To effect an evolutionary change, an innovation must be known, widely accepted, and then transmitted from generation to generation. The most distinctive aspect of the argument is the claim that these are processes in which the influence of other people is critical. Individuals are viewed as largely free to accept or reject a practice, but it is assumed that their decisions are strongly influenced by the examples set by parents and peers.

The study that aroused my interest in this view of culture change is described by its authors—Shuzhuo Li, Marcus W. Feldman, and Nan Li (2000: 159)—as "a study of the cultural transmission of uxorilocal marriage in Lueyang, China." An uxorilocal marriage is a marriage in which the newly married couple lives with the bride's family rather than with the groom's family. Because marriages of this kind run against the grain of the Chinese kinship system, men who marry in this fashion are commonly accused of having "deserted their parents," and are viewed as outsiders in their wife's community. Informants in Sichuan told me that when an uxorilocally married man passed, his wife's relatives would sniff the ground and say, "Smells bad. Must be an outsider." Thus the cultural trait in this study is a trait with a bad odor. I think it is correct to say that no Chinese man would marry uxorilocally if he had any other way of getting a wife.

Lueyang is the name of a county (xian) located in the Qing Mountains in southern Shaanxi Province, close to its borders with Sichuan and Gansu. The county is rich in natural resources but is "relatively underdeveloped," with a per-capita income "far less" than that for farmers elsewhere in Shaanxi and in China generally (Li et al. 2000: 161). The area was once the home of Ti and Qiang peoples, but most of them were driven out or absorbed by Han Chinese more than a thousand years ago. Two facts about Lueyang are important to the research at hand. One is that most of the ancestors of the present residents were "immigrants from other areas." "The result," according to Li et al. (2000: 161), "is that large family clans in Lueyang are few and unimportant in village life." This, in their view, has weakened "the patriarchal family system" typical of rural China. The other fact is that Lueyang's rich natural resources—minerals, forests, and arable land—offer opportunities not available elsewhere, but only for those willing and able to do heavy manual labor. As Li et al. (2000: 170) put it, the county "can absorb more laborers and people," but "the difficult environment necessitates hard labor, mostly by men."

Despite their bad odor, uxorilocal marriages were popular in Lueyang. Li et al.'s team interviewed 1,581 married couples living in forty-six natural villages. The 1,463 couples who provided complete data included 458 (31.3 percent of the total population) who had married uxorilocally. The proportion who had married uxorilocally was lower among the older couples than among the younger couples, but it was still substantial. Of the 175 marriages initiated before 1960, 22.9 were uxorilocal. That uxorilocal marriage was an established custom is also indicated by a high proportion of uxorilocal marriages among the subjects' parents. The husbands' parents included 16.7 percent who had married uxorilocally, and the wives' parents 20.2 percent (Li et al. 2000: 164 tab. 2).

Li et al. suggest several reasons for the high frequency of uxorilocal marriages in Lueyang. These include the absence of village regulations discouraging such marriages, the lack of strong patrilineal descent groups, the large bride price required for a virilocal marriage, and, most importantly, the conditions created by the county's rich natural resources. On one hand, these attract ambitious, able-bodied men from neighboring counties; on the other, they mean that every family must raise or recruit one or more sons. The result is a marriage economy with a ready supply of and a strong demand for men willing to marry uxorilocally. The men from outside need to find wives, and local families who fail to raise a son need to recruit a son-in-law.

The most interesting and most important of Li et al.'s findings concerns the way the custom of uxorilocal marriages is perpetuated. As Cavalli-Sforza and Feldman's model predicts, parental influence is critical. Young people whose parents married uxorilocally are far more likely to marry uxorilocally themselves than those who parents married virilocally.

> Before 1978, the odds of being in a uxorilocal marriage for a son whose parents were in a uxorilocal marriage are about 189 percent those for a son whose parents are in a nonuxorilocal marriage. For a daughter whose parents are in a uxorilocal marriage, the odds are 271 percent those for a daughter whose parents are in a uxorilocal marriage. After 1978, these two figures decline to 59 percent and 146 percent, respectively. (Li et al. 2000: 167)

I do not doubt these results, but I find them very surprising. One reason is the very large number of uxorilocal marriages in Lueyang. A brief item from an early twentieth century survey of customary law notes the occurrence of uxo-rilocal marriages in Lueyang and four of its neighbors (MXDB 1918),[1] but nothing suggests that uxorilocal marriages accounted for percentages as high as those marriages reported by Li et al. They are surprising figures because Li et al. (2000: 170) noted that "more than 99 percent" of their informants were Mandarin-speaking Han Chinese. In my experience, uxorilocal marriages rarely if ever occur as first marriages in Mandarin-speaking regions. They are largely limited to Wu-, Gan-, Xiang-, Hakka-, and Min-speaking areas of South China. In 1979–80 I reconstructed from interviews the marriages of 553 women aged sixty-five or over in seven communities in seven different provinces. I found twenty-two uxorilocal marriages among 240 women in Jiangsu, Zhejiang, and Fujian, but none at all among 313 women in Beijing, Shandong, Shaanxi, and Sichuan (A. Wolf 1989: 245 tab. 1). The cadres who arranged my interviews were the local representatives of the Women's Federation. Two of these at Mandarin-speaking sites had married uxorilocally, but insisted that theirs were the first uxorilocal marriages in their communities.

Entries in the Gansu and Shaanxi sections of the 1918 survey suggests that uxorilocal marriages were common outside the Great Wall, but this was clearly an exceptional area. The sociologist Li Jinghan (1929: 17, 93) found only one uxorilocally married man among 164 families in two Beijing suburbs, and researchers working for the South Manchurian Railway at midcentury did not turn up even one such marriage in their field sites in Hebei and Shandong (CNKCK 1957). More impressive still is the fact that Li Jinghan's survey of 5,255 families in Ting xian discovered only one married woman who was liv-ing in her parents' home (Gamble 1954: 29).[2] This was not because there was no demand for uxorilocal marriages in Mandarin-speaking North China. They were simply not an institutional option.

My second reason for surprise is Li et al.'s informants' willingness to admit to their parents having married uxorilocally. In my experience, people never admit to an uxorilocal marriage if it is easily concealed. We will see below (in figure 11.1) that when the marriages of Taiwanese born between 1850 and 1925 are arranged by birth cohort, the frequency of uxorilocal marriage appears to rise dramatically among people born before 1885, peaks among those born between 1886 and 1900, and then gradually declines among those born after 1900. The reason for the dramatic rise is simply that at the time the household registers were established in 1905, older people whose parents were dead found it easy to conceal how they had married, the result being that uxorilocal marriages are underreported for the cohorts born before 1880 (A. Wolf 1995: 42–45, 51–52). Underreporting is probably also the reason that John Lossing Buck's (1930: 321–22 tab. 2) 1921–23 surveys turned up only two uxorilocally married men in 3,456 households. His informants would have concealed what they considered a malodorous marriage.

But what I find most surprising about Li et al.'s results is not the high frequency of uxorilocal marriages or people's admitting their existence. It is their evidence arguing that having parents who married uxorilocally inclined people to marry uxorilocally themselves. Why should this be? It cannot be because this made young people aware of the possibility of marrying uxorilocally. Every mentally competent Chinese person knows what an uxorilocal marriage is and how it is arranged. He learns this as surely and effortlessly as he learns his native language. It must be, then, that having a parent who married uxorilocally makes this form of marriage more acceptable. But why? Why would having a father who was abused as a smelly outsider make someone see marrying uxorilocally as more acceptable? As I hope to demonstrate below, marrying uxorilocally was not a cultural option like eating or not eating beef. It was more analogous to the options open to a wage worker in nineteenth-century Manchester. He could either accept a starvation wage or starve. I find it difficult to think of uxorilocal marriage as a cultural trait that may or not be transmitted. It is, rather, an unhappy condition that may or may not be avoided.

Gabriel Tarde might have argued that young people in Lueyang were simply victims of an innate tendency to imitate.[3] They married uxorilocally for no other reason than their parents' having married uxorilocally. For Clark Wissler, such a tendency was not only innate but indispensable. He could not imagine how man could have culture at all without such an imitative function: to exist, culture "must be perpetuated by the imitation of the older by the younger members of the group" (Wissler 1923: 208). Certainly human beings do imitate one another, but surely Floyd Allport was right to insist that "there is no general instinctive drive to imitate." Behind such complex activity there is always "some personal and prepotent interest other than the desire to imitate" (Allport 1924: 241). In other words, imitation is an instrumental act. But what could be the end or purpose of imitation in the case at hand? Why would people undertake what is generally regarded as a disreputable form of marriage when they did not have to?

It could be that uxorilocal marriages were not held in such low regard in Lueyang as elsewhere in China. In this case, parental example would not have had a negative effect, but it is hard to imagine uxorilocal marriages being held in such high regard that parental example would have a positive effect. The ideology of patrilineal kinship always discourages marriages that threaten or contravene patriliny. Another possibility is that poor people married uxorilocally and had children who also married uxorilocally because they were also poor. Because Li et al.'s models do not control for socioeconomic status, this is a logical possibility, but it is not a likely one. We will see below that although poverty favored uxorilocality in Taiwan, this did not produce the results reported by Li et al. The only other possibility I can think of is that couples who had married uxorilocally were less likely to conceal their parents' having married uxorilocally. Unable to conceal the odor of their own marriage, they had nothing to lose—and perhaps a little to gain—by admitting that their parents had done no better.

I cannot explain Li et al.'s results, but I can show that I have reason to be surprised. My evidence comes from household registers compiled by the Japanese colonial government in Taiwan during the years 1905–45. I have reconstructed from these registers the marriages of 5,814 men and 5,153 women born in the years 1850–1925. They included all the men and women living in eleven villages and two small towns in Haishan, in the southwestern corner of the Taibei Basin. They are all the descendants of eighteenth- and early nineteenth-century immigrants from Anxi xian in Fujian. They therefore represent a culture area in which uxorilocal marriages were as common as they were among the older couples included in Li et al.'s study (A. Wolf and Huang 1980: 1–15, 326–39).

Table 11.1 assesses the extent of parental influence on marriage choices in Haishan. The sample is limited to people born after 1890 because there is no way to determine the marriage choices of many of the parents of persons born earlier. It should also be noted that the information used to determine the parents' choices is not entirely reliable. These people were all married before the household registers were created in 1905. This means that their form of marriage was recorded in 1905 from their responses to questions put by the Japanese police. Thus the older people among them had an opportunity to

Table 11.1. Percent of Haishan men and women who married uxorilocally, by whether or not their parents married uxorilocally

|  | Number of men | Percent of men married uxorilocally | Number of women | Percent of women married uxorilocally |
|---|---|---|---|---|
| *Father's form of marriage* | | | | |
| Uxorilocal | 511 | 3.7 | 571 | 8.1 |
| Virilocal | 3,576 | 4.2 | 3,333 | 8.9 |
| *Mother's form of marriage* | | | | |
| Uxorilocal | 474 | 4.6 | 530 | 7.4 |
| Virilocal | 4,165 | 4.5 | 3,799 | 9.9 |

NOTE: Includes only persons who were born between 1890 and 1925 and survived to age fifteen.

conceal the fact if they had married uxorilocally. This is an important limitation, but is not unique to the Haishan data. The data reported by Li et al. (2000: 175) for Lueyang have the same limitation. Their information on parental marriage type was reported by the married children, and must in many cases refer to people long since deceased. The children could have easily concealed an embarrassing uxorilocal marriage.

The data in table 11.1 clearly contradict the findings that Li et al. report for Lueyang. Among Haishan men, their mother's marrying uxorilocally had no influence on their chances of marrying uxorilocally, and their father's marrying uxorilocally had a slightly negative influence. Among Haishan women, their father's marrying uxorilocally had a slightly negative influence, and their mother's marrying uxorilocally had what appears to have been a strong negative influence. The negative influence of the father's marriage among men is what we might expect given the way uxorilocally men were usually treated. The only anomalous finding is what appears to be the strongly negative influence among women of their mother's marriage. Women who married uxorilocally were not stigmatized as unfilial and were not ostracized as outsiders. They were the only women who escaped that fate. Why, then, would a woman's marrying uxorilocally reduce her daughter's chance of marrying uxorilocally?

Table 11.2. Percent of Haishan women who married uxorilocally or bore a child before marriage, by whether or not their parents married uxorilocally

| | Number of women | Percent marrying uxorilocally or bearing a child before marriage |
|---|---|---|
| *Father's form of marriage* | | |
| Uxorilocal | 571 | 21.4 |
| Virilocal | 3,333 | 19.7 |
| *Mother's form of marriage* | | |
| Uxorilocal | 530 | 21.5 |
| Virilocal | 3,799 | 20.2 |

NOTE: Includes only women who were born between 1890 and 1925 and survived to age fifteen.

The answer to this question is implied by the figures displayed in table 11.2. They show that women whose parents married uxorilocally were more likely to bear an illegitimate child than women whose parents married virilocally. The result is that the influence of their parents' form of marriage disappears when women who married uxorilocally or bore an illegitimate child are classified together. The ultimate reason for this is that even before the railway was completed in 1905, Haishan was an integral part of a social and economic world centered on Taibei city (Wang 1996). It produced much of the tea processed in Taibei and provided much of the required labor (Davidson 1903: 371–96). In this setting, sending a daughter to work as a prostitute was an option for parents who had no son or whose sons were still young. Many of these women never married, but those who did commonly married uxorilocally because they were still needed by their parents. The result was that women whose mothers married uxorilocally were often the daughters of former prostitutes and consequently became prostitutes themselves. They were less likely than other women to marry uxorilocally because they were less likely to marry at all.

I conclude that in Haishan, having a parent who married uxorilocally did not increase a person's own chances of marrying uxorilocally. If there was any influence, it was negative.

Why, then, did many people marry uxorilocally? What were the critical conditions? My view of uxorilocal marriages is that they were the products of conditions that gave people no choice. These were conditions of the kind that make people work for a minimum wage. Thus an obvious prediction is that people who married uxorilocally were poor people: poor men married uxorilocally because it was the only way they could get a wife, and poor women married uxorilocally because their parents had no son and could not afford to buy one.[4] Many of the latter cases were women who worked as prostitutes and eventually married one of their clients.

The data for testing these predictions come from land-tax registers compiled by the Japanese colonial government. They provide a robust index of socioeconomic status in an agricultural economy of the kind found in Haishan. The "no land tax" class in table 11.3 includes tenant farmers, farm laborers, and what in China were commonly called "coolies"—from the word ku-li, "strenuous effort." The "twenty or more yen" class at the other end of the scale includes what the Chinese communists called "middle" and "rich" peasants, together with a few shopkeepers and a few country landlords. The elite of Taiwanese society—the people who lived in walled compounds with gardens—are not represented in the table. They all lived in Taibei City or on the Chinese mainland.

The figures in table 11.3 confirm the prediction outlined above. The poorer people were, the more likely they were to marry uxorilocally. The poorest men in Haishan were 3.6 times as likely to marry uxorilocally as were the wealthiest men, and the poorest women, 2.0 times as likely as the wealthiest women. The fact that poverty influenced men more than women is primarily the result of a sex ratio that favored women. Poor men commonly had trouble finding wives, and married uxorilocally because it was their only option. Women never had trouble finding husbands and were therefore never forced into an uxorilocal marriage as their only option. What forced women into uxorilocal marriages were poor parents who had to hold on to a daughter because they had no one else to support them in their old age.

Anthropologists have often noted that men with brothers are more likely to marry uxorilocally than men without brothers. For one thing, families with several sons are less willing and often less able to invest in the marriage of any one son; in addition, men with several brothers have less chance of inheriting

Table 11.3. Percent of Haishan men and women married uxorilocally, by amount of land tax paid by natal family

| Amount of land tax | Number of men | Percent of men married uxorilocally | Number of women | Percent of women married uxorilocally |
|---|---|---|---|---|
| No land tax | 4,092 | 4.7 | 3,199 | 11.3 |
| Less than one yen | 425 | 4.2 | 426 | 8.2 |
| One to nine yen | 872 | 3.1 | 752 | 9.3 |
| Ten to nineteen yen | 319 | 2.2 | 258 | 6.2 |
| Twenty or more yen | 401 | 1.3 | 393 | 5.6 |

NOTE: Includes only persons who were born between 1890 and 1925 and survived to age fifteen.

enough land to support a family than men with without brothers. Thus it is not surprising to find that in Lueyang "the odds of occurrence of an uxorilocal marriage for a husband with no brothers are only 50 percent those of a husband with brothers" (Li et al. 2000: 167).

Table 11.4 shows that Haishan men also behaved as expected. Among men who survived to age fifteen, those with brothers were far more likely to marry uxorilocally than those without brothers. There is, however, an interesting complication. Although men with older brothers were far more likely to marry uxorilocally than those with no older brothers, men with younger brothers— but younger brothers only—were less likely to marry uxorilocally than those with no brothers. This is puzzling because it contradicts the claim that the more brothers a man has, the more likely he is to marry uxorilocally. I found the solution to the puzzle only after reexamining the histories of the men who married uxorilocally despite being only sons. They turned out to be almost exclusively men whose parents had died before their only son was old enough to marry. This was the reason they had no younger brothers, and was also the reason they married uxorilocally. They had no one to help them find a bride and no one to help them pay a bride price.

The benefits of this discovery are evident in tables 11.5 and 11.6. Table 11.5 shows that a man's losing one or both of his parents before he reached age

Table 11.4. Percent of Haishan men who married uxorilocally, by number of older and younger brothers present at age fifteen

| | Number of men | Percent married uxorilocally | Average percent |
|---|---|---|---|
| No brothers | 623 | 3.9 | |
| No older brothers, one younger brother | 950 | 3.1 | 3.0 |
| No older brothers, two or more younger brothers | 1,223 | 2.5 | |
| One older brother, no younger brothers | 540 | 6.9 | |
| One older brother, one younger brother | 376 | 5.6 | 6.3 |
| One older brother, two or more younger brothers | 429 | 6.1 | |
| Two or more older brothers, no younger brothers | 319 | 6.0 | |
| Two or more older brothers, one younger brother | 198 | 9.6 | 7.8 |
| Two or more older brothers, two or more younger brothers | 589 | 8.2 | |

NOTE: Includes only men who were born between 1886 and 1926 and survived to age fifteen.

fifteen strongly enhanced the odds of his marrying uxorilocally. A man who had lost both of his parents by age fifteen was 4.6 times as likely to marry uxorilocally as a man whose parents both survived until he was fifteen. The death of even one parent made a substantial difference. Men who lost their fathers by age fifteen were 2.3 times as likely to marry uxorilocally as those whose parents both survived, and men who lost their mothers were 2.1 times as likely to marry uxorilocally as those fortunate enough to have both parents present when they reached marriageable age. Readers who doubt the influence of women in

Table 11.5. Percent of Haishan men who married uxorilocally, by whether, at age fifteen, they had living parents

|  | Number of men | Percent married uxorilocally |
|---|---|---|
| Both parents dead | 90 | 16.7 |
| Father dead, mother alive | 548 | 8.2 |
| Father alive, mother dead | 672 | 7.7 |
| Both parents alive | 3,936 | 3.6 |

NOTE: Includes only men who were born between 1886 and 1926 and survived to age fifteen.

Table 11.6. Percent of men who married uxorilocally, by number of brothers and presence of parents at age fifteen

|  | Both parents alive | | One or both parents dead | |
|---|---|---|---|---|
|  | Number of men | Percent uxorilocal | Number of men | Percent uxorilocal |
| No brothers | 292 | 0.3 | 306 | 7.5 |
| No older brothers, one younger brother | 746 | 2.0 | 205 | 7.8 |
| No older brothers, two or more younger brothers | 1,107 | 2.0 | 140 | 5.0 |
| One older brother, no younger brothers | 354 | 4.2 | 186 | 11.8 |
| One older brother, one younger brother | 289 | 4.5 | 87 | 9.2 |
| One older brother, two or more younger brothers | 368 | 5.7 | 61 | 4.9 |
| Two or more older brothers, no younger brothers | 200 | 5.5 | 119 | 7.7 |
| Two or more older brothers, one younger brother | 152 | 7.2 | 46 | 17.4 |
| Two or more older brothers, two or more younger brothers | 428 | 7.2 | 161 | 10.6 |

NOTE: Includes only men who were born between 1886 and 1926 and survived to age fifteen.

Chinese family life should note that the loss of a man's mother was almost as damaging to his chances of marrying respectably as the loss of his father.

Table 11.6 clarifies the influence of brothers by controlling for the influence exerted by the death of a parent. We now see what I had anticipated seeing in table 11.4. Under normal conditions, men with no surviving brothers at age fifteen almost never married uxorilocally. They only did so when they had lost one or both of their parents. We also see in table 11.6 that the chances of a man's marrying uxorilocally was not a function simply of how many brothers he had—whether his brothers were older or younger also mattered. Men with one brother who was older were 2.1 times as likely to marry uxorilocally as men with one brother who was younger. When we take into account the relative ages of men's brothers as well as the number of brothers, the result is a steady progression in the percentage who married uxorilocally as we move from men with no brothers to those with younger brothers only, and finally to those with one or more older brothers.

Anthropologists have also long noted that for women the influence of brothers is just the opposite of what it is for men. The fewer brothers a woman has, the more likely she is to marry uxorilocally. The reason for this is simply that if a family has no sons they have to use their daughter to attract a son-in-law to work their fields and support them in their old age. Lueyang appears to have been an exceptional place in many respects, but not in this respect. Li et al. (2000: 167) report that "the odds of occurrence of an uxorilocal marriage . . . for a wife with no brothers are about 345 percent those of a wife with brothers." One would like to know whether the brother's age makes a difference. Table 11.7 shows that this was critical in Haishan. It made little difference whether Haishan women had younger brothers or not. The deciding variable was whether or not they had older brothers. Those who had no older brothers were 1.8 times as likely to marry uxorilocally as those who had one older brother, and 2.9 times as likely to do so as those with two or more older brothers. Older brothers mattered and younger brothers did not because sons younger than a daughter of marriageable age were almost always the children of older couples who would soon need help supporting their families.

Tables 11.8 and 11.9 are the female equivalents of tables 11.5 and 11.6. The figures displayed here argue that number of brothers was not the only variable

Table 11.7. Percent of Haishan women who married uxorilocally, by number of older and younger brothers present at age fifteen

| | Number of women | Percent married uxorilocally | Average percent |
|---|---|---|---|
| No brothers | 904 | 11.4 | |
| No older brothers, one younger brother | 1,509 | 13.4 | 11.9 |
| No older brothers, two or more younger brothers | 1,199 | 10.2 | |
| One older brother, no younger brothers | 620 | 6.0 | |
| One older brother, one younger brother | 311 | 7.7 | 6.5 |
| One older brother, two or more younger brothers | 148 | 6.1 | |
| Two or more older brothers, no younger brothers | 185 | 4.9 | |
| Two or more older brothers, one younger brother | 83 | 4.8 | 4.1 |
| Two or more older brothers, two or more younger brothers | 102 | 2.0 | |

NOTE: Includes only women who were born between 1886 and 1926 and survived to age fifteen.

that had opposite effects for men and women: so also did death of a parent. Whereas table 11.5 shows that men who had lost both of their parents by age fifteen were 4.6 times as likely to marry uxorilocally as were men whose parents survived, table 11.8 shows that women whose parents *survived* were 3.1 times as likely to marry uxorilocally as were those whose parents were lost. This enormous difference highlights a fundamental contrast in marriage dynamics between men and women. Men married uxorilocally when it was the only way to get a wife. Women married uxorilocally when their parents made

Table 11.8. Percent of women who married uxorilocally, by whether, at age fifteen, they had living parents

|  | Number of women | Percent married uxorilocally |
|---|---|---|
| Both parents dead | 85 | 3.5 |
| Father dead, mother alive | 499 | 8.6 |
| Father alive, mother dead | 581 | 6.9 |
| Both parents alive | 3,896 | 11.0 |

NOTE: Includes only women who were born between 1886 and 1926 and survived to age fifteen.

them marry uxorilocally. Thus the death of a parent raised the odds that a man would marry uxorilocally, but lowered the odds that a woman would do so. The death deprived a man of the means to marry virilocally but gave a woman the chance to do so by freeing her of the need to serve as a substitute for a son. This and much of the preceding argument can be summed up in one sentence: An uxorilocal marriage was a practical arrangement uniting an indispensable daughter and an expendable son.

Figures 11.1 and 11.2 plot the evolution of uxorilocal marriages for men and women in Haishan. The solid lines trace the frequency of uxorilocal marriages among all men or women in the sample who survived to age fifteen, and the broken lines the frequency among all men or women who eventually married. Their trajectories are much the same. Among both sexes the frequency rises among persons born before 1880, peaks for those born between 1880 and 1900, and then declines. The rise among persons born before 1880 is significant only as evidence that people did not like to admit to marrying uxorilocally. The interesting change is the decline in the frequency of uxorilocal marriages among people born after 1900. The implication of Cavalli-Sforza and Feldman's conception of cultural evolution is that change of this kind is somehow the result of the transmission process. In their view, "the evolution of traits that are cultural depends ultimately on the way in which such traits are transmitted among individuals within a generation and between generations." My view of the matter is very different. I believe that a social transformation, the source of which was not visible to most people and not controllable by

Table 11.9. Percent of women who married uxorilocally, by number of brothers and presence of parents at age fifteen.

| | Both parents alive | | One or both parents dead | |
|---|---|---|---|---|
| | Number of women | Percent uxorilocal | Number of women | Percent uxorilocal |
| No brothers | 521 | 10.8 | 363 | 11.9 |
| No older brothers, one younger brother | 1,250 | 14.8 | 264 | 8.0 |
| No older brothers, two or more younger brothers | 1,086 | 10.7 | 128 | 5.5 |
| One older brother, no younger brothers | 408 | 7.1 | 212 | 3.8 |
| One older brother, one younger brother | 247 | 9.3 | 64 | 1.6 |
| One older brother, two or more younger brothers | 126 | 5.6 | 22 | 9.1 |
| Two or more older brothers, no younger brothers | 117 | 5.1 | 68 | 4.4 |
| Two or more older brothers, one younger brother | 63 | 4.7 | 20 | 5.0 |
| Two or more older brothers, two or more younger brothers | 78 | 2.6 | 24 | 0.0 |

NOTE: Includes only women who were born between 1886 and 1926 and survived to age fifteen.

them, changed irrevocably the conditions that forced some people into uxorilocal marriages.

We have seen above that men in Haishan rarely chose to marry uxorilocally. They did so only when circumstances deprived them of the possibility of a virilocal marriage. These circumstances were poverty, the death of one or both parents, and the existence of a number of older brothers. Thus it is easy to see why the frequency with which men married uxorilocally began declining among men born in the first decade of the twentieth century. These men

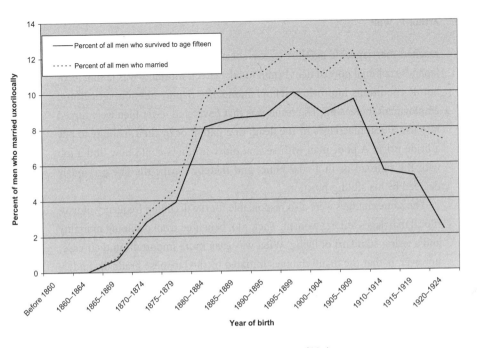

11.1.  Percent of Haishan men who married uxorilocally, by year of birth.

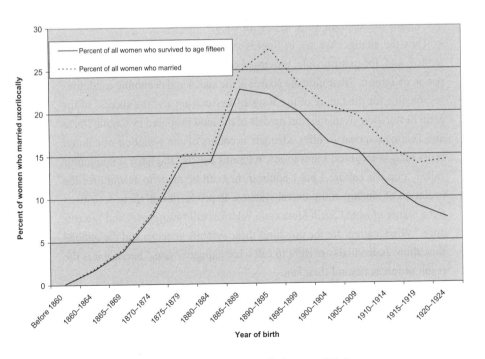

11.2.  Percent of Haishan women who married uxorilocally, by year of birth.

reached marriageable age just as the social and economic initiatives of the Japanese colonial government began to take effect. Mortality was falling, economic conditions improving, school attendance rising, and non-agricultural employment increasing. The result was that fewer and fewer men were forced into uxorilocal marriages by the death of a parent or by their inability to pay a bride price. A poor man with several older brothers could now find a job with the railroad or in a coal mine, and thereby escape the fate previously decreed by his sibling position.

The same changes were also responsible for the declining frequency of uxorilocal marriages among women. They, too, benefited from declining mortality and a rising standard of living. What was even more important in their case, however, is that the many changes initiated by the Japanese gradually undermined parental authority. This was critical for women because—as the effect of the death of a parent demonstrates—most women married uxorilocally only because their parents required it. Where men were victims of circumstances, women were victims of parental authority. The admission of women to the new primary schools and the creation of new employment opportunities changed all this. Parents gradually lost their ability to use their daughters as a means to substitute sons-in-law for sons.

Thus the changes we see in figures 11.1 and 11.2 are the indirect, unintended, and, for many people, unwanted consequences of the Japanese occupation of Taiwan. Their ultimate source is the social and economic conditions responsible for the failure of the Tongzhi Restoration and the success of the Meiji Restoration. Transmission models of the kind proposed by Cavalli-Sforza and Feldman appear to offer adequate accounts of the evolution of cultural traits such as technical innovations, religious beliefs, and what is commonly called "popular culture," but I doubt if they can be made to account for the evolution of kinship and political systems. Reproduction in these systems is not a matter of what Cavalli-Sforza and Feldman call "awareness" and "acceptance." What matters are the pressures these systems generate and the options they allow. Individuals are right to call what happens "fate," because it is the result of forces beyond their ken.

# APPENDIX 11.A: DATA POINTS GRAPHED IN FIGURES 11.1 AND 11.2

Table 11.A1. Persons who survived to age fifteen and who married uxorilocally, by year of birth

| Year of birth | Number of men | Percent of men married uxorilocally | Number of women | Percent of women married uxorilocally |
|---|---|---|---|---|
| Before 1860 | 666 | 0.0 | 434 | 0.1 |
| 1860–64 | 227 | 0.0 | 119 | 1.7 |
| 1865–69 | 307 | 0.7 | 155 | 3.9 |
| 1870–74 | 355 | 2.8 | 211 | 8.1 |
| 1875–79 | 484 | 3.9 | 269 | 14.1 |
| 1880–84 | 545 | 8.1 | 392 | 14.3 |
| 1885–89 | 477 | 8.6 | 353 | 22.7 |
| 1890–95 | 472 | 8.7 | 439 | 22.1 |
| 1895–99 | 588 | 10.0 | 585 | 20.0 |
| 1900–04 | 726 | 8.8 | 619 | 16.5 |
| 1905–09 | 739 | 9.6 | 658 | 15.4 |
| 1910–14 | 713 | 5.6 | 719 | 11.4 |
| 1915–19 | 797 | 5.3 | 732 | 9.0 |
| 1920–24 | 919 | 2.3 | 970 | 7.6 |

Table 11.A2. Persons, of those who married at all, who married uxorilocally, by year of birth

| Year of birth | Number of men | Percent of men married uxorilocally | Number of women | Percent of women married uxorilocally |
|---|---|---|---|---|
| Before 1860 | 477 | 0.0 | 388 | 0.1 |
| 1860–64 | 182 | 0.0 | 109 | 1.8 |
| 1865–69 | 250 | 0.8 | 145 | 4.1 |
| 1870–74 | 300 | 3.3 | 203 | 8.4 |
| 1875–79 | 414 | 4.6 | 251 | 15.1 |
| 1880–84 | 452 | 9.7 | 367 | 15.3 |
| 1885–89 | 379 | 10.8 | 323 | 24.8 |
| 1890–95 | 365 | 11.2 | 353 | 27.4 |
| 1895–99 | 473 | 12.5 | 502 | 23.3 |
| 1900–04 | 582 | 11.0 | 492 | 20.7 |
| 1905–09 | 579 | 12.3 | 517 | 19.5 |
| 1910–14 | 551 | 7.3 | 514 | 16.0 |
| 1915–19 | 523 | 8.0 | 474 | 13.9 |
| 1920–24 | 287 | 7.3 | 515 | 14.4 |

## NOTES

1. The item reads: "Among the common people a family that has no sons and only one daughter feels that if their daughter marries out they will have no one to depend on. They therefore call in for their daughter a man who changes his name and on whom they can rely for support. Colloquially, this is known as zhi nü bao xu, "appoint a daughter and adopt a son" (MXDB 1918, vol. 2, chap. 15, para. 7).

2. This book was published in English by Sidney Gamble, but the research was done by Li Jinghan and published in Chinese under Li's name (Li Jinghan 1933).

3. Tarde's views and those psychologists with similar views are summarized and briefly evaluated by Neal E. Miller and John Dollard (1941: 289–318).

4. There were two forms of male adoption in Taiwan at this time—adoption from a close agnate and adoption from a distant relative or a stranger. The latter always required compensation in cash (A. Wolf and Huang 1980: 108–13).

# 12

## When Theory Is Data

### Coming to Terms with "Culture" as a Way of Life

GREGORY STARRETT

Anthropology and sociology were originally conceived as reformers' sciences, battling superstition and ideology for the hearts and minds of modern men. Enlightenment consisted in part of recognizing, articulating, and changing, when necessary, the true forces shaping our actions. On this both the royalist Auguste Comte and the radical Karl Marx might agree with the bourgeois Edward Burnett Tylor (Tylor 1871: 453). If we scientists could use new modes of thought to free ourselves from the shackles of tradition, surely the rest of the world, if properly educated and encouraged, could do the same. This heritage runs deep in European history. During the Renaissance and the Enlightenment, progress in the nascent social sciences was at least as instrumental as that in the natural sciences in undermining European beliefs in astrology, fear of witchcraft, and confidence in divine providence. Attributing social misfortune to the inhumanity of economic forces, and attributing personal misery to poor upbringing, came gradually to replace their attribution to the movements of the stars, to the punishment of an angry god, or to malevolent neighbors with dark magical powers (Thomas 1971: 654–55).

But the optimism that fueled anthropology's growth through most of the last century has faded. One symptom of this is that a number of anthropologists have called for the abandonment of central concepts in the discipline, including the seemingly irreplaceable concept of "culture" (Kuper 1999; Trouillot 2002; R. Wilson 2002). The strategic proposal to abandon "culture" is supported by four types of arguments: that the concept is trivial, that it is incoherent, that it is incomplete, and that it is politically contaminated. The first argument holds that culture is an epiphenomenon without behavioral influence. It is the caused rather than the causal, and we should attend to tangibles such as social structure, economy, politics, or neuropsychology if we wish to understand our behavior (see the chapters by Brown and by Wolf in this volume). The second argument claims that culture is not a thing at all, but rather a congeries of disparate objects, a junk drawer into which we have thrown symbolic systems, social groups, codes of law, observations of behavior, bodies of folklore, and a colorful pile of house types, culinary traditions, pottery, baskets, weapons, and textiles. The third argument charges that the concept of "culture" ignores powerful political and socioeconomic inequalities that shape human experience. And the final argument points out that, however useful it was in anthropology's infancy, the term "culture" has since been tainted by its spread into popular usage. Because non-anthropologists have proved incapable of using the culture concept responsibly, and as we cannot take it back from them, we should leave it to be mutilated, and find another.

But replacing the term "culture" would be a pointless exercise. Let us look first at the argument that points to incoherence. Here we can proceed from Adam Kuper's critique:

> Complex notions like culture . . . inhibit an analysis of the relationships among the variables they pack together. Even in sophisticated modern formulations, culture . . . tends to be represented as a single system, though one shot through with arguments and inconsistencies. However, to understand culture, we must first deconstruct it. Religious beliefs, rituals, knowledge, moral values, the arts, rhetorical genres, and so on should be separated from each other rather than bound together into a single bundle labeled culture. . . . Separating out these elements, one is led on to explore the changing configurations in which language, knowledge, techniques, political ideologies, rituals, commodities, and so on are related to one another. (Kuper 1999: 245)

Kuper's stereotype of American cultural anthropology, drawn from the work of Clifford Geertz, David Schneider, and Marshall Sahlins, is unconvincing because these high theorists are so unrepresentative of anthropology as it is normally practiced. It's like arguing that one can understand contemporary Christian practice in the roadside churches of South Carolina by reading the works of existentialist theologians Søren Kierkegaard and Hans Küng. Kuper's failure is particularly sharp because on the penultimate page of his book he discusses alternatives developed by Roy D'Andrade and Eric Wolf, who have been influential in portraying culture as complex and partible, "a series of processes that construct, reconstruct, and dismantle cultural material, in response to identifiable determinants" (E. Wolf, quoted in Kuper 1999: 246). Contra Kuper, there has never been a single hegemonic concept of culture in anthropology (Kroeber and Kluckhohn 1952). Culture is an essentially contested concept (Gallie 1956; see also Borofsky, this volume), the object of argument precisely because of its complexity and centrality to our being, both as humans and as scholars. Were we to choose another concept to be central to our field, we would eventually complain that it, too, was impossibly muddled, the result of our using the concept as a focus for working through manifold concerns.

The argument of incompleteness holds that culture, "launched [at the beginning of the discipline] as the negation of race . . . became the negation of class and history. Launched as a shield against some of the manifestations of racial power, culture . . . protected anthropology from all conceptual fields and apparatuses that spoke of power and inequality" (Trouillot 2002: 41). Culture, in this view, has been treated as an ethereal and self-contained phenomenon sealed off from the realities of force and helplessness, wealth and poverty, stratification and marginality. But neither "history" nor "class" is a more natural object than culture. And the idea that either has been excluded from anthropological attention is impossible to sustain. By the late 1920s Margaret Mead was comparing the cruelties of power in traditional Samoa and their "counterparts in producing misery" in modern Western culture (M. Mead 1928, appx 3). Earlier, Robert Lowie surveyed systems of inequality, power, and punishment worldwide (Lowie 1920, chap. 12–14). Franz Boas discussed socioeconomic class (Boas 1928, chap. 8), and Ruth Benedict's 1946 *The Chrysanthemum and the Sword* was a book-length meditation on the nature of stratification. Edward Sapir's famous 1924 essay "Culture, Genuine and Spurious" (1924) was a reflection on Marx's concept of alienation, and Paul Radin's *Primitive Religion* (1937) was a cheerfully vulgar Marxist analysis of religious systems as structures of economic exploitation. Boas's students, coming of intellectual age in the 1920s and 1930s at Columbia University and the New School, were well acquainted with Marxist concepts and used them freely—albeit without citing Marx—in their critiques of American capitalism.

Early and mid-twentieth-century anthropological interest in race did not eclipse concern with economic and military inequalities, either on a domestic or an international scale (D. Price 2004; D. Price and Peace 2003). Suspicion that anthropologists were *too* interested in class and gender as well as racial and international inequality earned some scholars unwanted and career-destroying attention from their universities as well as from the FBI and congressional anticommunists. That some anthropologists might have taken refuge in a more ethereal culture concept implies that they were all too aware of the operations of power. Trouillout and other critics fail to apply to anthropology the materialist analysis they accuse it of lacking.

Probably the most interesting objection to the concept of culture is that it has become irredeemably spoiled by non-anthropological use. Marketers use it to hawk their products. Political scientists deploy it to explain global conflict or lack of development. In other contexts it has become a euphemism for "race" among groups as diverse as the apartheid regime of South Africa, grassroots ethnic movements, and corporate trainers who advise American executives on managing the "diverse, multicultural workforce of the twenty-first century." Expressed forcefully particularly with respect to the role of culture in South Africa, this view holds that "An essentialized version of culture . . . was central to the doomed delusions of Afrikaner nationalism. In this historical context, one wonders whether culture should ever be used again in an *analytical* capacity. Social and cultural anthropologists have generally rejected the term 'race' as an analytical category . . . and 'culture' seems to be in an analogous position in the aftermath of apartheid" (R. Wilson 2002: 229). Encouraging us to think in cosmopolitan and "postcultural" terms, such critics point out that culture is not merely an analytical category, but that "it is the bricks and mortar of nation-builders. . . . 'Culture' is above all a political principle that asserts the congruence between a social group and a set of beliefs and practices in order to legitimate the strategies of social actors. In short, it is an ideology of political contestation that motivates social behavior but provides no independent theoretical insights into actual social behavior" (R. Wilson 2002: 229–30; see also Kuper 1999).

But when we cast about for alternatives to culture, a problem presents itself. What concept has *not* been used as the brick and mortar of nationalists? Tradition? History? Custom? Society? Ethnos? One might argue that this critique itself is ethnocentric for decrying the contamination conveyed by the politics of the West and North but not the contamination conveyed by the politics of the East and South. Somalis, Tanzanians, North Koreans, Chinese, Cambodians, and Siberians might equally have objected at various times to the contamination of terms such as "science," "materialism," "inequality," "history," and "ideology" by Marxist regimes. If "culture" is disposable on the grounds that it has been dirtied, then what of presumably cosmopolitan, postcultural concepts such as "class" or "power"?

Let's not abandon poor, beleaguered culture when it needs us most, uncertain of its identity and tossed about by entrepreneurs and politicians whose very touch is defiling. Physicists have not abandoned the concepts of space, time, and matter merely because they are difficult to understand at very large and very small scales, or because there are fundamental arguments about their interrelations. Nor have biologists abandoned the concept of life merely because their entrepreneurial colleagues aim to patent, reproduce, and sell it, or because the assassins of abortion providers claim to be in favor of it.

If concepts such as culture, ideology, class, and even science have been such successful replicators that we have lost control over their use, we should recognize this as an interesting anthropological problem in itself. It is what these concepts were designed to do, for how else to spread enlightenment if not by spreading enlightening concepts? What does it mean for anthropology when our analytical vocabulary is fastened on to by people who deploy it to build their own institutions, goals, understandings, and experiences of the world? Can we use these concepts to understand institutions organized around these concepts themselves? In exploring this issue, I use evidence from the Middle East to look at the careers of two cultural objects: the practice of mass popular schooling, and the idea of "function." The spread of the idea and the transformation of local and national societies wrought by schooling were parts of an integrated process (see Wolf, this volume). I argue that contemporary unease about the concept of culture flows from the fact that life and thought among the global middle classes has become so saturated with social science perspectives that we have lost sight of the fact that these perspectives are cultural ones, ones that we hold and use to help organize our own lives.

## FUNCTIONALISM AS MODERNIST THEOLOGY

Religions fulfill many social and psychological needs. They offer continuity of existence beyond death and enable individuals to transcend their arduous earthly existence and attain . . . spiritual satisfaction. Socially, religions reinforce group norms and contribute to . . . moral sanctions for individual conduct. They also provide common goals and values, which in turn give a sense of stability and solidarity to human societies and contribute to their security and equilibrium. Thus, it is only logical that a people who find themselves unable to deal with problems that threaten their society, culture, and spiritual and psychological well-being should turn to religion for . . . solutions. Indeed, in societies in which secular systems offer no escape from daily frustrations caused by cultural and political alienation, identity crisis, and economic deprivation, religion has proven a refuge offering social and psychological support. (Sonbol 1988: 23)

This is the first paragraph of an article on Islamic fundamentalism by an Egyptian historian. But it could just as well have come from the "Religion" chapter in any introductory anthropology or sociology textbook. It is the most common explanation that educated, secular Egyptians—not to mention most foreign analysts—provide for the success of Islamist movements. It's common sense. But it pays to examine the taken-for-grantedness of this functionalist perspective in the minds of educated people, because there was a time, not so very long ago, when this sort of explanation was revolutionary rather than commonsensical, and a time, not so long before that, when this sort of explanation would have been incomprehensible. Before the late nineteenth century, most scholars had very different ways of perceiving, contextualizing, and explaining religious phenomena.

This perspective is not restricted to secular intellectuals. A functionalist reading of religion forms the backbone of Islamic instruction in Egyptian public schools, and in both the secular and the religious press. God has designed the universe to provide for human life and to instruct mankind about His own nature and majesty. So social forms like marriage and hierarchy, ideas like equality and duty, and ritual requirements like prayer, fasting, and ablution, all fulfill specific and multiple functions. Take, for example, the following passage from the fifth-grade Islamic-studies text used in the late 1980s in Egyptian schools:

> Islam had preceded all advanced nations by ages in its call for cleanliness, and it made "cleanliness next to Godliness" when [the Prophet Muhammad] . . . says "God is pleasant [*tayyib*] and loves pleasant things." Perhaps the *wudu'* [the ritual ablution before prayer] clarifies best the scope of Islam's interest in cleanliness, since it is part of prayer. . . . There's no doubt it cleans man's body, and the modern physician has established that the *wudu'* a number of times a day brings health and keeps away skin diseases, just as he has proved that the *istinshaq* [the inhalation of water through the nostrils during the *wudu'*] protects people from various respiratory diseases, and just as in rinsing there is a cleaning of the teeth guaranteed to freshen the breath; this had been mentioned in the noble Hadith: "Not to burden my people, I ordered them to use the *siwak* [a short stick for cleaning the teeth] at every prayer." . . . Among the manifestations of Islam's concern with cleanliness: that it calls on us to bathe for prayer on Fridays and on the two feast days. . . . And the modern physician agrees with Islam in this, for doctors call on us to bathe at least once a week, guarding the body's cleanliness and freeing it from diseases. . . . The conclusion is that whoever wants to maintain the teachings of his religion looks after cleanliness. And whoever wants people to love and respect him is neat and clean, for Islam is a religion of cleanliness, and therefore it's a religion of advancement and civilization. (Yunis et. al. 1987–88: 73–75)

This passage equates the ritual purity of the *wudu'* with the physical purity of a bath. Although it describes cleanliness as a sacred requirement, the requirement has been functionalized, implying that the reason for the prescription is its presumed effect on health and well-being, rather than to mark a separation between sacred and profane (Douglas 1966: 29). Furthermore, linking cleanliness and civilization creates a hierarchy of peoples in which the Islamic community is historically first, and places the tradition of the Prophet in the domain of the urban planner and the public health official, a quite recent development (Greene 1885: 78). Max Weber noted similar changes in Judaism and Zoroastrianism, as modern elites sought to rescue their faiths from accusations of irrationality (Weber 1963: 93, Luhrmann 1996: 107–8).

Egyptian religious studies textbooks commonly present moral lessons in the context of activities, stories, or skits, encouraging the child to remember to apply them in everyday situations. Morality and hygiene are linked in two ways: the text itself links them explicitly; and they are structurally similar as systems of rules that are intended, by those disseminating them, both to provide meaning and to generate behavior. Both are applied subjects in which the real test is the conduct of life rather than performance on exams. A 1929 British evaluation of Egyptian schools, commissioned by the Egyptian Ministry of Education, complained that the

> process [of examination and cramming] is objectionable in itself but most of all when applied to such subjects as hygiene and morals. Not only is examination in these subjects apt to confuse the essential issue but it attempts to test what obviously cannot be tested by the simplicities of question and answer. The dirtiest little boy ever born might easily get full marks in a written examination in hygiene, and the most doubtful juvenile ever conceived the first place in morality by sheer capacity for the reproduction of platitudes, in the one case physiological, in the other, ethical. (F. Mann 1932: 21)

Schools are one of the institutions in which children are supposed to learn the skills and discipline needed to transform the rules of cultural elites into the public good of virtuous performance.

This functionalization is a two-step process. First, social functions (increased health, cleanliness, order) are attributed to Islamic practices. Then these functions are interpreted not only as effects, but as the primary intent of given practices, and therefore divinely sanctioned themselves. Textbooks for higher grades also emphasize this natural theology, developing logical proofs from natural models for the necessity of the division of labor and the order of society. Just as communities of ants, bees, and humans have their leaders, so must the cosmos have a supreme authority in God, who orders life in such a way that acts of worship benefit both the worshiper and society at large (al-Duwwa et al. 1986–87: 40, 83, 87–88, 156). Daily prayers, for example, not only "invigorate the body," but "accustom the Muslim to organization, and respect for appointed times" (al-Naqa 1988–89: 29). Prayer is incumbent upon Muslims not only as a link between the divine and the created, but because in prayer "there is rising and bowing and prostration, all actions that invigorate the body, and the Muslim devotes himself to work with zeal and energy, and increases production and spreads the good, and promotes [the progress of] the nation. . . . [P]rayer accustoms us to order, and the keeping of appointments, and the binding together of Muslims with cooperative ties and love and harmony" (al-Duwwa et al. 1986–87: 158).

Both in government-issued textbooks and in children's books produced by private companies, moral behavior is linked closely not only to social order—the idea that "Islam is a religion of order and discipline" (Yusuf and Abduh 1988: 14)—but to economic development. The Ramadan fast reduces friction between the rich and poor by letting the wealthy experience hunger and privation, prompting them to give alms to the needy. This produces serenity in the hearts of the poor so that "everybody applies themselves to their work, and production increases, society becomes happy, and its economy develops" (al-Duwwa et al. 1987–88: 133). The fast also works indirectly by providing practice in willpower, helping free Muslims of "ugly habits like smoking, which takes its evil toll on the person's health, and then he can't do his work, and it reduces his productivity and reduces family income and causes the country's economy to slump" (al-Duwwa et al. 1987–88: 133). In summary, "The renaissance of Islamic society stands upon the faith of individuals, and on the effects of this faith on their behavior. . . . [The Muslim] balances the demands of religion and the world, and works for [this] world as he does for the next, is precise in his work, and increases production . . . until he has achieved prosperity, advancement and economic development for society" (al-Duwwa et al. 1987–88: 140).

Such functionalist readings of religious duty have three particularly significant aspects. First, the duties are portrayed as signs. We are expected to learn from them something about God's intentions for other parts of our lives (God's requirement of regular prayer shows that he wants us to live in an orderly manner). Second, the duties are portrayed as actually effective (regular prayer helps us build the self-discipline that will allow us to live orderly lives, and the resulting social solidarity spurs the economy). And third, such readings demonstrate to the student, in a more general sense, that religious duties and ideas have not only meaning at a personal level, but purpose on a social level as well. God has designed a universe in which personal habit and social tranquility are of a piece.

## THE SOCIAL TRANSMISSION OF FUNCTIONALIST THEORY

*Islam is an evolutionary religion that is constantly striving to attain the ideal in spiritual and material life alike. Hence, its defenders and propagators must be equally flexible and idealistic.*

—TAHA HUSAYN, The Future of Culture in Egypt

Contemporary textbook form in Egypt is the result of a long series of intellectual and institutional developments over the last century and a half. Muslim reformers and European colonial administrators in the late 1800s began to seek new ways of conceptualizing Islamic practice. Critics argued that Islam's scholarly custodians had become intellectually narrow and rigid, martinets attending to ritual niceties rather than spiritual guides disseminating the fundamental moral principles of a once-vibrant civilization. But popular religious concerns, too, had largely to do with the conduct of ritual life rather than the sorts of issues that engaged the country's secular urban elite: nationalism, modernity, science and technology. As late as the 1940s, University of London–trained Egyptian anthropologist Hamed Ammar remarked on the sorts of questions that Upper Egyptian villagers addressed to a visiting religious scholar during his annual visit. Most concerned the details of prayers and prostration. "Other questions that arose were queries about whether certain sayings or deeds were considered to be religiously approved or disapproved, how and to whom one would make a sacrifice oath, whether visiting saints' tombs is a good thing or not, and so forth" (Ammar 1954:78).

A less neutral view from a British traveler a century earlier noted the difference between a "modern" appreciation of the purpose of religion, and that institutionalized in Egypt at the time. Religious instruction at the renowned mosque-university of al-Azhar turns principally upon the religious observances required by the Koran, and degenerates into extreme frivolity. Rarely is any lesson of morality given, and the passages of the Koran, which teach the cultivation of the virtues, are much less introduced and commented on than those which bear upon the ceremonials of the Mussulman faith. Inquiries as to the quantity of adulteration, which makes water improper for ablution—into the grammatical turn of the language of prayer—into the cases in which the obligations to fast may be modified—into the gestures in adoration most acceptable to Allah—into the comparative sanctity of different localities, and similar points—are the controversies which are deemed of the highest importance, and the settlement of which is supposed to confer a paramount reputation upon the Ulema [religious scholars]. (Bowring 1840: 137)

The sole hope for overcoming this backwardness was that "the Koran might be made, like the Bible, a means of imparting moral truth combined with instructive history" (Cunynghame 1887: 232), instead of a ritual guide or a text to be memorized. Muslim reformers agreed that Islam as taught in indigenous institutions was "dry, dead, ritualistic, and irrelevant to the needs of living Muslims" (R. Mitchell 1969: 213). European critics of Egypt's Coptic Christians made the same point, faulting the church for encouraging thoughtless ritual rather than spiritual development, and calling for the revitalization of Egyptian Christianity through the Protestant "evangelical ethos" of industry, discipline, and order (Sedra 2001). Living Egyptians needed a new type of religion emphasizing morality rather than ritual, virtuous political life rather than doctrinal minutiae, and national integration rather than village-level particularism (Starrett 1998). Labeling common religious practices as superstitions to be purged from popular life, Egypt's secular elites sought to rescue Islam from the past, and from its marginalization on the international scene. The solution, according to twentieth-century Islamic activist Hasan al-Banna, "is the production of a state of mind imbued with the Islamic theory of life, to give permanence to external forces leading to this form of life. . . . And the natural method of establishing that philosophy is by education" (quoted in R. Mitchell 1969: 284).

Al-Banna was hardly alone in making this prescription. The spread of mass popular schooling in the nineteenth century was connected with changing economic and political forces consequent to the Industrial Revolution. Rapid urbanization, changing technology, new labor demands, and the intensified power of states that were centralizing bureaucratically and expanding geographically, encouraged European countries, and then their colonial dependencies and foreign rivals, to establish modern schools for the training of a modern workforce. In Egypt and elsewhere, states sought to borrow the "'secret wisdom' of the West" by establishing European-style educational systems (Fortna 2002). Science was the poetry of the new age; among its luminaries were Charles Darwin and Herbert Spencer in England, and Gustave Le Bon, Edmond Demolins, and Emile Durkheim in France (Hanioglu 1995, 2001: 290, 294).

In Central Asia, Egypt, and Anatolia, the expansion of religious education out of traditional mosque study circles into modern schools during the late Victorian era "transform[ed] the notion of what constituted 'Islam' just as, conversely, the 'secular' nature of modern schooling was itself altered as it became a vehicle for religious education" (Fortna 2000: 371; see Khalid 1999). In each of these places, states increasingly used religious education for their own ends, a complex process I've called "functionalization" (Starrett 1998). Functionalization is a set of processes in which ideas from one cultural domain (or discourse) come to serve the strategic or utilitarian ends of another. This translation not only places ideas, practices, and institutions in new fields of significance, but radically shifts the meaning of their initial context. In Egypt, a whole series of existing religious discourses have been reified, systematized in novel fashion, and put to work fulfilling the ends of the modern secular discourse of "public policy." Traditions, customs, beliefs, institutions, and values that originally had a host of varied meanings and uses are consciously subsumed, by modern-educated elites, to the evaluative criteria of social and political utility. Institutionally, independent local scriptural study circles are brought under the control of central or district government bureaucracies to act as tools of mass socialization, for the family is now considered inadequate to the task of preparing children for social life (Husayn 1954: 29; Durkheim 1961: 18–19; Fortna 2000: 385; Salmoni 2001). Logistically, educators fashion formal religious studies curricula, and formal testing patrols the borders of class mobility. Philosophically, ancient rituals and beliefs as well as the facts of history are reinterpreted to underscore political legitimacy, or are brought to bear, as we saw above, on social concerns such as public health, economic productivity, and crime. In all of these processes, existing meanings are purposely altered, and control is shifted to a central authority or entrusted to groups other than those who traditionally set their terms. These changes were made possible both by the financial, technical, administrative, and military resources of the government, and by new intellectual resources provided by the nascent social sciences.

Early structural-functionalist theories of social organization helped European and Middle Eastern elites think explicitly about the phenomenon of "society" as an interdependent organism, and about how best to ensure its efficient operation. There is disagreement about the lines of influence between various scholars, in part because of contemporary citation practices. Whereas Mary Douglas (1966: 20), for example, sees Le Bon's influence on Durkheim, Steven Lukes (1985: 462) does not. M. Sukru Hanioglu, on the other hand, writes of both LeBon and Demolins as popularizers of Spencer and Durkheim (Hanioglu 2001: 308). Whatever the case, European sociology's impact on the Middle Eastern intellectuals who would go on to become the region's ideologues and technocrats is indisputable (Hourani 1983; Ibrahim 1997).

In Ottoman lands, reformers set out to present positivist concepts in Islamic terms (Hanioglu 1995: 200–202), simultaneously appealing to preexisting values, and emphasizing the importance of a new division of labor guided by a modern-educated elite (Hanioglu 2001: 309). As early as the 1820s, followers of the French utopian Claude Henri Saint-Simon—to whom Auguste Comte himself was disciple and secretary—emigrated to Egypt to help its ruler establish a modern military and economic infrastructure of irrigation works, schools, public-health facilities, and factories. Some settled permanently in Egypt, reinforcing the influence of a small but steady stream of Egyptian students returning from study in French universities with an appreciation of the intellectual currents of positivism (Ibrahim 1997).

These intellectual currents were not intended to be esoteric knowledge, but rather vital ideas whose effectiveness relied on their dissemination. In the central Ottoman lands, many members of the Committee of Union and Progress (CUP) (a reformist movement founded in 1889 which eventually staged a revolution against the sultan in 1908), worked in education. They taught, established schools, published newspapers, and translated works from the French *science sociale* tradition of Edmond Demolins and his circle, which held, according to CUP leader Ahmed Riza, that "society is a complex organism dependent solely upon natural laws . . . [and t]his body is subject to cyclic illnesses" curable only by scientific methodology (Hanioglu 1995: 208). The physicians to perform this healing were lawmakers, administrators, and politicians trained in modern subjects such as natural science and sociology. Even members of the royal family, such as Mehmet Sabahaddin Bey, were convinced that the empire's problems were rooted in its social structure and educational system. Their solution lay in educational theories arising from the scientific sociology of France and England (Hanioglu 2001: 292). Intellectual developments in the Middle East followed two strands of European sociological theory, both focused on the organismic analogy summarized above by Ahmed Riza. One theoretical strand, associated with Spencer and Demolins, stressed adaptability and evolutionary change on the part of the system's decentralized constituents; the other strand, associated with Durkheim and Le Bon, emphasized the totality and functional interdependence of the system as a whole.

Prince Sabahaddin Bey joined Demolins's Societé Internationale de Science Sociale in 1904, followed two years later by Fathi Zaghlul Pasha, brother of Saʿd Zaghlul, the latter of whom became Egypt's minister of education in 1906, and later its most prominent nationalist leader (Hanioglu 2001: 87). Saʿd Zaghlul boasted in 1908 of the utility of education as an engine of social progress: "The Government has found itself the means of developing general morality amongst the popular masses, in order to diminish the number of noxious and blameworthy acts due to the ignorance of true principles and of the exact rules of religion" (quoted in Salama 1939: 303). In 1899, Fathi Zaghlul translated Demolins's book *A Quai tent la superiority des Anglo-Saxons* into Arabic. Demolins asserted not only that a new style of education was key to forging and transmitting modern ways of life, but that "a moral or mental gulf had opened [in the world at large] between those whose minds were formed by the social sciences and the rest" (T. Mitchell 1988: 110). The former group understood that competition in the international political sphere required national citizens "trained by their families, in their schools, and indeed by all their social surroundings, with the idea that a man 'ought always to fall on his feet, like a cat.' They are not brought up for rest, and *dolce far niente*, but for the *Struggle for existence*, for *Self-help*; in short, they are taught to march forward and go ahead" (Demolins 1899: 273). Progress was seen as a Darwinian process in which group selection is the operative mechanism.

Along with his 1898 book *L'Education nouvelle*, Demolins's argument about Anglo-Saxon cultural superiority was widely read by Egyptian intellectuals and leaders (T. Mitchell 1988: III), as were Le Bon's discussions of crowd psychology and the evolution of national character, which influenced the Egyptian elite's vision of itself as the vanguard of a new national consciousness. "The totality of the ideas and sentiments," wrote Le Bon,

> that are, as it were, the birthright of all the individuals of a given country form the soul of the race. Invisible in its essence, this soul is very visible in its effects, since it determines in reality the entire evolution of a people. A race may be compared to the totality of the cells that constitute a living being. The existence of these milliards of cells is very short, whereas the existence of the being formed by their unition is relatively very long; they possess at once their own personal life and a collective life, that of the being of which they form the substance. In the same way each individual of a race has a very short individual life and a very long collective life. The latter life is that of the race of which he is sprung, which he helps to perpetuate, and on which he is always dependent. (1974: 10)

When Le Bon spoke of "the three fundamental bases of the soul of a people: common sentiments, common interests, and common beliefs" (Le Bon 1974: 13), Egyptian nationalists knew they had found the scientific charter for the future of Egypt. But this development, he stressed, was a modern phenomenon.

> The congeries of sentiments, ideas, traditions, and beliefs which form the soul of a collectivity of men has always existed more or less in the case of all peoples and at all ages, but its progressive extension has been slowly accomplished. Restricted at first to the family and gradually extended to the village, the city, and the province, the collective soul has only spread to all the inhabitants of a country in comparatively modern times. It was only when this last result had been achieved, that the notion of a native country, as we understand it to-day, came into existence. The notion is not possible until the national soul is formed. (Le Bon 1974: 13–14)

Organismic analogies of both Spencerian and Durkheimian spirit mixed freely in Egyptian and European intellectual life. Egyptian religious reformer Muhammad 'Abduh—another one of Sa'd Zaghlul's teachers—so admired the work of Herbert Spencer that he visited him on a journey to Brighton and later referred to him as "the chief of the philosophers on social questions" (C. Adams 1968: 167). Spencer's social theories likewise influenced Qasim Amin, who agitated at the turn of the twentieth century for the emancipation of women (Hourani 1983: 167), Lutfi al-Sayyid, newspaper editor and director of the national library (Hourani 1983: 171), and others. Lutfi al-Sayyid's newspaper *al-Jarida*, like the publications of the Ottoman CUP, presented early twentieth-century Egyptian readers with discussions of the philosophy of Montesquieu, Comte, Mill, and Spencer (Mahmoudi 1998: 44).

Durkheim's lectures on sociology at the Sorbonne were attended by several Egyptians, including Mustafa 'Abd al-Raziq, who would become rector of the ancient and world-renowned mosque/university of al-Azhar in 1945 (Hourani 1983: 163), and Taha Husayn, who became minister of culture and education in 1950 (Mahmoudi 1998; T. Mitchell 1988: 121). Durkheim supervised Husayn's dissertation on fourteenth-century Arab historian Ibn Khaldun, and one of Husayn's two oral examinations for his doctoral degree was on Comte's sociology (Mahmoudi 1998: 182 n.50). Durkheimian theory guided the development of Turkish nationalism in the 1920s through that movement's chief theorist, sociologist Ziya Gokalp (Gokalp 1958).

For both Spencer and Durkheim, the key to creating a progressive and self-conscious society was the process of education. The mechanism varied somewhat. For Spencer and Demolins, a proper modern education primed society for evolutionary change by ensuring the mental and practical adaptability of its citizens. Durkheimian pedagogy, on the other hand, sought to create a sense of moral obligation to the collective by desacralizing traditional religious symbols and then resacralizing them as social ones, allowing students

> to divest this conception of the mythical forms in which [moral authority] has been shrouded in the course of history, and to grasp the reality beneath the symbolism. This reality is society. In molding us morally, society has inculcated in us those feelings that prescribe our conduct so imperatively; and that

kick back with such force when we fail to abide by their injunctions. Our moral conscience is its product and reflects it. When our conscience speaks, it is society speaking within us. (Durkheim 1961: 90)

Once demystified, social imperatives become self-evident and self-justifying. Once the child is bound properly to the social group, supernatural justifications for behavior become superfluous. "To teach morality," Durkheim wrote, "is neither to preach nor to indoctrinate; it is to explain. If we refuse the child all explanation of this sort, if we do not try to help him understand the reasons for the rules he should abide by, we would be condemning him to an incomplete and inferior morality" (1961: 120–21).

As expressed by Durkheim's student Taha Husayn, elementary education should teach the Egyptian child to "be aware that he belongs to an entity called the nation which existed before he was born and will continue to exist after he dies. Accordingly, he must have . . . familiarity with his country's history, present conditions, and laws; he must in addition sense its hopes and conceive somehow its future" (Husayn 1954: 29). In the case of Egypt, the process Durkheim outlined above occurred by functionalizing rather than desacralizing the religious tradition. According to Husayn, "The reality that is the nation functions within the geographical limits set by God. Only here can its people live in full security, carry on their traditional way of life, and expect to realize their aspirations" (Husayn 1954: 29). Rational social processes exist within a supernatural framework, for Muslims perceive the two as non-contradictory. For if God is concerned with the welfare of his creation, the prescriptions of Islam must be not only beneficial, but manifestly rational. Furthermore, common slogans such as "Islam is a religion of cleanliness," "Islam is a religion of order and discipline," or "Islam is a religion of work," not only mark these practices with divine intent, but transform the Islamic tradition into a source of policy guidelines and symbols that can bridge the orders of tradition and change. Once religion is perceived as useful in achieving given ends, it is used by state bureaucracies, private voluntary organizations, and mass political movements in the prosecution of those ends, whether or not it does in fact achieve them. And that use becomes in turn one of religion's empirical features. In being used in this very conscious way it becomes, perforce, a functional element of social life, "an important or even essential part of the social machinery, as are morality and law, part of the complex system by which human beings are enabled to live together in an orderly arrangement of social relations" (Radcliffe-Brown 1945: 154).

## CONCLUSIONS

*There is no more reason for the sociologist to adopt for his thinking about society the terms used for thinking in society (to take, in other words, his analytical tools straight out of his data) than there is for a carpenter to use nothing but wooden saws.*

—RONALD PHILIP DORE, *"Function and Cause"*

Anthropology students of a certain age remember reading in their methodology classes James Spradley's The Ethnographic Interview (Spradley 1979), which provides useful guidelines for analyzing culture based on the careful and systematic elicitation of cultural texts. As a caution to students, Spradley tells the story of doing evaluation research in an alcohol rehabilitation facility in Seattle, interviewing "tramps"—homeless alcoholics—about their survival strategies and their experience of the jail and legal systems. He recalls being very excited one day, in reading through arrest records from the county jail, to find a subject who had graduated from Harvard and done some graduate work in anthropology. A trained insider! A key informant! What a resource for understanding the world of tramps! At a preliminary interview, he asked "Bob" to reflect on the variety of tramps he had met. Later, at a longer follow-up interview,

> We chatted casually for a few minutes, then I started asking him some ethnographic questions. "What kind of men go through the Seattle City Jail and end up at this alcoholism treatment center?" I asked. "I've been thinking about the men who are here," Bob said thoughtfully. "I would divide them up first in terms of race. There are Negroes, Indians, Caucasians, and a few Eskimo. Next I think I would divide them on the basis of their education. Some have almost none, a few have some college. Then some of the men are married and some are single." For the next fifteen minutes he proceeded to give me the standard analytic categories that many social scientists use. "Have you ever heard men referred to as tramps?" I asked. From numerous informants I knew this identity was the most important. "Oh, yes," Bob said, "some guys

use that term." "Are there different kinds of tramps?" I asked. "I suppose so, but I'm not up on what they would be." Bob then proceeded to talk about intelligence, education, race, and other categories that usually interested social scientists. In later interviews Bob tended to analyze the motives men had for drinking and other behavior, but his analysis always reflected his background in college. He had great difficulty recalling how most other tramps would refer to things. (Spradley 1979: 53)

Spradley concludes by warning novice ethnographers away from informants like Bob: "In our society, many persons draw from psychology and the social sciences to analyze their own behavior. They mistakenly believe they can assist the ethnographer by offering these analytic insights" (Spradley 1979: 53). Even experienced ethnographers need to take "special precautions" when dealing with people who use the ethnographer's own language in framing their experience.

Spradley had compelling reasons to try to insulate native texts from the contamination of social-science vocabulary. He assumed that the latter was part of an artificial language that could only confuse an analysis of the lived experience of his subjects. His own world and that of the tramps, though inter-connected, had to be kept analytically separate for the ethnographic analysis to work. Some ethnographers have taken this idea in a different direction, per-ceiving the analytical isolation of "cultures" as an unfortunate general feature of anthropology, and urging us to "'write against culture' by working against generalizations" (Abu-Lughod 1993: 6).

[O]ur goal as anthropologists is usually to use details and the particulars of individual lives to produce typifications. The drawback . . . for those working with people living in other societies is that generalization can make these "others" seem simultaneously more coherent, self-contained, and different from ourselves than they might be. Generalization . . . helps make concepts like "culture" and "cultures" seem sensible. This in turn allows for the fixing of boundaries between self and other. (Abu-Lughod 1993: 7)

Anthropological generalizations can, as shown in the critiques of culture by Borofsky (this volume) and others, tend to flatten out internal differentiation and erase change and conflict, making "others" appear homogenous, coherent, and timeless. Ironically, in the process of attempting to understand the experiences of others, "anthropology ends up also constructing, producing, and maintaining difference. Anthropological discourse helps give cultural difference (and the separation between groups of people that it implies) the air of the self-evident" (Abu-Lughod 1993: 9, 12).

Ignoring Bob may have helped Spradley construct a vision of "pure" tramp culture. But surely the "many persons" who use social sciences to understand themselves deserve our analytical attention as well. The separation of "native" and "analytical" categories of thought is nice in the abstract, but increasingly difficult in practice. For even if, in Dore's terms, we avoid depending on the language *of* life to think *about* life, and thus avoid wooden saws, we face the dilemma of people who snatch up our steel blades not only as tools for second-order analyses of social life, but as the very raw material for its construction. Such is the problematic nature of "culture" as well as "function": Do we have to develop new conceptual tools in order to analyze ways of life built in part out of the old ones?

The artificial analytical separation of cultures is a problem. But I think Abu-Lughod's approach to its solution—as well as the "solution" of abandoning the idea of culture altogether—is misplaced. Assuming the homogeneity, coherence, and timelessness of other cultures is surely a mistake, but ethnographies that "write against culture" and emphasize individual experience as a bridge to understanding our common humanity, make a different sort of mistake: they take for granted the reader's experience as an individual. For if we treat the Other as fully individual and unique, unmotivated and unshaped by culture (or history, or class, or environment), we must treat ourselves the same way. Our human experiences suddenly appear to have "the air of the self-evident," to seem fundamental, natural, and unanalyzable. But the social sciences were developed precisely to undo the faith that our lives and selves are independent of larger social, political, economic, and cultural forces and processes. "Good social science," Robert Redfield wrote in 1947, "provides categories in terms of which we come to understand ourselves. Our buying and selling, our praying, our hopes, prejudices, and fears, as well as the institutions which embody all these, turn out . . . to have form, perspective, rule. Shown the general, we are liberated from the tyranny of the particular. I am not merely I; I am an instance of a natural law" (Redfield 1963: 197).

The same point that applies to the unique "us" as individuals applies also to the restricted "us" of social scientists as well as to the expansive "us" of "we of the West." As Roxanne Euben has noted in her study of European and Middle Eastern social criticism, "it often seems as if ideas and beliefs we regard as reasonable—ones we often hold—frequently do not require social scientific explanation at all" (Euben 1999: 14). We appreciate our own society's intellectuals and cultural critics for their insight into the problems and potentials of modern life, but we analyze similar arguments made by Muslim cultural critics by referring to the author's nationality, class background, biography, and politics, looking at the social forces that shape their views. Our ideas have meaning while theirs—epiphenomena of real objective forces—have only function.

Clearly one of the many problems with studying human beings is maintaining a consistency between the way we seek to understand "them," and the way we seek to understand "us." Where our analytical vocabulary has been appropriated as bricks and mortar for the construction of a way of life, we encounter the difficulty that our analyses may be self-confirming. Is religion in contemporary Egypt functional in same sense as religion among Radcliffe-Brown's Andaman Islanders? Does it matter that Egyptians recognize and even revel in the purported functionality of their religion, as did the third-century B.C.E. Confucian scholars to whom Radcliffe-Brown (1965: 157–60) attributes the "original" insight about function? Or do we perceive religions everywhere as functional because, beginning in the latter part of the nineteenth century, European social theory developed new ways to apply organismic metaphors to human groups, which lent themselves, in concert with the development of new communications media, administrative practices, and institutions like mass popular schooling, to the self-conscious functionalization of religious ideas and practices whose functionality is now, as a result, real and self-evident?

Function, like other social-science concepts, is part of a set of culturally transmitted ideas whose historical origin and successful career have helped make the world around us what it is today. Those ideas have been transmitted in the context of a massive transformation of social and political institutions that have altered the life experiences and possibilities of both children and adults. Each of us has been shaped intellectually, morally, and behaviorally (whether or not deterministically, completely, or beneficially) by a system of schooling that was put in place by social elites whose evolving understanding of evolving social systems made mass schooling look like the key to national progress and the solution to a number of economic, political, and social problems (Starrett 1998). Whether we are Egyptian, Bolivian, or Japanese, we have in turn learned from those schools the rationalist perspectives of their founders, who were inspired by functionalist and evolutionary social theory. We are also enmeshed in new sorts of social relations defined by the use of schools as status-granting institutions, tools of social policy, employers, recipients of state and private investment, generators (at higher levels) of new technologies, centers for social networking, political agitation or military recruitment, and so on. These relations and infrastructures, too, and not merely the "ideas" schools teach, constitute parts of our culture, niches that elites and working classes have cooperated in constructing for very different reasons (see Feldman, this volume). Given the spread of education from elementary to graduate level in countries around the world, and the development and diffusion of technocratic forms of government in the nineteenth and twentieth centuries, control of the economic and political systems that generate power and inequality (or, from the perspective of the system's insiders, freedom and wealth) may be ultimately in the hands of natural law of one sort or another. But proximately it is in the hands of men and women who have been trained in economics, political science, sociology, management, and the other human sciences, to act as the technicians of the modern economic and political order (T. Mitchell 2002). It is in the hands of people with theories about the world and strategies for making it work (some of which actually make things worse rather than better, just as among the Fore—see Durham, this volume). It is in the hands, in other words, of cultured beings very much like us. Thus the very first rationalization for abandoning the culture concept in anthropology—the argument that culture doesn't really do anything—crumbles into dust. Culture influences behavior. But, to complicate the matter, so does "culture" (the concept), along with concepts like "function," "class," "history," and all the others that we've developed in order to enlighten ourselves. Since culture so often works to obscure its own operation, though, the culture of social scientists sometimes makes us among the first to forget it.

# 13

## Studying "Culture" Scientifically Is an Oxymoron

### The Interesting Question Is Why People Don't Accept This

ROBERT BOROFSKY

There is an edge to this chapter's title. In a book entitled *Explaining Culture Scientifically*, it is meant to challenge the status quo of our conceptions. We tend to use terms such as "science" and "culture" in seemingly self-evident ways. Yet, clearly, they are problematic concepts that need be handled with care. There is not one science, for example, but many sciences. And often science constitutes a way for credentializing results as much as, if not more than, a way for challenging them. In respect to culture, we frequently give short shrift to its conceptual complexities—a point Brown, Starrett, and D'Andrade also make in this volume. Without necessarily intending to, many writers lean toward simplistic phrasings of culture as shared beliefs, that skim over issues of critical public importance.

We seem to sleepwalk through issues that lie at the heart of this book. If we are to explain scientifically how culture influences behavior, we might well start by asking ourselves why we, as social scientists, have not developed the scientific tools to unmask the hegemons—surrounding science and culture—that we deal with on an ongoing basis, in our writing, in our behavior. We often seem to go with the flow in dealing with these concepts, fitting into how others use them. That is the reason for my title's edge. It is meant to shake us awake.

### SCIENCE: CHALLENGING KNOWLEDGE CLAIMS

Let me start with science. There are plenty of abstract statements about what science is, or is not. We have, for example, the Vienna circle's concern with logical positivism—with the assertion that "propositions should not be accepted as meaningful unless they are [empirically] verifiable" (Passmore 1967: 55). But abstract pronouncements of this sort leave us uncertain, hesitant. How do we apply them in concrete ways? Why should we accept one pronouncement rather than another?

It would be wiser, I would suggest, to move from *pronouncements* about science to *practices* of science. When we look closely at scientific practice, it becomes clear that science is conducted in a number of ways. If we take, as an example, Karin Knorr Cetina's *Epistemic Cultures: How The Sciences Make Knowledge*—which examines scientific practices in experimental high-energy physics (at CERN in Geneva, Switzerland) and molecular biology (at the Max Planck Institute in Göttingen and Heidelberg, Germany)—we perceive "the fragmentation of contemporary science; it displays different architectures of empirical approaches, specific constructions of the referent, particular ontologies of instruments, and different social machines. In others words . . . the diversity of epistemic cultures" (1999: 3).

That said there is, however, at least one unifying element. In looking at the practices of science, we can perceive this unity in what happens when scientific results are called into question: when the possibility of challenging results— a pervasive phenomena—turns into the actual reality of challenging them—a rarer event. Challenges are made, people respond, and the broader community works its way toward consensus on the issue of the cause (or causes) of the divergent results.

A scientist's claim that she did A with B to produce C is questioned by others who, following the same basic procedures, attempt to replicate her results. Because the dynamics of replication are often complicated, what becomes crucial—especially when, as not infrequently happens, the replicated results are different—are the conversations that seek to resolve the difficulty. As Peter Galison states in his important study of physics, *Image and Logic*, with respect to replicability:

It is surely true, as authors from Michael Polanyi, to Thomas Kuhn, to Harry Collins have insisted, that there are moments when individuals cannot spell out rules for replication. That should be occasion, not to stop the inquiry, but to ask: Why not? What pieces of practice will not fit the public discourse of science, and why? Sometimes the movement of machine knowledge may be impeded because no one knows what portion of a complicated procedure is efficacious and what is superfluous . . . [such] knowledge typically takes place not all of a piece; engineers from different laboratories meet to share tricks about gaskets and seals, about computer analyses and simulations, about chemistry, cryogenics, and optics. (1997: xx–xxi)

Credibility is established when various people with various perspectives come to a general consensus regarding a challenge. It is not so much the replicable data themselves that generate trust—because these may well not be fully replicated—as much as the negotiated conversations that ensue. These conversations analyze and explain the differing results that formed the basis for the challenge in the first place.

By way of illustration, take the case of Victor Ninov at the Lawrence Berkeley National Lab. A world-renowned expert on atomic particles, he was a prominent figure on a team that claimed to have discovered (or produced) a new atomic element, number 118. On October 15, 2002, the *New York Times* reported:

Over five days in early April 1999 . . . experiments bombarded a lead target with a beam of krypton nuclei. The debris from the tiny collisions passed through the gas-filled separator, and various detectors recorded the energy, position, and timing. . . . [the] result was an enormous amount of raw data that Dr. Ninov processed using software he had mastered at GSI [a competing German research institute that had taken the lead in discovering new atomic elements and for which Dr. Ninov had previously worked]. As the only one on the team familiar with the program, he was put in charge of the analysis.

Several days later, he began telling colleagues that he had observed three instances of what appeared to be the decay of a 118 nucleus . . . it was an extraordinary claim, but there was good reason to trust Dr. Ninov's instincts. "I had hired a world-recognized expert and was trusting him to do the job," Dr. Gregorich [the project leader] said. . . . After the group had closely reviewed his calculations—Dr. Loveland filled a binder with supporting evidence—one chain [out of four] was discarded. But that still left three good ones. Scientists at GSI also examined the results and agreed that something noteworthy might have occurred. Everyone was working with the numbers Dr.

Ninov had gleaned from his own analysis. No one felt a need to go back and examine the original data. The group submitted a report to *Physical Review Letters*, which published it on Aug. 9. (G. Johnson 2002)

For an element to be allocated a name (rather than just a number) the same results need to be produced by a different laboratory. Both GSI in Germany and the Riken Institute in Japan tried to do this but failed. As the *New York Times* piece notes, "these negative results were not necessarily fatal. Events like these are exceedingly rare, and it was possible that Dr. Ninov and his colleagues had just been luckier than the others." The first independent review committee to look at the results was able to rule out possible machine errors—relating to beam alignments, detector errors, or flaws in the processing of the data. A second independent review committee, however, could not find the decay chains (that others earlier had accepted) in the original raw data. A third committee confirmed the second's conclusions. "The initial suspect," the *New York Times* reported,

was the analysis software, nicknamed Goosy, a somewhat temperamental computer program known on occasion to randomly corrupt data. Over the years, users had developed tricks for dealing with Goosy's irregularities. . . . But a close look at the element 118 experiments found no signs that Goosy had seriously misread the data.

What turned out to be the smoking gun was a computer "log file"—a diary automatically generated by Goosy of everything that had occurred during the handling of the data. . . . According to this history, an analysis performed around noon on May 7 indeed showed what appeared to be an element 118 decay chain. But when the very same data was analyzed again, a few hours later, the chain was not there. A closer look showed that it was the earlier record that had been altered; page lengths were inconsistent, and the timing of some of the events was off. In fact, investigators discovered the events passed off as a 118 decay chain could be manufactured by cutting and pasting a few lines from elsewhere in the file and changing some of the numbers. Records from the 1999 run also indicated that at least one of the original three chains had been edited in a similar manner—by someone using the account Vninov. (G. Johnson 2002)

It was on this basis that the fourth, and final, investigative committee concluded there was "clear and convincing evidence" that Ninov had fabricated the data. Dr. Ninov maintained his innocence. Although he acknowledged that the previously claimed decay chains were apparently not in the data, and some files were seemingly tampered with, he seemed puzzled as to who might have done it. Why would he, a world-famous scientist with a world-famous reputation to protect, do such a thing, he asked. (He noted that others at the lab knew his password and used his account.) It was only in the negotiated conversations surrounding efforts to replicate earlier results that fraud was discovered. We might ask what one should make of the fourteen coauthors whose names were included in the original *Physical Review Letters* article announcing the discovery. A researcher quoted in the *New York Times* piece commented that it was fortunate that Glenn T. Seaborg—a world-famous scientist at the Berkeley lab who holds the Guinness world record for discovering the most elements—"died before this, because he would have been one of the co-authors too" (G. Johnson 2002).

What causes me concern is that even though many acknowledge the importance of challenges to the scientific endeavor, fewer practice it. Science is often used as a way for credentializing rather than challenging results. We can see this in how many authors added their names to the original article as well as in how the peer review process initially failed to ferret out fraud.

Let me support this point with a case that will be familiar to many anthropological readers: the changing color of peppered moths in industrializing England. Here is how Gaden Robinson, writing in the *Times Literary Supplement*, summarizes Judith Hooper's 2002 book, *Of Moths and Men: An Evolutionary Tale, The Untold Story of Science and the Peppered Moth*:

[The book] is an extraordinary . . . researched book that dissects the theories, warped personalities, feuds and chicanery that took the phenomenon of industrial melanism in the peppered moth [in which darker-colored moths seemingly flourished in areas where industrial soot darkened trees because such trees offered more protective camouflage] and turned it into a cornerstone of evolutionary theory . . . the universal exemplar of natural selection. Corners were cut, results were "selected" and scandal and tragedy eventually ensued. (Robinson 2002: 3)

As Hooper had pointed out:

> Majerus's book [that reexamined that the original research done by Kettlewell on the subject] left no doubt that the classic story [by Kettlewell] was wrong in almost every detail. Peppered moths, if left to their own devices, surely do not rest on tree trunks, bird vision is nothing like human vision, Kettlewell was wrong about how peppered moths choose their resting sites; the high densities of moths he used may have skewed the results; the method of release was faulty. . . . The various predation and survey studies conducted after Kettlewell have not replicated his results particularly well, and other "factors" kept having to be invoked to squeeze the data into the standard industrial melanism model. (Hooper 2002: 283)

And yet, intriguingly, "although there have been dozens of experiments since Kettlewell's, no one yet has tried to redo Kettlewell's original experiments in *toto*, using proper controls and statistical techniques as well as live moths. (While Kettlewell used live moths, all of his successors, except for Majerus, have used dead ones, for reasons of convenience)" (Hooper 2002: 294).

*Scientific credibility does not derive from the way research proceeds—how one behaves (or claims to behave) in getting results—as much as from the challenges to these results.* Scientific experts and appearances can be deceiving when data are not challenged. In the case of peppered moths, most scientists wanted to believe the story: it is a charmingly clear story of evolution in action. The problem is that, on closer inspection, there are doubts as to whether Kettlewell was measuring what he claimed to be measuring. What is intriguing, given the prominent nature of the case, is that no one has deemed the experiment worth replicating. As a result, we have shimmering hopes and much uncertainty as to exactly what was measured in this most classic of classic experiments. All we can say is: who knows?

This brings me back to my central concern. Though we talk about science, we are not always willing to challenge the intellectual hegemons that surround the way we work, the way we behave. Actually challenging certain data—not simply the possibility of challenging them—is central to science. It is what gives data their scientific validity. But many times we turn to the appearance of science—systematic procedures and/or the production of results that fit with what is accepted—rather than to challenges and the resolution of these challenges to credentialize our results. A good example of how this plays out, with serious consequences, is the way anthropologists generally discuss the concept of culture.

## CULTURE

Despite the fact that we talk of culture as something "real," as something that exists "out there," culture is, in fact, an intellectual construct used for describing (and explaining) a complex cluster of human behaviors, ideas, emotions, and artifacts. Scholars have been making this point for decades (and D'Andrade and Starrett make it in this volume).

Lowie, for example, wrote in 1937 that "culture is invariably an artificial unit segregated for purposes of expediency" (1937: 235). The point that culture is a concept—not an actual reality—can be illustrated by noting the varied ways people interpret the term. Hatch writes: "Even though the term has been discussed in countless books and articles, there is still a large degree of uncertainty in its use—anthropologists employ the notion in fundamentally different ways" (1973: 1). That is why Appiah (1997: Electronic Edition) writes, playing on Johnst's famous phrasing (often attributed to Goering), "when you hear the word 'culture', you reach for your dictionary."

Still, one element that is repeatedly highlighted in the anthropological definitions of culture—either explicitly or implicitly—is a sense of sharing. I must be careful here. Many would probably concur that people who constantly engage with one another, share certain interactive styles because they are interacting with one another with some degree of presumed effectiveness. (This, of course, does not necessarily mean they share the same understandings of their interactions, as a reading of Steinbeck's The Chrysanthemums makes clear. One might also cite Wallace's example of the tooth fairy in this regard—1961: 37). When I refer to cultural sharing I mean sharing that extends beyond the limited case of a few people effectively interacting together on a continuing basis. As is commonly used in anthropology, the group specified as sharing cultural traits often includes people with whom an individual does not tend to interact on a personal level. Moreover, the anthropological usage often extends sharing to traits beyond the traits directly related to the interaction itself, such as sharing a belief in certain origin myths or sharing certain ways of behaving toward strangers (we see this meaning in the B and C models of D'Andrade's chapter in this volume).

A good way to perceive how anthropologists focus on sharing as a central element in their definitions of culture is to refer to the discipline's introductory textbooks. In the way anthropologists go about inculcating disciplinary perspectives in novitiates, we can see how anthropologists wish to portray their discipline to others. In his best-selling textbook, Haviland defines culture as "the ideals, values, and beliefs shared by members of a society" (2002: 488). In response to the question "What is culture?" Kottak includes, in his list of characteristics, "culture is shared" (1997: 38). Robbins states that culture is "the system of meanings . . . that are shared by a people . . ." (2001: 275). Peoples and Bailey define culture as "the socially transmitted knowledge and behavior shared by a group of people" (2000: 366). Miller defines culture as "learned and shared human behaviors and ideas" (2002: 388). A slightly more implicit sense of sharing occurs in definitions of culture as a set of behaviors and beliefs characteristic of a group (if they are characteristic of a group, it seems reasonable to presume that they are generally shared by individuals in the group in some form or other). In this regard, Ember and Ember define culture as "the set of learned behaviors, beliefs, attitudes, values, and ideals that are characteristic of a particular society or population" (1996: 402). Nanda and Warms refer to culture as "the way of life characteristic of a particular human society" (2002: 420). If one moves slightly further afield and looks at the more than 150 definitions of culture cited in Kroeber and Kluckhohn's famous work on the topic, we learn that "the attribution of culture to a group or social group is the single element most often given explicit mention" (1952: 154). My point is that explicitly or implicitly sharing constitutes a significant element in anthropological conceptions of culture.

But there is a problem with viewing sharing as central to culture. If anthropologists have proven one fact, again and again, it is that human diversity—in respect to behaviors and beliefs—is pervasive. Why should this diversity stop at cultural borders, especially when these borders are often ambiguous and amorphous? (Recent history makes clear they are not necessarily coterminus with national boundaries.)

Anthropologists have been aware of diversity within cultural groups for decades—for more than a century if one includes Dorsey's 1884 *Omaha Sociology*. More than fifty years ago, Boas (1938: 683) noted that "most attempts to characterize the social life of peoples are hampered by the lack of uniform behavior of all individuals." And Sapir (1938: 570) commented "Two Crows, a perfectly good and authoritative [informant] . . . could presume to rule out of court the very existence of a custom or attitude or belief vouched for by some other [informant] . . . equally good and authoritative."

A host of interesting insights have accumulated through the years. Jane Schneider, referring to Boas's *Primitive Art*, writes, "Boas determined statistically that Northwest Coast Indians were as likely as foreigners to assign different meanings to the same motifs or utilize different motifs to convey the same meanings" (1987: 416). Sapir argued for a more sensitive accounting of the psychological factors behind individual variation (the reference to Two Crows, just cited, comes from a paper entitled "Why Cultural Anthropology Needs the Psychiatrist"). "Only through an analysis of variation . . . [and] a minute and sympathetic study of individual behavior," Sapir (1938: 576) states, will it "ultimately be possible to say things about . . . culture that are more than fairly convenient abstractions." Pelto and Pelto (1975: 1) suggest that "intragroup diversity . . . like genetic diversity within populations, is of great significance for on-going processes of human adaptation." They compare cultural evolution to biological evolution, and the role that diversity plays in each.

Minturn et al. (1964), in their analysis of data from the Whiting's "Six Cultures" project—one of the most careful studies ever done of cross-cultural variations in child rearing—emphasize the importance of not ignoring "the precise measurement of individuals that specifies variation of behavior among people who share a common culture" (293). And in a follow-up to this analysis, Whiting and Whiting, in a chapter entitled "Differences within Cultures," write, "variations in the socioeconomic system and the household structure at the cultural level have been shown to influence the learning environments of children, causing their social behavior to be predictably different" (1975: 130). (Brown, in this volume, shows how such diversity—in the form of classifications—changes with changing political/economic contexts.)

## PUKAPUKAN PERSPECTIVES

Notwithstanding the studies I have highlighted above, detailed studies of informant variation within a cultural unit tend to be a minority affair within anthropology. The studies that do exist tend to focus on relatively narrow topics or, in
the case of consensus theory, on mathematical abstract values (such as eigen-
values). Let me offer a detailed ethnographic sense of what is involved in informant variation. The following material is drawn from what would seem, from an
anthropological perspective, to be an ideal locale for cultural homogeneity.
Pukapuka, located in the northern Cook Islands of the South Pacific, had (at the
time of my forty-one-month study begun in 1977) some 780 inhabitants (see
Borofsky 1987). It is considered, by Cook Islanders, to be the most traditional
island in their multi-island nation and, by many beyond the Cook Islands, as one
of the most traditional islands in Polynesia. Boat visits—the main form of contact with the outside world aside from radio—average three to four per year. As
we shall see, there is considerable intracultural diversity on the atoll.

### The Tale of Wutu

The tale of Wutu has been described by Ernest Beaglehole (n.d.: 1021–23) and
by myself (Borofsky 1987: 118). Pukapukans view it as a popular tale on the
island, known by both adults and children. I asked ten individuals from each
of the "a" through "h" cohorts listed in table 13.1 to recite as much as they
knew of the tale. (Details regarding this and other surveys noted here are
discussed in Appendix 13.A.) Following Lévi-Strauss (1963: 206–31), I then
divided informants' versions of the story into constituent units, so as to
explore the degree to which the versions shared certain elements. The constituent units were (1) Wutu goes to an isolated spot; (1a) specification of
location; (1b) specification of reason why he goes there; (2) Wutu falls asleep;
(3) ghosts come to where Wutu is; (4) they make a plan to eat him; (5) they
carry him away in a large wooden bowl (kumete); (6) they sing a chant (or
chants); (6a) a chant centering around the phrase "ko wutu, ko wutu"; (6b) a
different chant centering around the phrase "tau laulau ma tau pala"; (7)
Wutu makes a plan of escape; (8) he defecates inside the wooden bowl (so the
bowl will be heavy when he climbs out of it); (9) Wutu climbs into a tree

**Table 13.1. Consensus among Pukapukans on the tale of Wutu**

| Category | 67% Consensus | 75% Consensus | 100% Consensus |
|---|---|---|---|
| a. Ten-year-olds | — | — | — |
| b. Twenty-year-olds | — | — | — |
| c. Thirty-year-olds | 3, 5, 6a | 6a | — |
| d. Forty-year-olds | 6a | 6a | — |
| e. Fifty-year-olds | 6a | 6a | — |
| f. Elderly men | 6a | — | — |
| g. Elderly women | 6a | 6a | — |
| h. Experts | 3, 4, 5, 6a, 8, 9, 10, 12, 14, 15, 15a | 3, 4, 5, 6a, 12 | 6a |
| All males | 3, 6a | 6a | — |
| All females | 6a | — | — |
| Total: all groups | 6a | 6a | — |

reaching across the path; (10) the ghosts continue on to their selected spot (10a) specification of location; (11) ghosts prepare to eat Wutu; (12) the ghosts throw down the wooden bowl; (13) the ghosts are covered with feces; (14) Wutu escapes from the ghosts; (15) Wutu runs away to another location; (15a) specification of location he runs to; (16) ghosts chase him; but (17) Wutu is saved; (17a) specification of why Wutu is saved. Table 13.1 shows the degree to which these constituent units were shared among the sample of informants.

Clearly, there was only limited consensus regarding the story. People mainly agreed on the 6a chant. It reached a 67-percent level of consensus for six of the eight "a" to "h" cohorts, and a 75-percent level for five of them. Other constituent units—such as 3, 4, 5, and 12—were shared to a lesser degree. And some units, such as 6b, were shared by only a few individuals. Even though the tale was reputed to be popular with children as well as adults, neither the ten-year-old nor twenty-year-old cohorts expressed any consensus regarding what transpired in the tale. Only those deemed by their peers to be experts on the subject—cohort "h"—were able to agree on anything near a coherent version of the tale. Clearly, then, the tale of Wutu, though supposedly popular, was told in a variety of ways.

Table 13.2. Pukapukans' identification of fish as present in local waters

| Category | 67% Consensus | 75% Consensus | 100% Consensus |
|---|---|---|---|
| Thirty-year-olds | (119/141) 84% | (100/141) 71% | (48/141) 34% |
| Forty-year-olds | (119/141) 84% | (96/141) 68% | (44/141) 31% |
| Fifty-year-olds | (116/141) 82% | (96/141) 68% | (39 /141) 28% |
| Elderly | (102/141) 72% | (74/141) 52% | (14/141) 10% |
| Experts | (103/141) 73% | (77/141) 55% | (34/141) 24% |
| Experienced fishermen | (660/846) 78% | (514/846) 61% | (173/846) 20% |

## Identifying Fish in Pukapukan Waters

In addition to the above data—whose collection reasonably paralleled how Pukapukans might discuss these matters (see Borofsky 1987: 74–130)—I systematically surveyed informants regarding their recognition of the area's fish. What made this second survey more formal than the previous one is that I used a set of pictures (from Fowler 1928), and asked informants to decide which of the fish represented in the pictures existed in the waters around the atoll. For those fish identified as living in the atoll's waters, I inquired as to their names. Although Pukapukans had little experience with a task such as this, they seemed to readily grasp what was requested. Most expressed confidence (and appeared, to me, quite comfortable) in making the identifications. Table 13.2 summarizes the degree of informant consensus regarding fish pictured in Fowler that were (and were not) present in Pukapukan waters.

Table 13.3 summarizes data on informants' names for fish that had been identified as present in these waters. (Readers interested in the subtleties of table 13.3 are encouraged to examine the appendix 13.A. The percentages along the top of each table refer to levels of agreement. If one focuses on 67 percent as an appropriate level of consensus (i.e., two-thirds of the sample concur), then the thirty-year-olds in table 13.2 agreed on 119 out of 141 identifications, that is, they agreed regarding 84 percent of the fish pictured. If one focuses on a 100-percent level of consensus, the thirty-year-olds concurred on only 48 out of 141, or 34 percent, of the fish pictured. The overall impression conveyed by the two tables is that, if one accepts a somewhat

Table 13.3. Pukapukans' naming of fish in local waters

| Category | 67% Consensus | 75% Consensus | 100% Consensus |
|---|---|---|---|
| Thirty-year-olds | (32/42) 76% | (26/42) 62% | (14/42) 33% |
| Forty-year-olds | (38/49) 78% | (28/49) 57% | (12/49) 24% |
| Fifty-year-olds | (41/56) 73% | (33/56) 59% | (13/56) 23% |
| Elderly | (32/47) 68% | (25/47) 53% | (4/47) 9% |
| Experts | (43/53) 81% | (30/53) 57% | (11/53) 21% |
| Experienced Fishermen | (207/288) 72% | (162/288) 56% | (51/288) 18% |

NOTE: Deciding which fish to focus on in requesting names presented somewhat of a problem. To ask the Pukapukan name for a fish that few Pukapukans had identified as living near the atoll seemed a bit absurd (one would be asking for the indigenous names of a fish that the particular informant might not know existed). For this survey, I used the fish identified at the 67-percent level of consensus in table 13.2. This allowed consideration of a range of fish and provided a clear sense of the extent to which various individuals' names for specific fish overlapped.

lower level of agreement (i.e., 67 percent), then there was a reasonable degree of consensus (68–84 percent). If one stresses a higher level of agreement (i.e. 100 percent), there was a relatively low level of consensus (9–34 percent). In other words, although some knowledge was shared, a reasonable amount apparently was not, even though it involved one of the most basic and important activities male Pukapukans perform—fishing.

## WHAT HAPPENS WHEN ANTHROPOLOGISTS STUDY DIVERSITY?

What we need, clearly, are more studies of what forms diversity takes in which settings. As Vayda notes, "whether specific beliefs and meanings are shared widely and persist in a given society is . . . a matter for empirical determination rather than for a priori postulation on the base of essentialist assumptions" (1994: 326).

What surprises is that there are few studies of intracultural diversity within anthropology. Aside from mathematically phrased work in consensus theory (see, e.g., Atran et al. 2002; and studies cited in D'Andrade 1995a: 212–16) and some studies cited in this volume, comparatively few anthropologists carry out

comprehensive studies to investigate systematically what kind of cultural sharing occurs in which contexts.

One might suspect that there would be massive studies, given that intracultural diversity—which most anthropologists admit exists—challenges a
basic anthropological concept: culture as sharing. But this does not seem to
be the case. Why are there so few of these studies? We might seek an answer
to this question in the work of specific people who have pursued this topic. Let
me use Anthony Wallace as an example. Wallace's The Modal Personality of the
Tuscarora Indians (1952) is a classic work on cultural diversity. Focusing on a
Tuscarora Indian reservation near Niagara Falls of roughly 600 people—a
seemingly small, homogeneous group—Wallace found significant variation in
personality traits (as judged by Rorschach protocols). Only 37 percent of the
seventy residents interviewed possessed what might be termed a modal personality—their collective responses (in terms of twenty-one identifiable
Rorschach categories) fell within a modal range. Another 23 percent of
responses fell within the modal range for some Rorschach categories but not
others. Wallace referred to these as "submodal." And a final 40 percent fell
completely outside the modal range. These he called "deviant."

Building on this work, Wallace made a critical distinction in his 1961 book,
Culture and Personality. He differentiated between two ways for conceptualizing
the "relation between cultural systems and personality systems." The first he
termed the "replication of uniformity": "the society may be regarded as culturally homogeneous and the individuals will be expected to share a uniform
nuclear character. . . . In this case, the interest of processual research lies . . .
in the mechanisms of socialization by which each generation becomes, culturally and characterologically, a replica of its predecessors" (1961: 27). In
the "organization of diversity" model, "the fact of diversity is emphasized
[leading to] the obvious question . . . how do . . . various individuals organize
themselves culturally into orderly, expanding, changing societies? When the
process of socialization is examined . . . it becomes apparent that . . . it is
not a perfectly reliable mechanism for replication" (1961: 27–28). Affirming
the second position, Wallace suggested that orderly relationships derived less
from shared behaviors than from a capacity for mutual prediction (1961: 28).
"Advocates of togetherness may argue, whether or not it is necessary that all
members of society share all cognitive maps, they must share at least one";

however, Wallace suggests that many social systems "simply will not 'work' if all participants share common knowledge of the system" (1961: 37, 40)—a point readily illustrated in politics and economics, where differential access to knowledge is frequently construed as power.

What did Wallace do with these insights? On the one hand, he sets them aside in a book entitled *Religion* (1966). In this book, he delineates four "Religious Culture Areas of the World." The "Communal" area, for instance, includes "communal cults which involve a pantheon of major deities controlling departments of nature, either presently or in the past. Rituals are performed in which many lay persons participate actively. . . . The mythology is rich and variegated, although the gods are not for the most part heroic figures" (1966: 97). He divides the "Communal" cultural area into four subcategories: American Indian, African, Australian, and Oceanic. In the Oceanic area (which includes Micronesia and Polynesia but not Melanesia), communal cults stress "fetishes (statuary residences of gods), taboo systems, age-grading, and subsistence" (1966: 99). Wallace, in other words, applies a broad cultural framework to make generalizations about a wide swath of seemingly related groups. To a lesser extent, he follows a similar pattern in *The Death and Rebirth of the Seneca* (1969). Here the generalizations are limited to a single group, the Seneca. In a chapter entitled "The Seneca Nation of Indians," he writes, "the basic ideal of manhood was that of 'the good hunter.' Such a man was self-disciplined, autonomous, responsible. He was a patient and efficient huntsman, a generous provider to his family and nation, and a loyal and thoughtful friend and clansman. He was also a stern and ruthless warrior in avenging any injury done to those under his care" (1969: 30).

On the other hand, when Wallace turns to writing history, he produces deeply nuanced, acclaimed accounts such as *Rockdale: The Growth of an American Village in the Early American Revolution* (1972) and *St. Clair: A Nineteenth-Century Coal Town's Experience with a Disaster-Prone Industry* (1981). In *Rockdale*, when Wallace discussed the period from 1825 to 1835 ("The Creating of a Way of Life"), we meet numerous individuals including William Martin Jr., John S. Phillips, Edward Darlington, and Richard and Elizabeth Smith. The analysis of the period is built around them. When we turn to the 1835–1850 period in a chapter subtitled "The Struggle for Control of a Way of Life," we again meet individually named people: Minshall and Jacob Painter, Lafayette and Fanny

Wright, Daniel Lamont, John Speakman. Reading Wallace's account, one gains a sense of how individuals wended their various ways through these times.

The irony here is that when Wallace stays with anthropology, he uses the framework of cultural sharing that he so effectively criticizes in his early work. We do not get a sense of cultural diversity in *Religion* or in *The Death and Rebirth of the Seneca*, but rather broad group formulations. When Wallace turns to history, on the other hand, he affirms his earlier position. We see individuals— we see cultural diversity, close up through particular personalities.

Let me take, as a second example, two anthropologists whose accounts make clear that Melanesian cultural dynamics do not fit the framework of neatly defined groups based on cultural sharing. De Lepervanche reports "one aspect of social life that is evident in the ethnographies [regarding New Guinea] is that New Guineans are flexible in conceptualizing their social relations" (1973: 5). Following Hogbin and Wedgwood, De Lepervanche uses the terms "phyle" for "cultural-linguistic units" and smaller "parishes" for "autonomous local groups" (1973: 1–2). She notes that it is unusual for there to be "any inclusive idea of social or territorial unity. It is rare for a people to be able to delineate their phyle boundaries. The notion of defined territory usually applies only to [localized] parishes" (1973: 30). Similarly, Strathern casts doubt on the relevance of our Western concept of "society" for understanding Melanesian social dynamics. She writes: "As I understand Melanesian concepts of sociality, there is no indigenous supposition of a society that lies over or above or is inclusive of individual acts and unique events . . . the imagined problems of social existence are not those of an exteriorized set of norms, values, or rules" (1988: 102).

Still, both of these distinguished anthropologists return to a sense of the cultural/social framework in writing for their anthropological peers. De Lepervanche, in discussing various ethnographic analyses, repeatedly refers back to named cultural-linguistic groups. We read such phrasings as: "among the Kuma," "the people of Tangu believe that," "the Motu use," "unlike Enga and Siane, Melpa have" (1973: 9, 15, 17). She frames her ethnographic examples within the very anthropological framework she is bent on challenging — cultural-linguistic "phyle" (e.g, 1973: 5). The same holds for Strathern: She uses such phrases as "Hagen women embody," "the Melanesian personification of objects," "the two significant axes of Muyuw male-female relations concern brother and sister, husband and wife" (1988: 87, 177, 193).

These two anthropologists, I would suggest, appear caught between their observations of fluid, heterogenous cultural formations and anthropological frameworks that focus on bounded, shared cultural systems. They seem to be using the traditional anthropological sense of culture (and society)—in explaining Melanesia to others—all the while challenging this perspective.

You have, I hope, gotten my point. Certainly some anthropologists consider the issues of cultural diversity. And if we include writings by anthropologists working with consensus theory (which contains ambiguous assumptions that still need be worked through—see, e.g., Aunger 1999), as well as work such as Feldman's in the volume, we might add more studies still. But for the majority of the discipline, the focus is on shared beliefs and behaviors. That is not to say that most anthropologists do not recognize that the world is a messy place, and that cultural sharing is less pervasive, is more problematic, than asserted. But intracultural diversity, although acknowledged, tends to be passed over. Cultural sharing remains the dominant disciplinary way for discussing collectivities of people.

## ACCEPTING THE NATION-STATE'S HEGEMON
## (AND ITS CONSEQUENCES)

Given the page limitations of a chapter such as this, I cannot really flesh out how we got ourselves into this difficulty and, more importantly, why we persist with it: Why do so few scholars seriously challenge the framework that frames our conception of culture? Still, let me briefly suggest what I suspect is the key dynamic, and you decide if it seems reasonable.

I perceive the nation-state—one of the world's dominant governing structures for more than a century—as shaping our conceptions of culture. Rather than challenging the nation-state's framework for perceiving collectivities of people, anthropologists seem to continually go with the flow, phrasing their conceptualizations in terms that support this hegemonic perspective. They do not attempt to speak truth to power, to challenge these conceptions with scientifically solid data. Rather they seem servants of the status quo—affirming the simplistic and often unproven, but politically potent, assumption of cultural sharing.

As Keesing (1994: 307) observes, our modern sense of "cultural identity has its origin in nineteenth-century cultural nationalism in Europe, expressed in an intense search for ethnic roots and folk origins, for primordiality and cultural tradition." Importantly, the nation-states that arose in Europe at this time were different in conception from earlier kingdoms on that continent. Anderson (1991: 19) points out that

> Kingship organizes everything around a high center. Its legitimacy derives from divinity, not from populations, who, after all, are subjects, not citizens. In the modern conception, state sovereignty is fully, flatly, and evenly operative over each square centimeter of a legally demarcated territory. But in the older imagining, where states were defined by centers, borders were porous and indistinct, and sovereignties faded imperceptibly into one another.

Modern nation-states, in order to support their new conceptions of statehood, developed, Anderson suggests, "imagined communities" (1991: 6)— "*imagined* because the members of even the smallest nation will never know most of their fellow-members, meet them, or even hear of them, yet in the minds of each lives the image of their communion." Language played a key role in these imagined communities. Anderson continues, "there is a special kind of contemporaneous community which language alone suggests—above all in the form of poetry and songs" (1991: 145). With the singing of national anthems, for example, "no matter how banal the words and mediocre the tunes, there is in this singing an experience of simultaneity. At precisely such moments, people wholly unknown to each other utter the same verses to the same melody. The image: unisonance" (1991: 145)

We can perceive this theme in our anthropological conceptions of culture. "'Culture' of the modern anthropological persuasion," Sahlins writes, "originated in Germany. . . . [It] defined the unity and demarcated the boundaries of a people" (1995: 11–12). Kuper adds that the German sense of *Kultur* "is bounded in time and space and is coterminous with a national identity" (1999: 30, see also 36). Or, as Elias writes, "the German concept of *Kultur* places special stress on national differences and the particular identity of groups. . . . [It] mirrors the self-consciousness of a nation which had constantly to seek out and constitute its boundaries anew, in a political as well as a spiritual sense" (1994: 5).

There is a dark side to this nationalistic framing of culture. Ethnic/cultural borders rarely match up neatly with political ones. To quote Connor:

> Of a total of 132 contemporary states, only 12 (9.1 percent) can be described as essentially homogeneous from an ethnic point of view . . . this portrait of ethnic diversity becomes more vivid when the number of distinct ethnic groups within states is considered. In some instances, the number of groups within a state runs into the hundreds, and in 53 states (40.2 percent of the total), the population is divided into more than *five* significant groups. (1994: 29–30)

What binds people together, then?

Modernization offers one means for overcoming internal divisiveness. It provides promises of progress toward broadly shared goals in health, education, and higher living standards. Cynically one might note that promises of progress mean a dependence on Western "development" agencies such as the World Bank. Independent nation-states, in emphasizing modernization/development as a way of establishing and maintaining their political legitimacy, often increase their dependency on Western aid (see Ludden 1992: 251; Chatterjee 1993: 203; F. Cooper 1997: 425).

At the same time as nation-states are caught up in this modernizing/dependency dynamic—or, perhaps, in an effort to moderate the process—newly formed nation-states often seek to differentiate themselves from the West and their neighbors by emphasizing their cultural independence: To be modern but different (to paraphrase Chakrabarty 1997: 373). The problem is that, in many cases, this national culture has to be constructed in a culturally diverse environment using rituals of affirmation—at national celebrations and public schools—that in drawing on diverse pasts attempt to override them. It is a contingent, constructed affirmation vulnerable to continual disruption.

The result is that many nation-states—especially outside Europe and North America—are fragile institutions. What keeps them together, ultimately, may be raw power. But raw power can undermine its own effectiveness if overused. Raw power works best when it remains in the background, is rarely directly called upon. What often binds people together (and we see this in "the West" as well as in "the Rest") are oppositions to those who seem (culturally) different—either within the nation-state or beyond the nation-state's borders. Phrased another way, think of the number of people in nation-states who are kept united by their common dislike of Americans, Jews, Arabs, Blacks, Bosnians, Serbs, Hutus, Tutsi, terrorists, Iraqis, Iranians, and maybe even anthropologists. Illusionary cultural units such as nation-states, I am saying, often need enemies—people to continually mobilize against—in order to cover over the ambiguities, diversities, divergences, and conflicts among their citizens. They need someone or something to fight against, to oppose, even to hate.

Surely, Milošević, Hitler—you name them—would not have stopped their genocidal acts if they were forced to used a more refined sense of human collectivities. Their actions were politically driven. But just as certainly, we need not offer ideological support for such actions by affirming the illusionary cultural homogeneity of particular groups. We become part of the problem rather than part of the solution by accepting, despite our doubts, the nation-state framing. We operate within its hegemonic perspective and get caught up in its violence. Take the case of Rwanda. Des Forges, reviewing three books on the subject writes:

> Prunier, Keane and McCullum all rightly reject the simplistic analysis of the genocide as a manifestation of age-old tribal hatreds. As Prunier writes, "Tutsi and Hutu have not been created by God as cats and dogs, predestined from all eternity to disembowel each other . . . " The three authors recognize that Hutu and Tutsi are not tribes, but social strata, speaking a common language, bound by shared customs, living interspersed in a nation that they created together. All conclude that the killing campaign was systematic and planned, the result of an organized and ruthless exploitation of fear and ethnic loyalty by leaders who were in danger of losing power. (1997: 27)

We might cite others in this regard as well. Here is Deák's review (under the title "The Crime of the Century") in the *New York Review of Books*: "ethnic violence is not the result of 'ancient hatreds,' . . . contemporary politics and the struggle for power . . . largely determine who will engage in violence and who, ultimately, will be killed" (2002: 48). Verghese writes in the *New York Times Book Review*, "Hedges [in *War Is a Force that Gives Us Meaning*] discounts the notion that wars are born out of pure ethnic or religious differences; most wars, he believes 'are manufactured wars, born out of the collapse of civil societies, perpetuated by fear, greed, and paranoia' . . . nationalism fuels wars. Nationalism warms the heart, unites a nation . . . but its danger . . . is that it can devour intellectuals and social critics as readily as it does the masses" (2002: 21).

Anthropologists of the scientific persuasion need not stay with old styles, perpetuate old patterns, hold on to old hegemons. They can pull the rug out from under the ideological foundations of such atrocities. What they need to illuminate, to follow Wallace, is the organization of diversity—how people of different perspectives effectively deal with one another without killing each other. A science worthy of its name speaks truth to power—challenges the status quo—showing that current justifications for mayhem and murder are politically, not culturally, driven.

So there you have it. You can see why there is an edge to the paper's title. Science can clearly be a good thing. It can be used to unmask the hegemons of power that foster politically destructive behaviors. But caught up in the appearance of science, we at times seem to lose a sense of its substance. If we are going to explain culture scientifically, we would do well to start with our own behaviors regarding culture (as Starrett suggests in his chapter) and see why we, as anthropologists, tend to affirm perspectives that have not been scientifically demonstrated, but rather belong to the political/economic structures that shape our own lives.

Anthropologists often behave as if sharing is a central, indeed critical, element of culture. But that remains to be proven in a host of contexts and a host of locales. Why we, as anthropologists, should accept such an assumption—based as it is on the nation-state model—and call it scientifically affirmed in so many cases seems strange, especially given the negative behaviors unintentionally condoned by such a perspective. It is behavior certainly deserving of further scientific investigation.

## APPENDIX 13.1

For the surveys described in tables 13.1, 13.2, and 13.3, I interviewed eighty individuals. The sample for the Wutu survey (table 13.1) involved both males and females, whereas the samples for the fish questions (tables 13.2 and 13.3) involved only men. I focused on males in the latter two surveys because men primarily do the fishing on the atoll and hence are more conversant with the details of such work. In the Wutu sample, there were five males and five females for each of the age cohorts labeled a, b, c, d, and e. In the fish identification survey there were ten males in each cohort.

I also interviewed two other groups. One was a cohort of "experts" (as identified by a survey of their peers). For the Wutu sample, this group involved ten adults. Selection was based on a survey in which all men and women over fifty on the island—ninety-one individuals—were questioned in regard to who they thought was knowledgeable about traditional matters. For the fish survey, the expert group consisted of fifteen adult males. Selection here was based on a survey in which all the men over fifty—fifty-seven individuals—were questioned in regard to people's knowledgeability on fishing matters. The other cohort interviewed involved elderly informants, sixty-four years of age or older, who were not part of the expert group. For the Wutu survey, this included ten elderly men and ten elderly women. For the fish survey, it included fifteen elderly men. Thus both the Wutu and fish surveys involved eighty informants.

The "experienced fishermen" cohort included all males in the sample thirty years of age and older—males who, on the basis of their age, should have had considerable experience with fishing. The younger age groups, although surveyed, are not included here because they were presumably less knowledgeable about fishing and hence would skew the results toward diversity. In the Wutu survey, these younger groups were included because the tale was reputed to be popular among them.

# Epilogue

*Future Considerations*

## MELISSA J. BROWN

Possibilities for future research on culture range at least as widely as the approaches and concerns represented in this book. Areas for research that I consider central to the development of a scientific approach to culture and human behavior, which are discussed below, include: intracultural variation, diffusion, motivation, social structure, interactions between systems, and adaptiveness. Bringing together insights from modeling and field research to approach these and other issues will both contribute to the development of a science of culture, as well as promoting that approach to a broader academic audience. Because the skills needed to conduct such research require a broad range of training—spanning quantitative and qualitative analyses and field methods—collaboration promises to facilitate integrative research.

## DEVELOPING A SCIENCE OF CULTURE

The challenges posed by Wolf, Starrett, and Borofsky in this volume, although serious, can be resolved, and resolution can demonstrate the benefits of a scientific approach to culture. It is good science to evaluate more than one model against a data set, to test a promising model against more than one data set, and to look beyond correlations for further evidence of causation. It is good science to consider whether there are biases in how we have set up our research—in definitions, in data collection, and in the identification of independent variables. It is good science to verify and critically evaluate research results, both our own results and those of colleagues.

We are establishing protocol for a scientific anthropology. The collection of empirical data with reasonable sampling of the population under consideration is a contribution, regardless of debate over conclusions drawn from that data. Moreover, disagreement with data-driven conclusions must be countered with data. Such debate is healthy for a field, not disloyal (see endnote 16 in the introduction to this volume), for it necessitates further data collection and allows refinement of conclusions and, in a pre-science phase, theory building.

The need for better integration of theoretical modeling and empirical data in the development of a scientific approach to human behavior cannot be over-emphasized. The challenge is to produce models and data sets that promote comparison, evaluation, dialogue, and cumulative knowledge. S. Li, Feldman, and Li's (2000, 2003) empirical data, gathered to test models (e.g., Laland et al. 1996, 2000), allowed Wolf's comparison (in this volume), which in turn identifies a need for additional data on number and ages of siblings as well as whether parents were alive when individuals married. Similarly, Brown's (1995, 2004) work on footbinding led to Richerson and Boyd's suggestion (this volume) that their prestige-biased transmission model (Boyd and Richerson 1985: chap. 8) is a promising explanation of the prevalence of this Han custom in late imperial China. The model identifies a need for better empirical data on actual frequencies of footbinding (which is extremely difficult to reconstruct, but see Gates, in progress; Bossen 2002; Brown 2004; Bossen et al., in progress). However, further models will also need to be considered, because some empirical data challenge the fit of this model, including evidence that footbinding among the Han was a form of female labor control (e.g., Gates 2001, in progress), and evidence that at ethnic borders, absence of footbinding served as an exclusionary marker of non-Han status (Brown 2004).

Likewise, empirical data compiled by the Centers for Disease Control and Prevention (e.g., 2001a) showing that the frequency of sexually transmitted infections (STIs) in the U.S. varies by ethnic group inspired application of quantitative models of social networks (Jones, this volume). Jones and his colleagues found that they could account for STI distribution patterns by considering the sociologically derived observation that individuals often have ethnic preferences in their sexual partners; this result challenged previous attribution of the variation to ethnic differences in promiscuity. These modeling studies suggest that future empirical research on STIs should ask study participants not only about the number of their sexual partners but also about ethnic categorizations of those partners and how they chose those partners.

This research could also be expanded to address the relationship between ideas and behavior. Empirical studies could be designed to investigate whether people choose sexual partners based, for example, on internalized stereotypes of different ethnic categories as "good" or "bad" partners, or perhaps based on an internalized rule to choose a partner of the same ethnic label. Such data could be compared to quantitative models on the relationship between ideational units and behavior. Additionally, Jones's conclusion that "social networks are . . . a vital substrate for the process of cultural transmission," as well as evidence that ethnic identity is uncorrelated to cultural ideas (Brown, this volume), suggest that it would be fruitful to consider a structural possibility. Empirical studies could examine constraints on social opportunities to meet potential partners of differing ethnic labels, and quantitative models could be developed to further examine such constraints.

The potential of combining critical evaluation with integration of theory and data is evident in these examples. Using empirical data to critically evaluate quantitative models requires comparing multiple models to the data as well as examining a range of empirical data to see if other factors than those captured by the models may be important. Quantitative models reveal the potential consequences of repeated interactions in human relationships, something most people cannot keep track of qualitatively. The application of quantitative modeling to specific empirical problems can also benefit theoretical modeling by indicating areas for further model development.

Such integration also raises methodological issues that will need to be thought through theoretically as well as investigated empirically. For example, which aspects of a society should be examined? We must decide—and justify the decision—whether studies should focus on characteristics considered "core" to a society or on nonessential ones, on functional or on stylistic/aesthetic ones, and so on. Also, which societies should be examined? To date, much evolutionary anthropology research has focused on foraging and pastoral peoples, but we must be cautious about assuming that some societies are more representative than others, of the past or of the present. We need more studies that are as well-grounded in their rationale for conducting experiments among foraging and pastoral peoples as are the experimental economics studies of Henrich, Bowles and Gintis and their colleagues (e.g., Henrich et al. 2004).

## SKILLS NEEDED

These examples also make clear that the ideal training for scholars working on scientific approaches to culture would include both in-depth field research and quantitative modeling. Although it may not be realistic to expect most individuals to be proficient in both areas, expectations of basic literacy in both areas would promote future integration.

Modelers could benefit from field experience that goes beyond advising on sampling, analyzing someone else's surveys, and conducting controlled experiments. In-depth life-history interviews and participant observation provide invaluable insights for modeling. Field collaboration is one way modelers could get such experience without necessarily having to develop the interviewing and participant-observation skills themselves. Restudies of previous field sites are another potential way to benefit from a colleague's in-depth knowledge (restudies also provide valuable longitudinal data).

Modelers need sufficient appreciation of the kind of information that interviewing and participant observation can obtain to take field data seriously. For example, a life-history interview method (from as large a population sample as possible) can get at cultural ideas shared in the past, despite memory bias (e.g., Agar 1980). A large sample provides an estimate of the distribution of ideas across individuals in the population as well as indicating whether particular ideas are linked to class or other local means of social differentiation. Using a life-history method, one can take an interview subject back to earlier periods of her life, by initially asking about a major personal event—such as the death of a grandparent or the subject's own marriage—then shifting discussion to more mundane aspects of life at the time of that event. Because such recollections unconsciously bring back earlier emotions, it is assumed the discussion also captures shared cultural ideas of that earlier time (e.g., Mandler and Johnson 1977; Moss and Goldstein 1979; D'Andrade 1984; Tagg 1985; Strauss 1992b). Taking field data seriously includes not ignoring evidence from such field sources, which can yield, for example, information about market influence on historical and contemporary foraging peoples.

By the same token, field researchers could benefit from quantitative training that goes beyond running standard statistical tests. Unfortunately, finding anthropological field researchers with even sufficient mathematical skills to run statistical tests can be a challenge. This ignorance of basic mathematical reasoning skills simply must change. Quantitative modeling tailored to a specific set of empirical data—which can include both mathematical recursion models and computer simulation models—provides valuable insights for empirical research.

Collaboration, again, is one way field researchers could get tailored models without necessarily having to develop the modeling skills themselves. However, field researchers need sufficient quantitative skills to appreciate the insights quantitative models can provide, even when the models are based on unrealistic assumptions. For example, ecologists and demographers often use a stationary population model that assumes, among other things, that the population is closed to migration (or the net migration is zero). Despite this unrealistic assumption, the model yields insights about population growth and selection forces by allowing calculation of the total fertility rate (the number of children an average non-contracepting female would have if she survived all her reproductive years). Appreciating the insights of quantitative models includes using conclusions—such as patterns of population increase or decline over time—from such sources to develop hypotheses for empirical testing. "The validation of a model is not that it is 'true' but that it generates good testable hypotheses relevant to important problems" (Levins 1966: 430, see also chapter 2 in Hilborn and Mangel 1997).

With some effort at dialogue, and good foundations in the basic skills necessary to communicate, better integration of theoretically derived models and empirical data will be quite productive.

## SOME AREAS FOR RESEARCH

Contributors to this volume raise many promising areas for research, including comparison and modeling of cultural and social systems, consideration of units of transmission, exploration of the adaptiveness of culture, and cross-species comparisons. Some of these topics are crucial to the development of a scientific approach to culture and human behavior.

Documentation of intracultural variation, advocated by Borofsky in this volume, is particularly important because it provides a platform for promoting a scientific anthropology to a broader academic audience. Such documentation will likely appeal to anthropologists who see themselves as humanists rather than scientists. It acknowledges the well-known (if rarely discussed) empirical fact that intracultural variation exists. It also identifies—and thereby gives voice to—views held by population minorities. In addition, it can resolve the indeterminate character of the concept of culture as widely shared, if it also identifies how many people within a population need to share an idea before it is perceived as cultural. In this way, documentation of intracultural variation can link the widely shared concept of culture that is still held broadly across anthropology with the frequency-based concept of culture used in modeling.

Further connection of these different concepts of culture raises a number of interesting theoretical and empirical problems. For example, diffusion across a population—whether of an idea or a behavior—means that distributions will change within the same population over time. If we apply a threshold, or tipping-point, model to the diffusion process, it leads us to ask how the tipping point occurs—that is, what causes cascading subsequent change. The notion of culture as a shared idea suggests that the tipping point would occur once some percentage of the society shares an idea, because it causes people to perceive that idea as being shared by most, if not all, of the society (or section of society), and therefore view the idea as cultural (M. Brown 1997a, 1997b, 2004: 216, 2007). Connecting such a model with an interpretive concept of culture sets the stage for ethnographic work to investigate tipping points: What is the critical percentage of people who must share an idea? And what social conditions can affect people's perceptions—for example, modern mass media, or the orthopraxy in public rituals demanded by the late-imperial Chinese state? Do different social conditions—across societies or across time—yield different tipping points? Does diffusion across a population differ depending on whether what is spreading is a practice or the cultural ideas behind a practice (see Brown 2007)?

A major challenge here is how to model ideas that only potentially trigger behavior. Feldman (this volume) suggests that it does not matter for modeling purposes whether units are ideas, behavior, or customs, as long as they are identifiable and quantifiable (although evolutionary archaeologists debate the units of transmission—see, e.g., Dunnell 1995; Lipo et al. 1997; Ramenofsky and Steffen 1998; O'Brien and Lyman 2000). Feldman suggests that modeling culture as behavior works as a simplification because such models capture the social nature of transmission—that is, they capture how learning a behavior socially might affect its spread across a population (they also allow cross-species comparisons). But can models using behavioral units approximate models using ideational units in explaining empirical data? Does the accuracy of models with different units vary across empirical cases? If so, factors may have varying importance in different contexts. Answering these questions requires development of transmission models using ideational units, which complicates models considerably. The best current methods for getting at ideational units—such as linguistic discourse analysis and psychological assessment tests—are used in the cognitive sciences (see D'Andrade 1995a for an introduction).

We also need to better understand motivation as proximate causation. What triggers an idea—either an idiosyncratic or a cultural one—to become motivational? What else motivates behaviors besides cultural ideas? Brown (this volume) suggests that social structure is motivational, but how is social structure identified and internalized in the brain? Are mechanisms dependent on cultural context, social context, or some other (e.g., environmental) context? This volume's contributors explore how one idea can influence another idea and its subsequent behavioral enactment (see the chapters by D'Andrade; Feldman; Richerson and Boyd; Aoki et al.; Durham; Wolf), and examine how the content of an idea can influence its own transmission (see the chapters by Durham; Starrett). However, we need to know more about variation in motivation across different individuals in a population and in the same individuals over time, as well as whether structural influences can account for some of that variation.

We also have to examine our basic assumptions about another issue related to motivation: intentionality may vary in its impact on the consequences of human behavior. "The problem of the unintended consequences of our actions . . . is the fundamental problem of the social scientist" (Popper 1994: 128). At the micro-level, as with intracultural variation (discussed by Borofsky, this volume), many ethnographers discover actions carried out with no conscious motivation or intention but fail to realize their theoretical significance. Field research yields many reports by individuals of enacting a behavior for no other reason than that it is "customary" (see also Brown 2007: 117). Such actions are not motivated by any conscious intention to perpetuate the custom; they are simply the result of habit (see also Bourdieu, e.g., 1977: 89, 120, 1990: 102–3). People can have a similar lack of intentionality in the demise of a custom. The sole motivation articulated by elderly Chinese women who experienced the end of footbinding was "Society changed!" (Gates, in progress). Moreover, despite our best intentions, we may not be able to enact behaviors, as in the case of the last woman spirit medium in Toushe (Brown, this volume). Intentionality may, of course, be important at times, as in the case of the first Han bride in Toushe (Brown, this volume), but we need to better understand the variation in its significance.

Such variation has been explored at the macro-level by evolutionary archaeologists. For example, archaeological bet-hedging models suggest that expensive artifacts or behaviors, such as pyramid building, are more likely to persist in highly variable environments. Production of these monuments creates a buffer against the potential impact of famine or other catastrophic events by diverting energy expenditure into "wasteful" activities that can be curtailed or eliminated when energy is needed elsewhere (Dunnell 1999; Dunnell and Greenlee 1999; Sterling 1999; Allen 2004; Truncer, in progress). Neutral-trait or random-drift models (e.g., Neiman 1995, O'Brien and Lyman 2000, Shennan and Wilkinson 2001, Shennan 2002) also downplay intention by showing that some archaeological changes—such as stylistic changes in pottery—are due only to stochastic processes (i.e., historically contingent random processes). The idea that human intentionality may have little explanatory relevance is unsettling—even supporters of Bourdieu criticize him for reducing people to mindlessness (e.g., Comaroff 1985: 5, see also Brown 2004: 220–22). Nevertheless, we need to consider the possibility that the intentions of individuals may not play as large a role in cultural change as previously assumed.

For those who want to examine other higher levels of causation, consideration of a social inheritance system is just beginning. One question to ask is whether the presence of another potentially competing, potentially interacting system in coevolutionary models can resolve some of the discrepancies between current quantitative models and empirical data, while still incorporating the successes of current models (Richerson and Boyd, this volume). Also important is what social units might be, how they might relate to an individual's dynamic position in the social structure, and by what mechanism they relate to behavior (M. Brown 1995, 1997b, in progress).

In addition, we need to understand how to deal with interactions between systems (Durham, this volume), which complicate the analysis of competing influences on human behavior. Institutions, for example, crosscut the distinction between culture and social structure (D'Andrade, this volume), and some categories of behavior—e.g., ethnic intermarriage, migration of large populations—lead to interactions between systems (Brown 2004). Cognition is also thought to mediate between different inheritance systems (see, e.g., the chapters in this volume by Richerson and Boyd; Durham; Henrich). Do these interactions work in a patterned, predictable way?

Adaptiveness of culture, another potential higher-level causation, is often linked to cognition, via the argument that the huge costs of developing and maintaining the human brain—both phylogenetically in the evolutionary past and ontogenetically today—could only be borne if the brain is adaptive (e.g., Tooby and Cosmides 1992; Smith and Winterhalder 1992). The implication is that natural selection plays an important role in the relationship between culture and human behavior. Empirical data suggest both potential adaptive benefits and maladaptive or adaptively neutral consequences of socially learned behavioral (discussed in the chapters by Boesch; Richerson and Boyd; Durham; Henrich).

A major issue yet to be resolved is how to test the hypothesis that a particular behavior is, or ever was, adaptive. For example, studies that demonstrate correlations of certain current behaviors (often based on surveys of college students) with reconstructed historic and prehistoric fertility need critical evaluation and independent replication. College students may not be a representative sample of the societies from which they come (as shown in the chapters by Henrich; Bowles and Gintis). We need more data on historic and prehistoric fertility, including comparison of estimates based on modeling, historical data, and archaeological data. Additionally, we must consider whether other models or explanatory concepts fit the correlations as well as or better than models of adaptiveness. Field research on the occurrence and dynamics of the correlated behavior in a non-experimental social setting is crucial, in order to consider whether important factors have not been surveyed. For example, footbinding is often assumed to have been maladaptive, given possible complications such as gangrene. However, in areas with high local frequencies of footbinding, women without bound feet were likely to end up as bond servants or prostitutes, and thus produce few if any children (see, e.g., Gates 1996, 2001, in progress). In such areas, having bound feet may have improved an individual's fitness, given the social system at the time.

Another major issue related to the adaptiveness of culture is whether, and to what extent, the concept of culture can be applied to nonhuman species. Is such application limited by use of ideational units? Leaf-clipping chimpanzees appear to have something reasonably labeled as culture according to the ideational criterion (Boesch, this volume). Further research on whether non-adaptive (but not necessarily maladaptive) behavior with symbolic meaning is part of a general pattern in nonhuman species has important implications for consideration of the adaptiveness of culture in humans' evolutionary past.

There are many more potential avenues of inquiry raised in this volume. The contributing authors continue to conduct research in these areas, and we hope that our work will spur further debate and research toward the development of a scientific approach to culture and to anthropology more broadly.

# References

Abu-Lughod, Lila. 1991. Writing against culture. In *Recapturing Anthropology: Working in the Present*, ed. R. G. Fox, 137–62. Santa Fe, NM: School of American Research Press.

―――. 1993. *Writing Women's Worlds: Bedouin Stories*. Berkeley: University of California Press.

Adams, Charles C. 1968. *Islam and Modernism in Egypt*. New York: Russell and Russell.

Adams, Morton S., and James V. Neel. 1967. Children of incest. *Pediatrics* 40:55–62.

Adimora, Adoara A., Victor J. Schoenbach, Dana M. Bonas, Francis E. A. Martinson, Kathryn H. Donaldson, and Tonya R. Stancil. 2002. Concurrent sexual partnerships among women in the United States. *Epidemiology* 13 (3): 320–27.

Agar, Michael. 1980. Stories, background knowledge and themes: Problems in the analysis of life history narrative. *American Ethnologist* 7:223–39.

Agresti, Alan. 1990. *Categorical Data Analysis*. New York: Wiley.

Alexander, Jeffrey C., Bernhard Giesen, Richard Münch, and Neil J. Smelser. 1987. *The Micro-Macro Link*. Berkeley: University of California Press.

Alexander, Richard D. 1979. *Darwinism and Human Affairs*. Seattle: University of Washington Press.

―――. 1987. *The Biology of Moral Systems*. Hawthorne, NY: A. de Gruyter.

Allen, Melinda S. 2004. Bet-hedging strategies, agricultural change, and unpredictable environments: historical development of dryland agriculture in Kona, Hawaii. *Journal of Anthropological Archaeology* 23 (2): 196–224.

Alley, Richard B. 2000. *The Two-mile Time Machine: Ice Cores, Abrupt Climate Change, and Our Future.* Princeton, NJ: Princeton University Press.

Allport, Floyd Henry. 1924. *Social Psychology.* Boston: Houghton Mifflin.

Alpers, Michael. 1965. Epidemiological changes in kuru, 1957 to 1963. In *Slow, Latent, and Temperate Virus Infections*, ed. D. Gajdusek, C. Gibbs, and M. Alpers, 65–82. Washington, DC: US Government Printing Office.

———. 1979. Epidemiology and ecology of kuru. In *Prions: Novel Infections Pathogens Causing Scrapie and Creutzfeldt-Jakob Disease*, ed. S. Prusiner and M. McKinley, 451–65. San Diego: Academic Press.

———. 1992. Kuru. In *Human Biology in Papua New Guinea: The Small Cosmos*, ed. R. Attenborough and M. Alpers, 313–34. Oxford: Clarendon.

Altmann, M., and M. Morris. 1994. Clarification of the phi-mixing model. *Mathematical Biosciences* 124 (1): 1–7.

Ammar, Hamed. 1954. *Growing Up in an Egyptian Village.* London: Routledge and Kegan Paul.

Ammerman, Albert J., and Luigi L. Cavalli-Sforza. 1973. A population model for the diffusion of early farming in Europe. In *The Explanation of Culture Change*, ed. Colin Renfrew, 343–59. Duckworth, London.

Anderson, Benedict. 1991. *Imagined Communities: Reflections on the Origin and Spread of Nationalism*, 2nd ed. London: Verso.

Anderson, Elijah. 1999. *The Code of the Street.* New York: Norton.

Anderson, Roy M., and Robert M. May. 1991. *Infectious Diseases of Humans: Dynamics and Control.* Oxford: Oxford University Press.

Andersson, Malte. 1994. *Sexual Selection.* Princeton, NJ: Princeton University Press.

Aoki, Kenichi. 2001. Theoretical and empirical aspects of gene-culture coevolution. *Theoretical Population Biology* 59:253–61.

———. 2005. Avoidance and prohibition of brother-sister sex in humans. *Population Ecology* 47:13–19.

Aoki, Kenichi, and Marcus W. Feldman. 1997. A gene-culture coevolutionary model for brother-sister mating. *Proceedings of the National Academy of Sciences (US)* 94:13046–50.

Appiah, Kwame Anthony. 1997. The multiculturalist misunderstanding. *The New York Review of Books* 44 (15), http://www.nybooks.com/articles/1057.

Arens, William. 1979. *The Man-Eating Myth: Anthropology and Anthropophagy.* New York: Oxford University Press.

Atran, Scott, Douglas Medin, Norbert Ross, Elizabeth Lynch, Valentina Vapnarsky, Edilberto Ucan Ek' John Coley, Christopher Timura, and Michael Baran. 2002. Folkecology, cultural epidemiology, and the spirit of the commons. *Current Anthropology* 43:421–50.

Aunger, Robert. 1992. An Ethnography of Variation: Food Avoidances among Horticulturalists and Foragers in the Ituri Forest, Zaire. PhD diss., University of California, Los Angeles.

———. 1994. Are food avoidances maladaptive in the Ituri Forest of Zaire? *Journal of Anthropological Research* 50(3): 277–310.

———. 1995. On ethnography: Story-telling or science? *Current Anthropology* 36 (1): 97–130.

———. 1999. Against idealism / Contra consensus. *Current Anthropology* 40, suppl. (Feb.): S93–S101.

Avital, Eytan, and Eva Jablonka. 2000. *Animal Traditions: Behavioural Inheritance in Evolution.* Cambridge: Cambridge University Press.

Axelrod, Robert. 1984. *The Evolution of Cooperation.* New York: Basic.

Bagnall, Robert S., and Bruce W. Frier. 1994. *The Demography of Roman Egypt.* Cambridge: Cambridge University Press.

Bailey, Norman T. J. 1975. *The Mathematical Theory of Infectious Disease.* New York: Hafner.

Bandura, Albert, and Richard H. Walters. 1963. *Social Learning and Personality Development.* New York: Holt, Rinehart, and Winston.

Bannister, Robert C. 1979. *Social Darwinism: Science and Myth in Anglo-American Social Thought, American civilization.* Philadelphia: Temple University Press.

Barkow, Jerome H. 1992. Beneath new culture is old psychology: Gossip and social stratification. In Barkow et al. 1992, 627–37.

Barkow, Jerome H., Leda Cosmides, and John Tooby, eds. 1992. *The Adapted Mind: Evolutionary, Psychology and the Generation of Culture.* New York: Oxford University.

Barnard, Alan. 2000. *History and Theory in Anthropology.* Cambridge: Cambridge University Press.

Barrett, Louise, Robin Dunbar, and John Lycett. 2002. *Human Evolutionary Psychology.* Princeton, NJ: Princeton University Press.

Barry, Herbert, III, Irvin L. Child, and Margaret K. Bacon. 1959. Relation of child training to subsistence economy. *American Anthropologist* 61:51–63.

Barth, Fredrik. 2002. Toward a richer description and analysis of cultural phenomena. In *Anthropology beyond Culture*, ed. R. G. Fox and B. J. King, 23–36. New York: Berg.

Bartram, Laurence E., Jr. 1997. A Comparison of Kua (Botswana) and Hadza (Tanzania) Bow and Arrow Hunting. In *Projectile Technology*, ed. Heidi Knecht, 321–44. New York: Plenum.

Bateson, Gregory. 1972. *Steps to an Ecology of Mind.* New York: Ballantine.

Baum, William M. 1994. *Understanding Behaviorism: Science, Behavior, and Culture.* New York: HarperCollins.

Beaglehole, Ernest. n.d. Myths, Stories and Chants from Pukapuka. Ms. Honolulu: Bernice P. Bishop Museum Library.

Becker, Gary S. 1996. *Accounting for Tastes.* Cambridge, MA: Harvard University Press.

Becker, Gary S., and George J. Stigler. 1977. De gustibus non est disputandum. *American Economic Review* 67 (2): 76–90.

Behrman, Jerer R., Hans-Peter Kohler, and Susan Cotts Watkins. 2002. Social networks and changes in contraceptive use over time: Evidence from a longitudinal study in rural Kenya. *Demography* 39:713–38.

Benedict, Ruth. 1943. *Race: Science and Politics.* Rev. ed. New York: Viking.

Bendix, Reinhard. 1988. *Embattled Reason: Essays on Social Knowledge.* New Brunswick, NJ: Transaction.

Berndt, Ronald M. 1962. *Excess and Restraint: Social Control among a New Guinea Mountain People.* Chicago: University of Chicago Press.

Bettinger, Robert L., and Jelmer W. Eerkens. 1997. Evolutionary implications of metrical variation in Great Basin projectile points. In *Rediscovering Darwin: Evolutionary Theory and Archaeological Explanation*, ed. C. M. Barton and G. A. Clark, 177–191. Arlington, VA: American Anthropological Association.

Bevc, I., and I. Silverman. 1993. Early proximity and intimacy between siblings and incestuous behavior: A test of the Westermarck theory. *Ethology and Sociobiology* 14:171–81.

Binford, L. 1963. Red ochre cadres from the Michigan area: A possible case of cultural drift. *Southwestern Journal of Anthropology* 19:89–108.

Bittles, Alan H., and James V. Neel. 1994. The costs of human inbreeding and their implications for variations at the DNA level. *Nature Genetics* 8:117–21.

Bittles, Alan H., and Wendy N. Erber. 2005. Genetics and population health. *Annals of Human Biology* 32:113–16.

Blake, Judith. 1989. *Family Size and Achievement*. Berkeley: University of California Press.

Bliege Bird, Rebecca L., and Eric A. Smith. 2005. Signaling theory, strategic interaction, and symbolic capital. *Current Anthropology* 46 (2): 221–48.

Bliege Bird, Rebecca L., Eric A. Smith, and Douglas W. Bird. 2001. The hunting handicap: Costly signaling in human foraging strategies. *Behavioral Ecology and Sociobiology* 50: 9–19.

Bloch, Maurice. 1991. Language, anthropology, and cognitive science. *Man*, n.s., 26 (2): 183–98.

Bloom, Paul. 2000. *How Children Learn the Meanings of Words*. Cambridge, MA: MIT Press.

Blythe, S., C. Castillo-Chavez, J. S. Palmer, and M. Cheng. 1991. Toward a unified theory of sexual mixing and pair formation. *Mathematical Biosciences* 107:379–407.

Boas, Franz. 1887. The Study of Geography. In *Volksgeist as Method and Ethic*, ed. G. W. Stocking Jr., 9–16. Madison: University of Wisconsin Press, 1996.

———. 1898. The Jesup North Pacific expedition. In *The Shaping of American Anthropology, 1883–1911: A Franz Boas Reader*, ed. G. W. Stocking Jr., 107–116. Chicago: University of Chicago Press, 1974.

———. 1928. *Anthropology and Modern Life*. New York: Dover, 1962.

———, ed. 1938. *General Anthropology*. Boston: D. C. Heath.

Boehm, Christopher. 1996. Emergency decisions, cultural-selection mechanics, and group selection. *Current Anthropology* 37 (5): 763–93.

———. 2000. Conflict and the evolution of social control. *Journal of Consciousness Studies* 7:79–101.

Boesch, Christophe. 1991. Teaching in wild chimpanzees. *Animal Behaviour*, 41 (3): 530–32.

———. 1996. The emergence of cultures among wild chimpanzees. In *Evolution of Social Behaviour Patterns in Primates and Man*, ed. W. Runciman, J. Maynard-Smith, and R. Dunbar, 251–68. London: British Academy.

———. 2003. Is culture a golden barrier between human and chimpanzee? *Evolutionary Anthropology*, 12:82–91.

Boesch, Christophe, and Hedwige Boesch. 1981. Sex differences in the use of natural hammers by wild chimpanzees: A preliminary report. *Journal of Human Evolution* 10:585–93.

———. 1983. Optimization of nut-cracking with natural hammers by wild chimpanzees. *Behaviour* 83:265–86.

———. 1984. Mental map in wild chimpanzees: An analysis of hammer transports for nut cracking. *Primates* 25:160–70.

———. 1990. Tool use and tool making in wild chimpanzees. *Folia primatologica* 54:86–99.

Boesch, Christophe, and Hedwige Boesch-Achermann. 2000. *The Chimpanzees of the Taï Forest: Behavioural Ecology and Evolution.* Oxford: Oxford University Press.

Boesch, Christophe, Paul Marchesi, Nathalie Marchesi, Barbara Fruth, and Frédéric Joulian. 1994. Is nut cracking in wild chimpanzees a cultural behaviour? *Journal of Human Evolution* 26:325–38.

Boesch, Christophe, and Michael Tomasello. 1998. Chimpanzee and human cultures. *Current Anthropology* 39 (5): 591–614.

Boni, Maciej F., and Marcus W. Feldman. 2005. Evolution of antibiotic resistance by human and bacterial niche construction. *Evolution* 59:477–91.

Bonner, John Tyler. 1980. *The Evolution of Culture in Animals.* Princeton, NJ: Princeton University Press.

Borofsky, Robert. 1987. *Making History: Pukapukan and Anthropological Constructions of Knowledge.* New York: Cambridge University Press.

———, ed. 1994. *Assessing Cultural Anthropology.* New York: McGraw-Hill.

Borofsky, Robert, and Bruce Albert, with Raymond Hames, Kim Hill, Lêda Leitão Martins, John Peters, and Terence Turner. 2005. *Yanomami: The Fierce Controversy and What We Can Learn from It.* Berkeley: University of California Press.

Borofsky, Robert, Fredrik Barth, Richard A. Shweder, Lars Rodseth, and Nomi Maya Stolzenberg. 2001. WHEN: A conversation about culture. *Current Anthropology* 103:432–46.

Boserup, Ester. 1981. *Population and Technological Change: A Study of Long-term Trends.* Chicago: University of Chicago Press.

Bossen, Laurel. 2002. *Chinese Women and Rural Development: Sixty Years of Change in Lu Village, Yunnan.* Lanham, MD: Rowman and Littlefield.

Bossen, Laurel, Melissa J. Brown, and Hill Gates. In progress. Chinese women's labor: The silent transformation (book manuscript).

Bourdieu, Pierre. 1977. *Outline of a Theory of Practice.* Cambridge: Cambridge University Press.

———. 1990. *The Logic of Practice.* Stanford, CA: Stanford University Press.

Bowles, Samuel. 1998. Endogenous preferences: The cultural consequences of markets and other economic institutions. *Journal of Economic Literature* 36 (1): 75–111.

———. 2004. *Microeconomics: Behavior, Institutions, and Evolution.* Princeton, NJ: Princeton University Press.

Bowles, Samuel, Jung-Kyoo Choi, and Astrid Hopfensitz. 2003. The coevolution of individual behaviors and group level institutions. *Journal of Theoretical Biology* 223 (2): 135–47.

Bowring, John. 1840. Report on Egypt and Candia. *Sessional Papers of the British House of Commons,* vol. 21.

Boyd, Robert, and Peter J. Richerson. 1982. Cultural transmission and the evolution of cooperative behavior. *Human Ecology* 10:325–51.

———. 1985. *Culture and the Evolutionary Process.* Chicago: University of Chicago Press.

———. 1989. Social learning as an adaptation. *Lectures on Mathematics in the Life Sciences* 20:1–26.

———. 1996. Why culture is common and cultural evolution is rare. In *Evolution of Social Behaviour Patterns in Primates and Man,* ed. W. G. Runciman, John Maynard-Smith, and R. I. M. Dunbar, 77–94. Oxford: Oxford University Press for the British Academy.

Boyer, Pascal. 1990. *Tradition as Truth and Communication: A Cognitive Description of Traditional Discourse.* Cambridge: Cambridge University Press.

———. 2001. *Religion Explained: The Evolutionary Origins of Religious Thought.* New York: Basic.

Brightman, Robert. 1995. Forget culture: Replacement, transcendence, relexification. *Cultural Anthropology* 10 (4): 509–46.

Brown, Melissa J. 1995. We Savages Didn't Bind Feet—The Implications of Cultural Contact and Change in Southwestern Taiwan for an Evolutionary Anthropology. PhD diss., University of Washington.

———. 1997a. Articulating collectivism and individualism: Choices at the border to Han. Paper presented at the 96th Annual American Anthropological Association Meeting (November).

———. 1997b. Meaning, power, cognition and evolution: A multiple systems synthesis (unpublished manuscript).

———. 2001a. Reconstructing ethnicity: Recorded and remembered identity in Taiwan. Ethnology 40 (2): 153–64.

———. 2001b. Ethnic classification and culture: The case of the Tujia in Hubei, China. Asian Ethnicity 2 (1): 55–72.

———. 2002. Local government agency: Manipulating Tujia identity. Modern China 28 (3): 362–95.

———. 2003. The cultural impact of gendered social roles and ethnicity: Changing religious practices in Taiwan. Journal of Anthropological Research 59 (1): 47–67.

———. 2004. Is Taiwan Chinese? The Impact of Culture, Power and Migration on Changing Identities. Berkeley: University of California Press.

———. 2007. Ethnic identity, cultural variation, and processes of change: Rethinking the insights of standardization and orthopraxy. Modern China 33:91–124.

———. In progress. Why Not China? A Unified Social Theory Explanation of the Eur-Asian Divergence (book manuscript).

Brown, Paula, and Donald Tuzin, eds. 1983. The Ethnography of Cannibalism. Washington, DC: Society for Psychological Anthropology.

Brown, Ryan A., and George J. Armelagos. 2001. Apportionment of racial diversity: A review. Evolutionary Anthropology 10:34–40.

Brumann, Christoph. 1999. Writing for culture: Why a successful concept should not be discarded. Current Anthropology 40, suppl. (Feb.): S1–S27.

Buck, John Lossing. 1930. Chinese Farm Economy: A Study of 2,866 Farms in Seventeen Localities and Seven Provinces in China. Chicago: University of Chicago Press.

Bulmer, Michael. 2003. Francis Galton: Pioneer of Heredity and Biometry. Baltimore, MD: Johns Hopkins University Press.

Bunzl, Matti. 1996. Franz Boas and the Humboldtian Tradition. In Volksgeist as Method and Ethic, ed. G. W. Stocking Jr., 17–78. History of Anthropology Series 8. Madison: University of Wisconsin Press.

Burnham, Kenneth.P., and David R. Anderson. 2002. Model Selection and Multi-

*model Inference: A Practical Information-Theoretical Approach*. 2nd ed., New York: Springer.

Byrne, Richard W. 1995. *The Thinking Ape*. Oxford: Oxford University Press.

Byrne, Richard W., and Andrew Whiten. 1988. *Machiavellian Intelligence: Social Intelligence and the Evolution of Intellect in Monkeys, Apes, and Humans*. Oxford: Clarendon.

Calvin, William H. 2002. *A Brain for All Seasons: Human Evolution and Abrupt Climate Change*. Chicago: University of Chicago Press.

Camerer, Colin F. 2003. *Behavioral Game Theory: Experiments in Strategic Interaction*. Princeton, NJ: Princeton University Press.

Camerer, Colin F., and Ernst Fehr. 2004. Measuring social norms and preferences using experimental games: A guide for social scientists. In *Foundations of Human Sociality: Economic Experiments and Ethnographic Evidence from Fifteen Small-Scale Societies*, ed. Joseph Henrich, Robert Boyd, Samuel Bowles, Ernst Fehr, and Herbert Gintis, 55–95. Oxford: Oxford University Press.

Camerer, Colin F., and Robin M. Hogarth. 1999. The effects of financial incentives in experiments: A review of capital-labor-production framework. *Journal of Risk and Uncertainty*, 18:7–42.

Cameron, Jessica A., Jeanette M. Alvarez, Diane N. Ruble, and Andrew J. Fuligni. 2001. Children's lay theories about ingroups and outgroups: Reconceptualizing research on prejudice. *Personality and Social Psychology Review* 5 (2): 118–28.

Cavalli-Sforza, Luigi L., and Marcus W. Feldman. 1973a. Cultural versus biological inheritance: Phenotypic transmission from parents to children (A theory of the effect of parental phenotypes on children's phenotypes). *American Journal of Human Genetics* 25:618–37.

———. 1973b. Models for cultural inheritance I. Group mean and within-group variation. *Theoretical Population Biology* 4:42–55.

———. 1981. *Cultural Transmission and Evolution: A Quantitative Approach*. Princeton, NJ: Princeton University Press.

Cavalli-Sforza, Luigi L., Marcus W. Feldman, K. H. Chen, and S. M. Dornbusch. 1982. Theory and observation of cultural transmission. *Science* 218 (4567): 19–27.

Celnarova, Xenia. 1997. The religious ideas of the early Turks from the point of view of Ziya Gokalp. *Asian and African Studies* 6 (1): 103–8.

Centers for Disease Control and Prevention. 2001a. *Sexually Transmitted Disease Surveillance, 2000.* Atlanta, GA: U.S. Department of Health and Human Services, Centers for Disease Control and Prevention.

———. 2001b. *Sexually Transmitted Disease Surveillance, 2000. Supplement: Gonococcal Isolate Surveillance Project (GISP).* Atlanta, GA: U.S. Department of Health and Human Services, Center for Disease Control and Prevention.

Cetina, Karin Knorr. 1999. *Epistemic Cultures: How The Sciences Make Knowledge.* Cambridge, MA: Harvard University Press.

Chakrabarty, Dipesh. 1997. The difference—deferral of a colonial modernity: Public debates on domesticity in British Bengal. In *Tensions of Empire*, ed. Frederick Cooper and Anne Stoler, 373–405. Berkeley: University of California Press.

Chang Mao-kuei. 1994. Toward an understanding of the shen-chi wen-ti in Taiwan: Focusing on changes after political liberalization. In *Ethnicity in Taiwan: Social, Historical, and Cultural Perspectives*, ed. Chen Chung-min, Chuang Ying-chang, and Huang Shu-min, 93–150. Taibei: Institute of Ethnology, Academia Sinica.

———. 2000. On the origins and transformation of Taiwanese national identity. *China Perspectives* 28 (March–April): 51–70.

Chatterjee, Partha. 1993. *The Nation and Its Fragments.* Princeton, NJ: Princeton University Press.

Chevalier-Skolnikoff, Suzanne. 1988. Spontaneous tool use and sensorimotor intelligence in Cebus compared with other monkeys and apes. *Behavioral and Brain Sciences* 12:561–627.

Chiou, Howard. 2006. More than a Mouthful: Transmissible Spongiform Encephalopathies. Master's thesis, Department of Anthropological Sciences, Stanford University.

Chun, Allen. 1994. From nationalism to nationalizing: Cultural imagination and state formation in postwar Taiwan. *Australian Journal of Chinese Affairs* 30 (January): 49–69.

CNKCK (Chugoku Noson Kanko Chosa Kankokai), ed. 1957. *Chugoku Noson Kanko Chosa* (Survey of Customs in North China). Tokyo: Iwanami Shoten.

Cohen, Dov, Richard E. Nisbett, Brian F. Bowdle, and Norbert Schwarz. 1996. Insult, aggression, and the southern culture of honor: An experimental ethnography. *Journal of Personality and Social Psychology* 70:945–60.

Cohen, Dov, Joe Vandello, and Adrian K. Rantilla. 1998. The sacred and the social: Honor and violence in cultural context. In *Shame: Interpersonal Behavior, Psychopathology, and Culture*, ed. P. Gilbert and B. Andrews, 261–82. New York: Oxford University Press.

Cohen, Mark Nathan. 1977. *The Food Crisis in Prehistory: Overpopulation and the Origins of Agriculture*. New Haven, CT: Yale University Press.

Collinge, John, and Mark S. Palmer, eds. 1997. *Prion Diseases*. Oxford: Oxford University Press.

Comaroff, Jean. 1985. *Body of Power, Spirit of Resistance: The Culture and History of a South African People*. Chicago: University of Chicago Press.

Connor, Walker. 1994. *Ethnonationalism: The Quest for Understanding*. Princeton, NJ: Princeton University Press.

Cooper, Frederick. 1997. The dialectics of decolonization: Nationalism and labor movements in postwar French Africa. In *Tensions of Empire: Colonial Cultures in a Bourgeois World*, ed. Frederick Cooper and Anne Stoler, 406–35. Berkeley: University of California Press.

Cooper, Richard S., Joan F. Kennelly, Ramon Durazo-Arvizu, Hyun-Joo Oh, George Kaplan, and John Lynch. 2001. Relationship between premature mortality and socioeconomic factors in black and white populations of US metropolitan areas. *Public Health Reports* 116 (5): 464–73.

Corcuff, Stéphane. 2000. Taiwan's "Mainlanders": A New Ethnic Category. *China Perspectives* 28 (March-April): 71–81.

Cosmides, Leda, and John Tooby. 1989. Evolutionary psychology and the generation of culture 2. Case study: A computational theory of social exchange. *Ethology and Sociobiology* 10:51–97.

Cready, Cynthia M., Mark A. Fossett, and K. Jill Kiecolt. 1997. Mate availability and African American family structure in the US nonmetropolitan South, 1960–1990. *Journal of Marriage and the Family* 59 (1): 192–203.

Cronk, Lee. 1995. Is there a role for culture in human behavioral ecology? *Ethology and Sociobiology* 16 (3): 181–205.

Cronk, Lee, Napoleon A. Chagnon, and William Irons, eds. 2000. *Adaptation and Human Behavior: An Anthropological Perspective*. New York: Aldine de Gruyter.

Crowder, K. D., and S. E. Tolnay. 2000. A new marriage squeeze for black women: The role of racial intermarriage by black men. *Journal of Marriage and the Family* 62 (3): 792–807.

Cunynghame, Henry. 1887. The present state of education in Egypt. *Journal of the Royal Asiatic Society of Great Britain and Ireland*, n.s. 19:223–37.

Dagg, Anne Innis. 1998. Infanticide by male lions hypothesis: A fallacy influencing research into human behavior. *American Anthropologist* 100 (4): 940–50.

Daly, Martin, and Margo Wilson. 1988. *Homicide*. New York: Aldine de Gruyter.

———. 1999. *The Truth about Cinderella: A Darwinian View of Parental Love*. New Haven, CT: Yale University Press.

D'Andrade, Roy G. 1984. Cultural meaning systems. In *Culture Theory: Essays on Mind, Self, and Emotion*, ed. R. A. Shweder and R. A. LeVine, 88–119. Cambridge: Cambridge University Press.

———. 1987. A folk model of the mind. In *Cultural Models in Language and Thought*, ed. D. Holland and N. Quinn, 113–47. Cambridge: Cambridge University Press.

———. 1995a. *The Development of Cognitive Anthropology*. Cambridge: Cambridge University Press.

———. 1995b. Moral models in anthropology. *Current Anthropology* 36 (3): 399–408.

———. 1999. Comment (on Brumann 1999). *Current Anthropology* 40, suppl. (Feb.): S16–S17.

———. 2002a. Cultural Darwinism and language. *American Anthropologist* 104:223–32.

———. 2002b. Violence without honor in the American South. In *Tournaments of Power*, ed. Tor Aase, 61–73. Burlington, VT: Ashgate.

D'Andrade, Roy, and Claudia Strauss, eds. 1992. *Human Motives and Cultural Models*. Cambridge: Cambridge University Press.

Darwin, Charles. 1874. *The Descent of Man and Selection in Relation to Sex*. 2nd ed., 2 vols. New York: American Home Library.

Davidson, James, W. 1903. *The Island of Formosa: Past and Present*. Shanghai: Kelly and Walsh.

Dawkins, Richard. 1976. *The Selfish Gene*. New York: Oxford University Press.

Deák, István. 2002. The crime of the century. *New York Review of Books* 45 (14): 48–51.

DeBello, William M., Daniel E. Feldman, and Eric J. Knudsen. 2001. Adaptive axonal remodeling in the midbrain auditory space map. *Journal of Neuroscience* 21 (9): 3161–74.

De Lepervanche, Marie. 1973. Social structure. In *Anthropology in Papua New Guinea*, ed. Ian Hogbin, 1–60. Melbourne: Melbourne University Press.

Demolins, Edmond. 1899. *Anglo-Saxon Superiority: To What It Is Due*. Trans. Louis Bert. Lavigne. New York: R. F. Fenno.

Des Forges, Alison. 1997. No more corpses on the road. *Times Literary Supplement* August 15:26–27.

Deutscher, Guy. 2005. *The Unfolding of Language: An Evolutionary Tour of Mankind's Greatest Invention*. New York: Henry Holt.

Diamond, Jared. 1978. The Tasmanians: The longest isolation, the simplest technology. *Nature* 273:185–86.

———. 1997. *Guns, Germs, and Steel: The Fates of Human Societies*. New York: W.W. Norton.

Diekmann, O., K. Dietz, and J. A. P. Heesterbeek. 1991. The basic reproduction ratio for sexually transmitted diseases: I. Theoretical considerations. *Mathematical Biosciences* 107 (2): 325–39.

Diekmann, O., and J. A. P. Heesterbeek. 2000. *Mathematical Epidemiology of Infectious Diseases: Model Building, Analysis, and Interpretation*. New York: Wiley.

Dietz, K., J. A. P. Heesterbeek, and D. W. Tudor. 1993. The basic reproduction ratio for sexually transmitted diseases: II. Effects of variable HIV infectivity. *Mathematical Biosciences* 117 (1–2): 35–47.

Dirks, Nicholas B., Geoff Eley, and Sherry B. Ortner, eds. 1994. *Culture/Power/History: A Reader in Contemporary Social Theory*. Princeton, NJ: Princeton University Press.

Donald, Merlin. 1991. *Origins of the Modern Mind: Three Stages in the Evolution of Culture and Cognition*. Cambridge, MA: Harvard University Press.

Donham, Donald L. 1990. *History, Power, Ideology: Central Issues in Marxism and Anthropology*. New York: Cambridge University Press.

Dore, Ronald Philip. 1968. Function and cause. In *Theory in Anthropology*, ed. Robert A. Manners and David Kaplan, 212–20. Chicago: Aldine.

Dorsey, James O. 1884. Omaha sociology. *Annual Report, Bureau of Ethnology, Smithsonian Institution* 3:322–70.

Douglas, Mary. 1966. *Purity and Danger: An Analysis of the Concepts of Pollution and Taboo*. London: Routledge and Kegan Paul.

Dunbar, Robin. 1988. *Primate Social Systems*. Ithaca, NY: Cornell University Press.

Dunnell, Robert C. 1995. What is it that actually evolves? In *Evolutionary Archaeology: Methodological Issues*, ed. Patrice A. Teltser, 32–50. Tucson: University of Arizona Press.

———. 1999. The concept of waste in an evolutionary archaeology. *Journal of Anthropological Archaeology* 18 (3): 243–250.

Dunnell, Robert C., and Diana M. Greenlee. 1999. Late Woodland Period "waste" reduction in the Ohio River Valley. *Journal of Anthropological Archaeology* 18 (3): 376–395.

Durham, William H. 1991. *Coevolution: Genes, Culture, and Human Diversity*. Stanford, CA: Stanford University Press.

———. 1992. Applications of evolutionary culture theory. *Annual Review of Anthropology* 21:331–55.

———. 2002. Cultural variation in time and space: The case for a populational theory of culture. In *Anthropology Beyond Culture*, ed. R. Fox and B. King, 193–206. New York: Berg.

Durham, William H., and Peter Weingart. 1997. Units of culture. In *Human By Nature: Between Biology and the Social Sciences*, ed. Peter Weingart, Sandra D. Mitchell, Peter J. Richerson, and Sabine Maasen, 300–313. Mahwah NJ: Lawrence Erlbaum.

Durkheim, Emile. 1912. *The Elementary Forms of Religious Life*. Trans. Karen E. Fields. New York: Free Press, 1995.

———. 1961. *Moral Education: A Study in the Theory and Application of the Sociology of Education*. New York: Free Press.

al-Duwwa, Mahmud al-Sayyid, et al. 1986–87. *Al-tarbiya al-islamiyya*. Seventh Grade. Cairo: Al-Jihaz al-Markazi lil-Kutub al-Jamiʿiyya wa al-Madrasiyya wa al-Wasaʾil al-Taʿlimiyya.

———. 1987–88. *Al-tarbiya al-islamiyya*. Eighth Grade. Cairo: Al-Jihaz al-Markazi lil-Kutub al-Jamiʿiyya wa al-Madrasiyya wa al-Wasʾil al-Taʿlimiyya.

Dworkin, Dennis. 1997. *Cultural Marxism in Postwar Britain: History, the New Left, and the Origins of Cultural Studies*. Durham, NC: Duke University Press.

Ebrey, Patricia. 1996. Surnames and Han Chinese identity. In *Negotiating Ethnicities in China and Taiwan*, ed. Melissa J. Brown, 19–36. Berkeley, CA: Institute of East Asian Studies, University of California.

Edgerton, Robert B. 1971. *The Individual in Cultural Adaptation: A Study of Four East African Peoples*. Berkeley: University of California Press.

————. 1992. *Sick Societies: Challenging the Myth of Primitive Harmony*. New York: Free Press.

Edmonds, Bruce. 2002. Three challenges for the survival of memetics. *Journal of Memetics—Evolutionary Models of Information Transmission* 6 (2), http://jom-emit.cfpm.org/2002/vol6/edmonds_b_letter.html.

————. 2005. The revealed poverty of the gene-meme analogy—Why memetics per se has failed to produce substantive results. *Journal of Memetics-Evolutionary Models of Information Transmission* 9 (1), http://jom-emit.cfpm.org/2005/vol9/edmonds_b.html.

Eerkens, Jelmer W., and Carl P. Lipo. 2005. Cultural transmission, copying errors, and the generation of variation in material culture and the archaeological record. *Journal of Anthropological Archaeology* 24:316–334.

Efferson, Charles, Peter J. Richerson, Richard McElreath, Mark Lubell, Ed Edsten, Timothy M. Waring, Brian Paciotti, and William Baum. 2007. Learning, productivity and noise: An experimental study of cultural transmission on the Bolivian Altiplano. *Evolution and Human Behavior* 28 (1):11–17.

Eibl-Eibesfeldt, Irenäus. 1982. Warfare, man's indoctrinability, and group selection. *Zeitschrift für Tierpsychologie* 67:177–98.

Elias, Norbert. 1994. *The Civilizing Process*. Trans. Edmund Jephcott. Cambridge, MA: Blackwell.

Ember, Carol, and Melvin Ember. 1996. *Cultural Anthropology*. 8th ed. Upper Saddle River, NJ: Prentice Hall.

Ensminger, Jean. 1998. Anthropology and the new institutionalism. *Journal of Institutional and Theoretical Economics* 127(4): 774–89.

Euben, Roxanne L. 1999. *Enemy in the Mirror: Islamic Fundamentalism and the Limits of Modern Rationalism*. Princeton, NJ: Princeton University Press.

Evans-Pritchard, Edward E. 1940. *The Nuer: A Description of the Modes of Livelihood and Political Institutions of a Nilotic People*. London: Oxford University Press.

Falk, Armin, Ernst Fehr, and Uris Fischbacher. 2005. Driving forces behind informal sanctions. *Econometrica* 73 (6): 2017–70.

Fehr, Ernst, and Simon Gächter. 2002. Altruistic punishment in humans. *Nature* 415:137–40.

Fehr, Ernst, and Joseph Henrich. 2003. Is strong reciprocity a maladaptation? In *Genetic and Cultural Evolution of Cooperation*, ed. P. Hammerstein, 55–82. Cambridge, MA: MIT Press.

Feldman, Daniel E., and Eric I. Knudsen. 1997. An anatomical basis for visual calibration of the auditory space map in the barn owl's midbrain. *Journal of Neuroscience* 17 (17): 6820–37.

Feldman, Marcus W., and Luigi L. Cavalli-Sforza. 1976. Cultural and biological evolutionary processes, selection for a trait under complex transmission. *Theoretical Population Biology* 9: 239–59.

———. 1989. On the theory of evolution under genetic and cultural transmission with application to the lactose absorption problem. In *Mathematical Evolutionary Theory*, ed. Marcus W. Feldman, 145–73. Princeton, NJ: Princeton University Press.

Feldman, Marcus W., and Kevin N. Laland. 1996. Gene-culture coevolutionary theory. *Trends in Ecology and Evolution* 11:453–57.

Fessler, Daniel M. T., and C. David Navarette. 2004. Third-party attitudes toward sibling incest. Evidence for Westermarck's hypotheses. *Evolution and Human Behavior* 25:277–94.

Fienberg, Stephen E., Matthew S. Johnson, and Brian W. Junker. 1999. Classical multilevel and Bayesian approaches to population size estimation using multiple lists. *Journal of the Royal Statistical Society*, Series A 162:383–405.

Finkelhor, David. 1980. Sex among siblings: A survey on prevalence, variety, and effects. *Archives of Sexual Behavior* 9:171–94.

Firth, Raymond. 1936. *We the Tikopia: A Sociological Study of Kinship in Primitive Polynesia*. Stanford, CA: Stanford University Press, 1983.

Fischl, M. A., G. M. Dickinson, G. B. Scott, N. Klimas, M. A. Fletcher, and W. Parks. 1987. Evaluation of heterosexual partners, children, and household contacts of adults with AIDS. *Journal of the American Medical Association* 257 (5): 640–44.

Fisher, R. A. 1918. The correlation between relatives on the supposition of Mendelian inheritance. *Transactions of the Royal Society of Edinburgh* 52 399–433.

Fiske, A. P. 1998. Learning Culture the Way Informants Do: Observing, Imitating, and Participating (article manuscript).

Flinn, Mark V. 1997. Culture and the evolution of social learning. *Evolution and Human Behavior* 18 (1): 23–67.

Fortes, Meyer. 1945. *The Dynamics of Clanship among the Tallensi*. London: Oxford University Press (for International African Institute).

————. 1949. *The Web of Kinship among the Tallensi*. London: Oxford University Press (for International African Institute).

Fortna, Benjamin. 2000. Islamic morality in late Ottoman "secular" schools. *International Journal of Middle East Studies* 32 (3): 369–93.

————. 2002. *Imperial Classroom: Islam, the State, and Education in the Late Ottoman Empire*. Oxford: Oxford University Press.

Foster, George M. 1960. *Culture and Conquest: America's Spanish Heritage*. Viking Fund Publications in Anthropology 27. New York: Wenner-Gren Foundation for Anthropological Research.

Foucault, Michel. 1965. *Madness and Civilization: A History of Insanity in the Age of Reason*. Trans. R. Howard. New York: Random House.

————. 1973. *The Birth of the Clinic: An Archaeology of Medical Perception*. Trans. A. M. Sheridan-Smith. New York: Vintage.

————. 1977. *Discipline and Punish: The Birth of the Prison*. New York: Pantheon.

————. 1980. *Power/knowledge: Selected Interviews and Other Writings, 1972–77*. New York: Pantheon.

————. 1983. The subject and power. In *Michel Foucault: Beyond Structuralism and Hermeneutics*, 2nd ed., ed. Hubert L. Dreyfus and Paul Rabinow, 208–28. Chicago: University of Chicago Press.

Fowler, Henry Weed. 1928. *Fishes in Oceania*. Bernice P. Bishop Museum Memoir 10. Honolulu: Bishop Museum Press.

Fox, Richard G., and Barbara J. King. 2002a. Introduction: Beyond culture worry. In Fox and King 2002b, 1–19.

————, eds. 2002b. *Anthropology beyond Culture*. New York: Berg.

Fox, Robin. 1962. Sibling incest. *British Journal of Sociology* 13 (2): 128–50.

————. 1980. *The Red Lamp of Incest*. New York: E. P. Dutton.

Fracchia, Joseph, and R. C. Lewontin. 1999. Does culture evolve? *History and Theory* 38:52–78.

Fragaszy, Dorothy M., and Susan Perry. 2003. *The Biology of Traditions: Models and Evidence*. Cambridge: Cambridge University Press.

Frank, Ove, and David Strauss. 1986. Markov graphs. *Journal of the American Statistical Association* 81 (395): 832–42.

Frazer, James George. 1910. *Totemism and Exogamy: A Treatise on Certain Early Forms of Superstition and Society*. 4 vols. Repr., London: Dawsons, 1968.

Freud, Sigmund. 1918. *Totem and Taboo*. New York: Vintage, 1946.

———. 1920. *Introductory Lectures on Psychoanalysis*. New York: W. W. Norton, 1977.

Furubotn, E. G., and R. Richter. 2000. *Institutions and Economic Theory*. Ann Arbor: University of Michigan Press.

Gajdusek, D. Carleton. 1977. Unconventional viruses and the origin and disappearance of kuru. *Science* 197 (4307): 943–960.

Galef, Bennett G., Jr. 1988. Imitation in animals. In *Social Learning: Psychological and Biological Perspectives*, ed. Thomas R. Zentall and Bennett G. Galef Jr., 3–28. Hillsdale, NJ: Lawrence Earlbaum.

Galison, Peter. 1997. *Image and Logic: A Material Culture of Microphysics*. Chicago: University of Chicago Press.

Gallie, Walter Bryce. 1956. Essentially contested concepts. *Proceedings of the Aristotelian Society* 56: 167–98.

Gamble, Sidney D. 1954. *Ting Hsien: A North China Rural Community*. New York: Institute of Pacific Relations.

Gates, Hill. 1981. Social Class and Ethnicity. In *The Anthropology of Taiwanese Society*, ed. Emily Martin Ahern and Hill Gates, 241–81. Stanford, CA: Stanford University Press.

———. 1987. *Chinese Working-Class Lives: Getting By in Taiwan*. Ithaca, NY: Cornell University Press.

———. 1996. *China's Motor: A Thousand Years of Petty Capitalism*. Ithaca, NY: Cornell University Press.

———. 2001. Footloose in Fujian: Economic correlates of footbinding. *Comparative Studies in Society and History* 43(1): 130–48.

———. 2004. Refining the incest taboo: With considerable help from Bronislaw Malinowski. In Wolf and Durham 2004, 139–60.

———. In progress. Hand and foot: Footbinding and women's labor (book manuscript).

Geertz, Clifford. 1957. Ritual and social change: A Javanese example. *American Anthropologist* 59 (1): 32–54.

———. 1973. *The Interpretation of Cultures*. New York: Basic.

———. 1983. *Local Knowledge: Further Essays in Interpretive Anthropology*. New York: Basic.

Geronimus, Arline T., John Bound, Timothy A. Waidmann, Marianna M. Hillemeier, and Patricia B. Burns. 1996. Excess mortality among blacks and whites in the United States. *New England Journal of Medicine* 335 (21): 1552–58.

Giedd, Jay N., Jonathan Blumenthal, Neil O. Jeffries, F. X. Castellanos, Hong Liu, Alex Zidjenbos, Tomás Paus, Alan C. Evans, and Judith L. Rapoport. 1999. Brain development during childhood and adolescence: A longitudinal MRI study. *Nature Neuroscience* 2 (10): 861–63.

Gilbert, Margaret. 2002. Belief and acceptance as features of groups. *Protosociology* 16:137–55.

Gillison, Gillian. 1980. Images of nature in Gimi thought. In *Nature, Culture and Gender*, ed. C. MacCormack and M. Strathern, 143–173. Cambridge: Cambridge University Press.

———. 1983. Cannibalism among women in the Eastern Highlands of Papua New Guinea. In Brown and Tuzin 1983, 33–50.

Gintis, Herbert, Samuel Bowles, Robert Boyd, and Ernst Fehr, eds. 2004. *Moral Sentiments and Material Interests: The Foundations of Cooperation in Economic Life.* Cambridge, MA: MIT Press.

Glasse, Shirley. 1964. The social effects of kuru. *Papua New Guinea Medical Journal* 7 (1): 36–47.

Glendon, Mary Ann. 1989. *The Transformation of Family Law.* Chicago: University of Chicago Press.

Godelier, Maurice. 1982. *The Making of Great Men: Male Domination and Power among the New Guinea Baruya.* Cambridge: Cambridge University Press, 1986.

Gokalp, Ziya. 1958. *Turkish Nationalism and Western Civilization.* New York: Columbia University Press.

Golden, Matthew R., Julia A. Schillinger, Lauri Markowitz, and Michael E. St. Louis. 2000. Duration of untreated genital infections with chlamydia trachomatis: a review of the literature. *Sexually Transmitted Diseases* 27 (6): 329–37.

Goldfarb, Lev G. 2002. Kuru: The old epidemic in a new mirror. *Microbes and Infection* 4:875–82.

Goodall, Jane. 1963. Feeding behaviour of wild chimpanzees: A preliminary report. *Symposium of the Zoological Society, London* 10:39–48.

———. 1968. Behaviour of free-living chimpanzees of the Gombe Stream area. *Animal Behaviour Monograph* 1: 163–311.

———. 1986. *The Chimpanzees of Gombe.* Cambridge, MA: Harvard University Press.

Goodenough, Ward H. 1957. Cultural anthropology and linguistics. In *Report of the Seventh Annual Round Table Meeting on Linguistics and Language Study*, ed. P. L. Garim.109–73. Washington, DC: Georgetown University Press.

Goodfield, June. 1985. *Quest for the Killers*. New York: Hill and Wang.

Goody, Jack. 1976. *Production and Reproduction: A Comparative Study of the Domestic Domain*. New York: Cambridge University Press.

———. 1994. Culture and its boundaries: A European view. In *Assessing Cultural Anthropology*, ed. R. Borofsky, 250–61. New York: McGraw Hill.

Gough, Kathleen. 1981. *Rural Society in Southeast India*. New York: Cambridge University Press.

———. 1990. *Political Economy in Vietnam*. Berkeley, CA: Folklore Institute.

Graebner, Fritz. 1909. *Ethnography of the Santa Cruz Islands*. New Haven, CT: Human Relations Area Files, 1962.

Gray, Jeremy R., Todd S. Braver, and Marcus E. Raichle. 2002. Integration of emotion and cognition in the lateral prefrontal cortex. *Proceedings of the National Academy of Science (US)* 99 (6): 4115–20.

Graziano, William G., Lauri A. Jensen-Campbell, Laura J. Shebilske, and Sharon R. Lundgren. 1993. Social influence, sex differences, and judgments of beauty: Putting the *interpersonal* back in interpersonal attraction. *Journal of Personality and Social Psychology* 65:522–31.

Greene, H. R. 1885. Report on the medical and sanitary administration of the Government of Egypt. Enclosure in item no. 19 in Egypt no. 15 (1885), *Reports on the State of Egypt and the Progress of Administrative Reforms, Sessional Papers of the British House of Commons*, vol. 89.

Guglielmino, C. R., C. Viganotti, B. Hewlett, and L. L. Cavalli-Sforza. 1995. Cultural variation in Africa: Role of mechanisms of transmission and adaptation. *Proceedings of the National Academy of Sciences (US)* 92:7585–89.

Gupta, S., R. M. Anderson, and R. M. May. 1989. Networks of sexual contacts: Implications for the pattern of spread of HIV. *AIDS* 3 (12): 807–17.

Guy, Gregory R., and William Labov. 1996. Variation and change in language and society. *Amsterdam Studies in the Theory and History of Linguistic Science*, series IV, *Current Issues in Linguistic Theory* 127. Amsterdam, PA: J. Benjamins.

Hamilton, William D. 1964. Genetic evolution of social behavior I, II. *Journal of Theoretical Biology* 7 (1): 1–52.

———. 1967. Extraordinary sex ratios. *Science* 156:477–88.

———. 1975. Innate social aptitudes of man: An approach from evolutionary genetics. In *Biosocial Anthropology*, ed. Robin Fox: 133–55. New York: Wiley.

Hammel, E. A., C. K. McDaniel, and K. W. Wachter. 1979. Demographic consequences of incest tabus: A microsimulation analysis. *Science* 205 (4410): 972–77.

Hammonds, Evelynn M. 2003. *The Logic of Difference: A History of Race in Science and Medicine in the United States*. Chapel Hill: University of North Carolina Press.

Hanioglu, M. Sukru. 1995. *The Young Turks in Opposition*. New York: Oxford University Press.

———. 2001. *Preparation for a Revolution: The Young Turks, 1902–1908*. New York: Oxford University Press.

Hankins, Thomas. 1985. *Science and the Enlightenment*. Cambridge: Cambridge University Press.

Harbaugh, William T., Kate Krause, and Steven G. Liday. 2002. Bargaining by Children (manuscript).

Hare, Brian, Josep Call, and Michael Tomasello. 2001. Do chimpanzees know what conspecifics know? *Animal Behaviour* 61 (1): 139–51.

Harrell, Stevan. 1977. Modes of Belief in Chinese Folk Religion. *Journal for Scientific Studies of Religion* 16 (1): 55–65.

Harris, Marvin. 1997. Comment (on O'Meara 1997). *Current Anthropology* 38 (3): 410–18.

———. 2001. *Cultural Materialism: The Struggle for a Science of Culture*. Updated ed. Walnut Creek, CA: AltaMira.

Harrison, Faye, ed. 1997. *Decolonizing Anthropology: Moving Further Toward an Anthropology for Liberation*. 2nd ed. Arlington, VA: Association of Black Anthropologists, American Anthropological Association.

Hatch, Elvin. 1973. *Theories of Man and Culture*. New York: Columbia University Press.

Hauert, Christoph. 2002. Effects of space in 2 x 2 games. *International Journal of Bifurcation and Chaos* 12 (7): 1531–48.

Hauser, Marc D. 2000. *Wild Minds: What Animals Really Think*. New York: Henry Holt.

Haviland, William. 2002. *Cultural Anthropology*. 10th ed. New York: Harcourt College Publishers.

Hawthorn, Geoffrey. 1987. *Enlightenment and Despair: A History of Social Theory.* Cambridge: Cambridge University Press.

Hayes, Cathy. 1951. *The Ape in Our House.* New York: Harper.

Heesterbeek, Johan Andre Peter. 2002. A brief history of R-0 and a recipe for its calculation. *Acta Biotheoretica* 50 (3): 189–204.

Henrich, Joseph. 2001. Cultural transmission and the diffusion of innovations: Adoption dynamics indicate that biased cultural transmission is the predominate force in behavioral change. *American Anthropologist* 103 (4): 992–1010.

————. 2002. Decision-making, cultural transmission and adaptation in economic anthropology. In *Theory in Economic Anthropology*, ed. J. Ensminger, 251–95. Rowman and Littlefield.

————. 2004a. Cultural group selection, coevolutionary processes and large-scale cooperation. *Journal of Economic Behavior and Organization* 53:3–35.

————. 2004b. Demography and cultural evolution: Why adaptive cultural processes produced maladaptive losses in Tasmania. *American Antiquity* 69 (2): 197–221.

Henrich, Joseph, and Robert Boyd. 1998. The evolution of conformist transmission and the emergence of between-group differences. *Evolution and Human Behavior* 19 (4): 215–241.

————. 2001. Why people punish defectors: Weak conformist transmission can stabilize costly enforcement of norms in cooperative dilemmas. *Journal of Theoretical Biology* 208: 79–89.

————. 2002. On modeling cognition and culture: Why replicators are not necessary for cultural evolution. *Culture and Cognition* 2 (2): 67–112.

Henrich, Joseph, Robert Boyd, Samuel Bowles, Colin Camerer, Ernst Fehr, and Herbert Gintis, eds. 2004. *Foundations of Human Sociality: Economic Experiments and Ethnographic Evidence from Fifteen Small-scale Societies.* Oxford: Oxford University Press.

Henrich, Joseph, Robert Boyd, Samuel Bowles, Colin Camerer, Ernst Fehr, Herbert Gintis, Richard McElreath, Michael Alvard, Abigail Barr, Jean Ensminger, Natalie Smith Henrich, Kim Hill, Francisco Gil-White, Michael Gurven, Frank Marlowe, John Patton, and David Tracer. 2005. "Economic man" in cross-cultural perspective: Behavioral experiments in 15 small-scale societies. *Behavioral and Brain Sciences* 28 (6): 795–855.

Henrich, Joseph, and Francisco J. Gil-White. 2001. The evolution of prestige: Freely conferred deference as a mechanism for enhancing the benefits of cultural transmission. *Evolution and Human Behavior* 22 (3): 165–96.

Henrich, Joseph, R. McElreath, A. Barr, J. Ensminger, C. Barrett, A. Bolyanatz, J.C. Cardenas, M. Gurven, E. Gwako, N. Henrich, C. Lesorogol, F. Marlowe, D. Tracer, and J. Ziker, 2006. Costly Punishment Across Human Societies. *Science* 312, 1767–1770.

Henrich, Joseph, and Richard McElreath. 2003. The evolution of cultural evolution. *Evolutionary Anthropology* 12 (3): 123–35.

Henrich, Joseph, and Natalie Smith Henrich. 2007. *Why Humans Cooperate: A Cultural and Evolutionary Explanation*. Oxford: Oxford University Press.

Herrnstein, Richard J., and Charles A. Murray. 1994. *The Bell Curve: Intelligence and Class Structure in American Life*. New York: Free Press.

Hethcote, Herbert W., and James A. Yorke. 1984. *Gonorrhea: Transmission Dynamics and Control. Lecture Notes in Biomathematics* vol. 56. New York: Springer.

Hewlett, Barry S., and Luigi L. Cavalli-Sforza. 1986. Cultural transmission among Aka pygmies. *American Anthropologist* 88:922–34.

Hewlett, Barry S., Annalisa De Silvestri, and C. Rosalba Guglielmino. 2002. Semes and genes in Africa. *Current Anthropology* 43:313–21.

Heyes, Cecilia M. 1994. Imitation, culture and cognition. *Animal Behaviour* 46:999–1010.

———. 1998. Theory of mind in nonhuman primates. *Behavioral and Brain Sciences* 21 (1): 101–34.

Heyes, Cecilia M., and Bennett G. Galef Jr., eds. 1996. *Social Learning in Animals: The Roots of Culture*. San Diego, CA: Academic Press.

Hill, Kim. 2001. Letter (reply to Mann 2001). *Science* 292 (5523): 1837.

Hirschfeld, Lawrence A., and Susan A. Gelman, eds. 1994. *Mapping the Mind: Domain Specificity in Cognition and Culture*. Cambridge: Cambridge University Press.

Hodgson, Geoffrey M. 2004. *The Evolution of Institutional Economics: Agency, Structure and Darwinism in American Institutionalism*. London: Routledge.

Hoffman, Elizabeth, Kevin A. McCabe, and Vernon L. Smith. 1998. Behavioral foundations of reciprocity: Experimental economics and evolutionary psychology. *Economic Inquiry* 36 (3): 335–52.

Hofstadter, Richard. 1945. *Social Darwinism in American Thought, 1860–1915.* Philadephia: University of Pennsylvania Press.

Hohmann, Gottfried, and Barbara Fruth. 2003. Culture in bonobos? Between-species and within-species variation in behavior. *Current Anthropology* 44 (4): 563–571.

Holden, Clare, and Ruth Mace. 1997. A phylogenetic analysis of the evolution of lactose digestion in adults. *Human Biology* 69:605–28.

Hooper, Judith. 2002. *Of Moths and Men: An Evolutionary Tale, The Untold Story of Science and the Peppered Moth.* New York: Norton.

Hopkins, Keith 1980. Brother-sister marriage in Roman Egypt. *Comparative Studies in Society and History* 22:303–54.

Hourani, Albert. 1983. *Arabic Thought in the Liberal Age, 1789–1939.* Cambridge: Cambridge University Press.

Hout, Michael, Andrew M. Greeley, and Melissa J. Wilde. 2001. The demographic imperative in religious change in the United States. *American Journal of Sociology* 107 (2): 468–86.

Hruschka, Daniel J., and Henrich, Joseph, 2006. Friendship, Cliquishness, and the Emergence of Cooperation. *Journal of Theoretical Biology* 239, 1–15.

Humle, Tatyana, and Tetsuro Matsuzawa. 2002. Ant-dipping among the chimpanzees of Boussou, Guinea, and some comparisons with other sites. *American Journal of Primatology* 58 (3): 133–148.

Hunt, Gavin R. 1996. Manufacture and use of hook-tools by New Caledonian crows. *Nature* 379:249–51.

Hunt, Gavin R. and Russel D. Gray. 2003. Diversification and cumulative evolution in New Caledonian crow tool manufacture. *Proceedings of the Royal Society of London, Series B* 270:867–874.

Husayn, Taha. 1954. *The Future of Culture in Egypt.* Trans. Sidney Glazer. New York: Octagon.

Hyde, Peter S., and Eric I. Knudsen 2002. The optic tectum controls visually guided adaptive plasticity in the owl's auditory space map. *Nature* 415 (6867): 73–76.

Ibrahim, Saad Eddin. 1997. Cross-Eyed Sociology in Egypt and the Arab World. *Contemporary Sociology* 25 (5): 547–51.

Ihara, Yasuo, and Marcus W. Feldman. 2004. Cultural niche construction and the evolution of small family size. *Theoretical Population Biology* 65:105–11.

Jacquez, John A. 1996. *Compartmental Analysis in Biology and Medicine*. 3rd ed. Ann Arbor, MI: Biomedware.

Jacquez, John A., Carl P. Simon, James S. Koopman, Lisa Sattenspiel, and Timothy Perry. 1988. Modeling and analyzing HIV transmission: The effect of contact patterns. *Mathematical Biosciences* 92 (2): 119–99.

Jensen, Arthur R. 1969. How much can we boost IQ and scholastic achievement? *Harvard Educational Review* 39 (1): 1–123.

———. 1970. IQ of identical twins reared apart. *Behavioral Genetics* 1:133–48.

Johnson, George. 2002. At Lawrence Berkeley, physicists say a colleague took them for a ride. *New York Times*, Oct. 15: F1 (accessed online at: http://sanacacio .net/118_saga/story.html).

Johnson, Kay M., Jorge Alarcón, Douglas M. Watts, Carlos Rodriguez, Carlos Velasquez, Jorge Sanchez, David Lockhart, Bradley P. Stoner, and King K. Holmes. 2003. Sexual networks of pregnant women with and without HIV infection. *AIDS* 17 (4): 605–12.

Kareiva, Peter. 1989. Renewing the dialogue between theory and experiments in population ecology. In *Perspectives in Ecological Theory*, ed. Jonathan Roughgarden, Robert M. May, and Simon A. Levin, 68–88. Princeton, NJ: Princeton University Press.

Karmiloff-Smith, Annette. 1994. Precis of beyond modularity: A developmental perspective on cognitive science. *Behavioral and Brain Sciences* 17 (4): 693–706.

Kasarda, John D., John O. G. Billy, and Kirsten West. 1986. *Status Enhancement and Fertility: Reproductive Responses to Social Mobility and Educational Opportunity*. Orlando, FL: Academic Press.

Katz, Barry P. 1992. Estimating transmission probabilities for chlamydial infection. *Statistics in Medicine* 11 (5): 565–77.

Katzner, Donald. In process. Culture and economic behavior (book manuscript).

Kavanagh, Etta, ed. 2006. Debating sexual selection and mating strategies (letters in response to Roughgarden et al. 2006). *Science* 312:689–97.

Kawai, Masao. 1965. Newly acquired precultural behavior of the natural troop of Japanese monkeys on Koshima islet. *Primates* 6:1–30.

Kawecki, Tadeusz J., and Robert D. Holt. 2002. Evolutionary consequences of asymmetric dispersal rates. *American Naturalist* 160 (3): 333–47.

Keesing, Roger. 1994. Theories of culture revisited. In *Assessing Cultural Anthropology*, ed. R. Borofsky, 301–12. New York: McGraw-Hill.

Kelly, Raymond C. 1985. *The Nuer Conquest: The Structure and Development of an Expansionist System*. Ann Arbor: University of Michigan Press.

Keyfitz, Nathan, and Wilhelm Flieger. 1990. *World Population Growth and Aging: Demographic Trends in the Late Twentieth Century*. Chicago: University of Chicago Press.

Khalid, Adeeb. 1999. *The Politics of Muslim Cultural Reform: Jadidism in Central Asia*. Berkeley: University of California Press.

Killingback, Timothy, and Etienne Studer. 2001. Spatial ultimatum games, collaborations and the evolution of fairness. *Proceedings of the Royal Society of London*, series B, Biological Sciences 268 (1478): 1797–1801.

Kirkpatrick, Mark, and Russell Lande. 1989. The evolution of maternal characters. *Evolution* 43: 485–503.

Klein, Richard G. 1999. *The Human Career: Human Biological and Cultural Origins*. 2nd ed. Chicago: University of Chicago Press.

Klitzman, Robert. 1998. *The Trembling Mountain: A Personal Account of Kuru, Cannibals, and Mad Cow Disease*. New York: Plenum.

Knauft, Bruce M. 1991. Violence and sociality in human evolution. *Current Anthropology* 32:391–428.

———. 1993. *South Coast New Guinea Cultures: History, Comparison, Dialectic*. Cambridge Studies in Social and Cultural Anthropology 89. Cambridge: Cambridge University Press.

Kohlberg, Lawrence. 1976. Moral stages and moralization. In *Moral Development and Behavior: Theory, Research, and Social Issues*, ed. T. Lickona, 31–53. New York: Holt, Rinehart, and Winston.

Kohn, Melvin, Atsushi Naoi, Carrie Schoenbach, Carmi Schooler, and Kazimierz Slomczynski. 1990. Position in the class structure and psychological functioning in the U.S., Japan, and Poland. *American Journal of Sociology* 95 (4): 964–1008.

Kottak, Conrad. 1997. *Anthropology: The Exploration of Human Diversity*. 7th ed. New York: McGraw-Hill.

Kowalski, Kurt, and Ya-Fen Lo. 2001. The influence of perceptual features, ethnic labels, and sociocultural information on the development of ethnic/racial bias in young children. *Journal of Cross-Cultural Psychology* 32 (4): 444–55.

Kraybill, Donald B., and Marc A. Olshan. 1994. *The Amish Struggle with Modernity*. Hanover, NH: University Press of New England.

Kroeber, Alfred L. 1923. *Anthropology*. New York: Harcourt, Brace.

———. 1925. *Handbook of the Indians of California*. New York: Dover, 1976.

———. 1945. The ancient Oikoumene as an historic culture aggregate. *Journal of the Royal Anthropological Institute* 75:9–20.

Kroeber, Alfred L., and Clyde Kluckhohn. 1952. *Culture: A Critical Review of Concepts and Definitions*. Papers of the Peabody Museum of American Archaeology and Ethnology, Harvard University, vol. 47, no. 1, Cambridge, MA: The Museum.

Kubler, Cornelius. 1985. *The Development of Mandarin in Taiwan: A Case Study of Language Contact*. Taipei: Student Book Co.

Kuhn, Thomas. 1962. *The Structure of Scientific Revolutions.*, 2nd ed. Chicago: University of Chicago Press, 1970.

Kumm, Jochen, Kevin N. Laland, and Marcus W. Feldman. 1994. Gene-culture coevolution and sex ratios—the effects of infanticide, sex-selective abortion, sex selection, and sex-biased parental investment on the evolution of sex ratios. *Theoretical Population Biology* 46 (3): 249–78.

Kuper, Adam. 1982. *Wives for Cattle*. Boston: Routledge and Kegan Paul.

———. 1999. *Culture: The Anthropologists' Account*. Cambridge, MA: Harvard University Press.

Labov, William. 2001. *Principles of Linguistic Change*, vol. 2, *Social Factors*. Language in Society 29. Malden, MA: Blackwell.

Laland, Kevin N., and Gillian R. Brown. 2002. *Sense and Nonsense: Evolutionary Perspectives on Human Behaviour*. Oxford: Oxford University Press.

Laland, Kevin N., Jochen Kumm, and Marcus W. Feldman. 1995. Gene-culture coevolutionary theory: A test case. *Current Anthropology* 36 (1): 131–56.

Laland, Kevin N., John Odling-Smee, and Marcus W. Feldman. 1996. The evolutionary consequences of niche construction: A theoretical investigation using two-locus theory. *Journal of Evolutionary Biology* 9:293–316.

———. 1999. Evolutionary consequences of niche construction and their implications for ecology. *Proceedings of the National Academy of Sciences (US)* 96:10242–47.

———. 2000. Niche construction, biological evolution, and cultural change. *Behavioral and Brain Sciences* 23 (1): 131–46.

Lamarck, Jean Baptiste Pierre Antoine de Monet de. 1984. *Zoological Philosophy: An Exposition with Regard to the Natural History of Animals*. Trans. Hugh Elliot;

intro. David L. Hull, Richard W. Burkhardt Jr. Chicago: University of Chicago Press.

Lancy, David F. 1996. *Playing on the Mother Ground: Cultural Routines for Children's Development*. London: Guilford.

Lande, Russell, and Douglas W. Schemske. 1985. The evolution of self-fertilization and inbreeding depression in plants. I. Genetic models. *Evolution* 39:24–40.

Lansing, J. Stephen, James N. Kremer, and Barbara B. Smuts. 1998. System-dependent selection, ecological feedback, and the emergence of functional structure in ecosystems. *Journal of Theoretical Biology* 192:377–91.

Lauden, Larry. 1990. *Science and Relativism: Some Key Controversies in the Philosophy of Science*. Chicago: University of Chicago Press.

Laumann, Edward O., John H. Gagnon, Robert T. Michael, and Stuart Michaels. 1994. *The Social Organization of Sexuality: Sexual Practices in the United States*. Chicago: University of Chicago Press.

Laumann, Edward O., and Yoosik Youm. 1999. Racial/ethnic group differences in the prevalence of sexually transmitted diseases in the United States: A network explanation. *Sexually Transmitted Diseases* 26 (5): 250–61.

Le Bon, Gustave. 1894. *The Psychology of Peoples*. New York: Arno Press, 1974.

Lee, Richard Borshay. 1979. *The !Kung San: Men, Women, and Work in a Foraging Society*. Cambridge, Cambridge University Press.

Levins, R. 1966. The strategy of model building in population biology. *American Scientist* 54: 421–31.

Lévi-Strauss, Claude. 1955. *Tristes Tropiques*. Trans. J. Weightman and D. Weightman. New York: Penguin, 1973.

———. 1964. *The Raw and the Cooked: Mythologiques Volume 1*. Trans. J. Weightman and D. Weightman. New York: Harper and Row, 1969.

———. 1969. *The Elementary Structures of Kinship*. Trans. J. H. Bell, J. R. von Sturmer, and R. Needham. Boston: Beacon.

Lewontin, Richard C. 1972. The apportionment of human genetic diversity. *Evolutionary Biology* 6:381–98.

———. 1974. *The Genetic Basis of Evolutionary Change*. New York: Columbia University Press.

———. 1983. Gene, organism, and environment. In *Evolution From Molecules to Men*, ed. D. S. Bendall, 273–85. Cambridge: Cambridge University Press.

————. 2005. The wars over evolution. *New York Review of Books* 52 (16), http://www.nybooks.com/articles/18363.

Li Jinghan. 1929. *Beiping jiaowai zhi xiangcun jiating* (Rural Families in Peri-urban Beijing). Beijing: Zhonghua Jiaoyu Wenhua Jijin Dongshihui, Shehui Diaocha Bu.

————. 1933. *Ding xian shehui gaikuang diaocha* (Ding Xian: A Social Survey). Beijing: Zhonghua Pingmin Jiaoyu Cujinhui.

Li Nan, Marcus W. Feldman, and Shripad Tuljapurkar. 1999. Nanhai pianhao he chusheng xingbie bi [Son preference and sex ratio at birth]. *Renkou yu jingji* [*Population and Economics*] 1999, supplemental issue, *Zhongguo nongcun nanhai pianhao wenhua de chuanbo yu yanhua* [*Cultural transmission and evolution of son preference in rural China*], 19–26.

Li Nan, Marcus W. Feldman, and Li Shuzhuo. 1999. Nanhai pianhao wenhua chuanbo: jiyu Zhongguo liangge xian diaocha de guji [Transmission of son preference: estimates from a survey in two counties of China]. *Renkou yu jingji* [*Population and Economics*] 1999, supplemental issue, *Zhongguo nongcun nanhai pianhao wenhua de chuanbo yu yanhua* [*Cultural transmission and evolution of son preference in rural China*], 48–58.

Li Shuzhuo, Marcus W. Feldman, and Nan Li. 2000. Cultural transmission of uxorilocal marriage in Lueyang, China. *Journal of Family History* 25 (2): 158–77.

————. 2003. Acceptance of two types of uxorilocal marriage in contemporary rural China: The case of Lueyang. *Journal of Family History* 28 (2): 314–33.

Liebenberg, Louis. 1990. *The Art of Tracking: The Origin of Science*. Cape Town, South Africa: David Philip Publishers.

Lieberman, Debra, John Tooby., and Leda Cosmides. 2003. Does morality have a biological basis? An empirical test of the factors governing moral sentiments relating to incest. *Proceedings of the Royal Society of London*, Series B, *Biological Sciences* 270:819–26.

Lindenbaum, Shirley. 1979. *Kuru Sorcery: Disease and Danger in the New Guinea Highlands*. Palo Alto, CA: Mayfield Publishing Co.

————. 2001. Kuru, prions, and human affairs: Thinking about epidemics. *Annual Review of Anthropology* 30:363–85.

————. 2004. Thinking about cannibalism. *Annual Review of Anthropology* 33:475–98.

Lipo, Carl P., Mark E. Madsen, Robert C. Dunnell, and Tim Hunt. 1997. Population structure, cultural transmission, and frequency seriation. *Journal of Anthropological Archaeology* 16:301–34.

Littlefield, Alice, and Hill Gates. 1991. *Marxist Approaches in Economic Anthropology*. Lanham, MD: University Press of America.

Longini, Ira M., Jr., Paul E. M. Fine, and Stephen B. Thacker. 1986. Predicting the global spread of new infectious agents. *American Journal of Epidemiology* 123 (3): 383–91.

Lowie, Robert H. 1920. *Primitive Society*. New York: Horace Liveright.

———. 1937. *The History of Ethnological Theory*. New York: Farrar and Rinehart.

Ludden, David. 1992. India's development regime. In *Colonialism and Culture*, ed. Nicholas Dirks, 247–87. Ann Arbor: University of Michigan Press.

Luhrmann, Tanya. 1996. *The Good Parsi: The Fate of a Colonial Elite in a Postcolonial Society*. Cambridge, MA: Harvard University Press.

Lukes, Steven. 1985. *Emile Durkheim: His Life and Work, a Historical and Critical Study*. Stanford, CA: Stanford University Press.

Lumsden, Charles J., and Edward O. Wilson. 1981. *Genes, Mind, and Culture: The Coevolutionary Process*. Cambridge, MA: Harvard University Press.

MacDonald, G. 1957. *The Epidemiology and Control of Malaria*. Oxford: Oxford University Press.

Mackie, Gerry. 1996. Ending footbinding and infibulation: A convention account. *American Sociological Review* 61 (6): 999–1017.

Mahajan, Vijay, and Robert A. Peterson. 1985. *Models for Innovation Diffusion*. Beverly Hills, CA: Sage.

Mahmoudi, Abdelrashid. 1998. *Taha Husain's Education: From the Azhar to the Sorbonne*. Surrey, UK: Curzon.

Malinowski, Bronislaw. 1922. *Argonauts of the Western Pacific*. New York: E.P. Dutton, 1961.

———. 1926. *Crime and Custom in Savage Society*. London: Kegan Paul, Trench, Truber.

———. 1929. *The Sexual Life of Savages in North-Western Melanesia: An Ethnographic Account of Courtship, Marriage, and Family Life Among the Natives of the Trobriand Islands, British New Guinea*. Boston: Beacon, 1987.

———. 1931. Culture. In *Selections from the Encyclopaedia of the Social Sciences* (7th ed.), 621–45. New York: Macmillan, 1937.

————. 1944. *A Scientific Theory of Culture and Other Essays*. London: Routledge, 2002.

Mandler, J. M., and N. S. Johnson. 1977. Remembrance of Things Parsed: Story Structure and Recall. *Cognitive Psychology* 9:111–51.

Mann, Charles C. 2001. Scientific community: Anthropological warfare. *Science* 291 (5503): 416–21.

Mann, F. O. 1932. *Report on Certain Aspects of Egyptian Education, Rendered to His Excellency, the Minister of Education at Cairo*. Cairo: Government Press.

Marks, John, and Edward Staski. 1988. Individuals and the evolution of biological and cultural systems. *Human Evolution* 3 (3): 147–61.

Martin, Emily. 1998. Anthropology and the cultural study of science. *Science, Technology and Human Values* 23 (1): 24–44.

Martrat, Belen, Joan O. Grimalt, Constancia Lopez-Martinez, Isabel Cacho, Francisco J. Sierro, Jose Abel Flores, Rainer Zahn, Miquel Canals, Jason H. Curtis, and David A. Hodell, 2004. Abrupt temperature changes in the Western Mediterranean over the past 250,000 years. *Science* 306 (5702): 1762–1765.

Marx, Karl. 2000. *Karl Marx: Selected Writings*. Ed. David McLellan. Oxford: Oxford University Press.

Maynard-Smith, John, and Eörs Szathmáry. 1995. *The Major Transitions in Evolution*. Oxford: Freeman.

McBrearty, Sally, and Alison S. Brooks. 2000. The revolution that wasn't: A new interpretation of the origin of modern human behavior. *Journal of Human Evolution* 39 (5): 453–563.

McCay, Bonnie J. 2002. Emergence of institutions for the commons: Contexts, situations, events. In *The Drama of the Commons*, ed. Elinor Ostrom, Thomas Dietz, Nives Dolšak, Paul C. Stern, Susan Stonich, Elke U. Weber, 361–402. Washington, DC: National Academy Press.

McCullagh, Peter John, and John A. Nelder. 1989. *Generalized Linear Models*. 2nd ed. London: Chapman and Hall.

McElreath, Richard. n.d. In the pastures and the fields: Ecology, community and cultural microevolution in Usangu, Tanzania (manuscript).

————. 2004. Social learning and the maintenance of cultural variation: An evolutionary model and data from East Africa. *American Anthropologist* 106 (2): 308–21.

McElreath, Richard, Robert Boyd, and Peter J. Richerson. 2003. Shared norms can lead to the evolution of ethnic markers. *Current Anthropology* 44 (1): 122–29.

McElreath, Richard, Mark Lubell, Peter J. Richerson, Timothy M. Waring, William Baum, Edward Edsten, Charles Efferson, and Brian Paciotti. 2005. Applying evolutionary models to the laboratory study of social learning. *Evolution and Human Behavior* 26:483–508.

McGrew, William C. 1974. Tool use by wild chimpanzees in feeding upon driver ants. *Journal of Human Evolution* 3:501–508.

———. 1992. *Chimpanzee Material Culture: Implications for Human Evolution.* Cambridge: Cambridge University Press.

McPherson, Miller, Lynn Smith-Lovin, and James M. Cook. 2001. Birds of a feather: Homophily in social networks. *Annual Review of Sociology* 27:415–44.

Mead, Margaret. 1928. *Coming of Age in Samoa.* New York: W. Morrow.

———. 1930. *Social Organization of Manu'a.* 2nd ed. Bernice P. Bishop Museum Bulletin 76. Honolulu: Bishop Museum Press, 1969.

Mead, Simone, Michael P. H. Stumpf, Jerome Whitfield, Jonathan A. Beck, Mark Poulter, Tracy Campbell, James B. Uphill, David Goldstein, Michael Alpers, Elizabeth M. C. Fisher, and John Collinge. 2003. Balancing selection at the prion protein gene consistent with prehistoric kurulike epidemics. *Science* 300 (5619): 640–43.

Meigs, Anna S. 1984. *Food, Sex, and Pollution: A New Guinea Religion.* New Brunswick, N.J.: Rutgers University Press.

Meltzoff, Andrew N. 2002. Elements of a developmental theory of imitation. In Meltzoff and Prinz 2002, 19–41.

Meltzoff, Andrew N., and Wolfgang Prinz, eds. 2002. *The Imitative Mind: Development, Evolution, and Brain Bases.* New York: Cambridge University Press.

Miller, Barbara. 2002. *Cultural Anthropology.* 2nd ed. Boston: Allyn and Bacon.

Miller, Neal E., and John Dollard. 1941. *Social Learning and Imitation.* New Haven, CT: Yale University Press.

Minturn, Leigh, and William Lambert, with John Fischer, Ann Fischer, Kimball Romney, Romaine Romney, William Nydegger, Corinne Nydegger, Thomas Maretzki, Hatsumi Maretzki, Robert LeVine, and Barbara LeVine. 1964. *Mothers of Six Cultures: Antecedents of Child Rearing.* New York: John Wiley and Sons.

Minturn, Leigh, and Lapporte, R. 1985. A new look at the universal incest taboo. In *A Different Perspective: Studies of Behavior across Cultures. Selected Papers from the Seventh Conference of International Association of Cross-Cultural Psychology*, ed. I. Reyes Lagunes and Y. H. Poorting, 159–74. Berwyn, PA: Swets North America.

Mintz, Sidney. 1985. *Sweetness and Power: The Place of Sugar in Modern History*. New York: Viking.

Mintz, Sidney, and Eric Wolf. 1989. Reply to Taussig. *Critique of Anthropology* 9(1), 25–31.

Mitchell, Richard. 1969. *The Society of the Muslim Brothers*. Oxford: Oxford University Press.

Mitchell, Timothy. 1988. *Colonising Egypt*. Cambridge: Cambridge University Press.

———. 2002. *Rule of Experts: Egypt, Techno-Politics, and Modernity*. Berkeley: University of California Press.

Mithen, Steven. 1996. *The Prehistory of Mind: The Cognitive Origin of Art and Science*. London: Thames and Hudson.

Montgomery, Mark R., Gebre-Egziabher Kiros, Dominic Agyeman, John B. Casterline, Peter Aglobitse, and Paul C. Hewett. 2001. Social networks and contraceptive dynamics in southern Ghana. Working paper 153, Policy Research Division, Population Council, New York.

Morris, Martina. 1991. A log-linear modeling framework for selective mixing. *Mathematical Biosciences* 107 (2): 349–77.

———. 1995. Data driven network models for the spread of disease. In *Epidemic Models: Their Structure and Relation to Data*, ed. D. Mollison, 302–22. Cambridge: Cambridge University Press.

———. 1996. Behaviour change and non-homogenous mixing. In *Models for Infectious Human Disseases*, ed. V. Isham and G. F. Medley, 239–52. Cambridge: Cambridge University Press.

Morris, Martina, and Laura Dean. 1994. Effect of sexual behavior change on long-term human immunodeficiency virus prevalence among homosexual men. *American Journal of Epidemiology* 140 (3): 217–32.

Morris, Martina, and Mirjam Kretzschmar. 1995. Concurrent partnerships and transmission dynamics in networks. *Social Networks* 17 (3–4): 299–318.

———. 1997. Concurrent partnerships and the spread of HIV. *AIDS* 11 (5): 641–48.

Morton, Newton E., James F. Crow, and H. J. Muller. 1956. An estimate of the mutational damage in man from data on consanguineous marriages. *Proceedings of the National Academy of Sciences (US)* 42:855–63.

Moss, Louis, and Harvey Goldstein. 1979. *The Recall Method in Social Surveys.* London: University of London Institute of Education.

Murdock, George Peter. 1949. *Social Structure.* London: Macmillan.

MXDB. 1918. *Minshangshi Xiguan Diaocha Baogaolu* (Report of research on commercial and customary law). Nanking: Supreme Court of China.

Nanda, Serena, and Richard Warms. 2002. *Cultural Anthropology.* 7th ed. Belmont, CA: Wadsworth/Thomson Learning.

al-Naqa, Mahmud Kamil, et al. 1988–89. *Al-tarbiya al-islamiyya.* Second grade. Cairo: Al-Jihaz al-Markazi lil-Kutub al-Jami'iyya wa al-Madrasiyya wa al-Wasa'il al-Ta'limiyya.

Neiman, Fraser D. 1995. Stylistic variation in evolutionary perspective: implications for Middle Woodland ceramic diversity. *American Antiquity* 60: 7–36.

Nelson, Cary, and Lawrence Grossberg. 1988. *Marxism and the Interpretation of Culture.* Urbana: University of Illinois Press.

Nettle, Daniel. 1997. On the status of methodological individualism. *Current Anthropology* 38 (2): 283–86.

Newman, M. E. J. 2002. Spread of epidemic disease on networks. *Physical Review E* 66 (1): art. no. 016128.

Newson, Lesley, Tom Postmes, S.E.G. Lea, Peter J. Richerson, and Richard McElreath. 2007. Influences on communication about reproduction: The cultural evolution of low fertility. *Evolution and Human Behavior* 28 (3): 199–210.

Nisbett, Richard E., and Dov Cohen. 1996. *The Culture of Honor: The Psychology of Violence in the South.* Boulder CO: Westview Press.

Nishida, Toshisada. 1987. Local traditions and cultural transmission. In *Primate Societies,* ed. S. S. Smuts, D. L. Cheney, R. M. Seyfarth, R. W. Wrangham, and T. T. Strusaker, 462–74. Chicago: University of Chicago Press.

Nowak, Martin A., and Karl Sigmund. 1998. Evolution of indirect reciprocity by image scoring. *Nature* 393:573–77.

Nowak, Martin A., Karen M. Page, and Karl Sigmund. 2000. Fairness versus reason in the ultimatum game. *Science* 289 (5485): 1773–75.

O'Brien, Michael J., and R. Lee Lyman. 2000. *Applying evolutionary archaeology: A systematic approach*. New York: Kluwer Academic/ Plenum.

———. 2003. *Cladistics and archaeology*. Salt Lake City: University of Utah Press.

Odling-Smee, F. John. 1988. Niche constructing phenotypes. In *The Role of Behavior in Evolution*, ed. H. C. Plotkin. Cambridge, MA: MIT Press.

Odling-Smee, F. John., Kevin N. Laland, and Marcus W. Feldman. 2003. *Niche Construction: The Neglected Process in Evolution*. Princeton, NJ: Princeton University Press.

O'Meara, Tim. 1997. Causation and the struggle for a science of culture. *Current Anthropology* 38 (3): 399–418.

Oota, Hiroki, Wannapa Settheetham-Ishida, Danai Tiwaweck, Takafumi Ishida, and Mark Stoneking. 2001. Human mtDNA and Y-chromosome variation is correlated with matrilocal versus patrilocal residence. *Nature Genetics* 29:20–21.

Ortner, Sherry B. 1997. Introduction. *Representations* 59, special issue, *The Fate of "Culture": Geertz and Beyond*, 1–13.

Ostrom, Elinor. 1998. A behavioral approach to the rational choice theory of collective action (presidential address, American Political Science Association, 1997). *American Political Science Review* 92 (1): 1–22.

Panger, Melissa A., Susan Perry, Lisa Rose, Julie Gros-Louis, Erin Vogel, Katherine C. MacKinnon, and Mary Baker. 2002. Cross-site differences in foraging behavior of white-faced capuchins (*Cebus capucinus*). *American Journal of Physical Anthropology*, 119:52–66.

Parsons, Talcott. 1951. *The Social System*. New York: Free Press.

———. 1961. On social structure. In *Theories of Society: Foundations of Modern Sociological Theory*, ed. T. Parsons, E. Shils, K. Naegele, J. Pitts, 102–34. New York: Free Press.

Passmore, John. 1967. Logical positivism. In *The Encyclopedia of Philosophy*, ed. Paul Edwards, vol. 5, 52–57. New York: Macmillan.

Pelto, Perti, and Gretel Pelto. 1975. Intra-cultural diversity: Some theoretical issues. *American Ethnologist* 2 (1): 1–18.

Pennisi, Elizabeth. 2003. Cannibalism and prion disease may have been rampant in ancient humans. *Science* 300:227–28.

Peoples, James, and Garrick Bailey. 2000. *Humanity: An Introduction to Cultural Anthropology*. 5th ed. Belmont, CA: Wadswoth/Thomson Learning.

Perry, Susan, Mary Baker, Linda Fedigan, Julie Gros-Louis, Katherine Jack, Katherine C. MacKinnon, Joseph H. Manson, Melissa Panger, Kendra Pyle, and Lisa Rose. 2003. Social conventions in wild white-faced capuchin monkeys: Evidence for traditions in a neotropical primate. *Current Anthropology* 44 (2): 241–68.

Perry, Susan, and Joseph H. Manson. 2003. Traditions in monkeys. *Evolutionary Anthropology* 12 (2): 71–81.

Peter, Karl A. 1987. *The Dynamics of Hutterite Society: An Analytical Approach.* Edmonton: University of Alberta Press.

Peterman, T. A., R. L. Stoneburner, J. R. Allen, H. W. Jaffe, and J. W. Curran. 1988. Risk of human immunodeficiency virus transmission from heterosexual adults with transfusion-associated infections. *Journal of the American Medical Association* 259 (1): 55–58.

Peters, Charles R. 1987. Nut-like oil seeds: Food for monkeys, chimpanzees, humans, and probably ape-man. *American Journal of Physical Anthropology* 73:333–63.

Petroski, Henry. 1992. *The Evolution of Useful Things.* New York: Vintage.

Pinker, Steven. 2002. *The Blank Slate: The Modern Denial of Human Nature.* New York: Viking.

Plummer, L., J. J. Potterat, S. Q. Muth, J. B. Muth, and W. W. Darrow. 1996. Providing support and assistance for low-income or homeless women. *Journal of the American Medical Association* 276 (23): 1874–75.

Polanich, Judith K. 1995. The origins of Western mono coiled basketry: A reconstruction of prehistoric change in material culture. *Museum Anthropology* 19 (3): 57–68.

Polsby, Nelson W. 2004. *How Congress Evolves: Social Bases of Institutional Change.* Oxford: Oxford University Press.

Popper, Karl. 1959. *The Logic of Scientific Discovery.* London: Routledge, 1992.
———. 1994. *The Myth of the Framework: In Defence of Science and Rationality.* London: Routledge.

Porac, Clare, and Stanley Coren. 1981. Life-span age trends in the perception of the Mueller-Lyer—Additional evidence for the existence of 2 illusions. *Canadian Journal of Psychology—Revue Canadienne de Psychologie* 35 (1): 58–62.

Poser, Charles M. 2002. Notes on the history of the prion diseases, Parts I and II. *Clinical Neurology and Neurosurgery* 104:1–9, 77–86.

Potts, Richard. 1996. Evolution and climate variability. *Science* 273 (5277): 922–23.

Price, David. 2004. *Threatening Anthropology: McCarthyism and the FBI's Surveillance of Activist Anthropologists.* Durham, NC: Duke University Press.

Price, David, and William J. Peace. 2003. Un-American anthropological thought: The Opler/Meggers exchange. *Journal of Anthropological Research* 59 (2): 183–203.

Price, George R. 1970. Selection and covariance. *Nature* 227:520–21.

Prusiner, Stanley B. 1998. Prions (Nobel lecture). *Proceedings of the National Academy of Sciences (US)* 95: 13363–83.

———. 2001a. Shattuck lecture—Neurodegenerative diseases and prions. *New England Journal of Medicine* 344 (2): 1516–26.

———. 2001b. Prion diseases. In *The Metabolic and Molecular Basis of Inherited Disease,* 8th ed, ed. C. Scriver, A. L. Beaudet, W. S. Sly, D. Valle, B. Childs, K. W. Kinzler, and B. Vogelstein, 5703–28. New York: McGraw-Hill.

Pulliam, H. Ronald, and Christopher Dunford. 1980. *Programmed to Learn: An Essay on the Evolution of Culture.* New York, Columbia University Press.

Pusey, Anne E. 1980. Inbreeding avoidance in chimpanzees. *Animal Behavior* 28:543–52.

Quartz, Steven R. 1999. The constructivist brain. *Trends in Cognitive Sciences* 3 (2): 48–57.

———. 2002. *Liars, Lovers, and Heroes: What the New Brain Science Reveals about How We Become Who We Are.* New York: William Morrow.

Quartz, Steven R., and Terrence J. Sejnowski. 1997. The neural basis of cognitive development: A constructivist manifesto. *Behavioral and Brain Sciences* 20 (4): 537–96.

———. 2000. Constraining constructivism: Cortical and sub-cortical constraints on learning in development. *Behavioral and Brain Sciences* 23 (5): 785–92.

Rabinow, Paul, and William M. Sullivan. 1987. The interpretive turn: A second look. In *Interpretive Social Science: A Second Look,* ed. P. Rabinow and W. M. Sullivan, 1–30. Berkeley, CA: University of California Press.

Radcliffe-Brown, Alfred R. 1945. Religion and society. In *Structure and Function in Primitive Society,* A.R. Radcliffe-Brown, 153 – 177. New York: Free Press, 1965.

————. 1952. *Structure and Function in Primitive Society.* New York: Free Press, 1965.

Radin, Paul. 1926. *Crashing Thunder: The Autobiography of an American Indian.* New York: Appleton.

————. 1933. *The Method and Theory of Ethnology.* New York: McGraw-Hill.

————. 1937. *Primitive Religion.* New York: Viking.

Ramenofsky, Ann, and Anastasia Steffen, eds. 1998. *Unit Issues in Archaeology: Measuring Time, Space, and Material.* Salt Lake City: University of Utah Press.

Redfield, Robert. 1963. The social uses of social science. In *The Social Uses of Social Science: The Papers of Robert Redfield,* vol. 2, ed. Margaret Park Redfield, ed., 191–98. Chicago: University of Chicago Press.

Rendell, Luke, and Hal Whitehead. 2001. Culture in whales and dolphins. *Behavioral and Brain Sciences* 24 (3): 309–24.

Reyna, Stephen P. 1994. Literary anthropology and the case against science. *Man* 29:555–81.

Rhodes, Richard. 1997. *Deadly Feasts: Tracking the Secrets of a Terrifying New Plague.* New York: Simon and Schuster.

Richards, Robert J. 1987. *Darwin and the Emergence of Evolutionary Theories of Mind and Behavior.* Chicago: University of Chicago Press.

Richerson, Peter J., and Robert Boyd. 1987. Simple models of complex phenomena: The case of cultural evolution. In *The Latest on the Best: Essays on Evolution and Optimality,* ed. J. Dupré. Cambridge, MA: MIT Press.

————. 1989. "The role of evolved predispositions in cultural evolution: Or, human sociobiology meets Pascal's wager." *Ethology and Sociobiology* 10: 195–219.

————. 1992. Cultural inheritance and evolutionary ecology. In Smith and Winterhalder 1992, 61–92. New York: Aldine De Gruyter.

————. 1998. The evolution of ultrasociality. In *Indoctrinability, Ideology and Warfare,* ed. I. Eibl-Eibesfeldt and F. K. Salte, 71–96. New York: Berghahn.

————. 1999. Complex societies: the evolutionary dynamics of a crude superorganism. *Human Nature* 10:253–89.

————. 2000. Built for speed: Pleistocene climate variation and the origin of human culture. In *Perspectives in Ethology 13: Evolution, Culture, and Behavior,* ed. F. Tonneau and N. S. Thompson, 1–45. New York: Kluwer Academic/Plenum.

————. 2001a. The evolution of subjective commitment to groups: A tribal instincts hypothesis. In *Evolution and the Capacity for Commitment*, ed. R. M. Nesse, 186–220. New York: Russell Sage Foundation.

————. 2001b. Institutional evolution in the Holocene: The rise of complex societies. In *The Origin of Human Social Institutions*, ed. W. G. Runciman, 197–234. Oxford: Oxford University Press.

————. 2005. *Not By Genes Alone: How Cultural Transformed Human Evolution.* Chicago: University of Chicago Press.

Richerson, Peter J., Robert Boyd, and Robert L. Bettinger. 2001. Was agriculture impossible during the Pleistocene but mandatory during the Holocene? A climate change hypothesis. *American Antiquity* 66 (3): 387–411.

Ridley, Rosalind, and Harry Baker. 1998. *Fatal Protein: The Story of CJD, BSE, and Other Prion Diseases.* Oxford: Oxford University Press.

Robbins, Joel. 1998. Becoming sinners, Christianity and desire among the Urapmin of Papua New Guinea. *Ethnology* 37 (4): 299–316.

Robbins, Richard H. 2001. *Cultural Anthropology: A Problem-Based Approach.* 3rd ed. Itasca, IL: F. E. Peacock.

Robinson, Gaden. 2002. How a moth lost its spots: Fantasy and fraud in the stories of evolution. *Times Literary Supplement.* July 19: 3–4.

Roemer, John E. 2002. The democratic dynamics of educational investment and income distribution. Departments of Economics and Political Science, Yale University 2002, http://www.nyu.edu/gsas/dept/politics/seminars/roemer.pdf.

Rogers, Everett M. 1995. *Diffusion of Innovations.* 4th ed. New York: Free Press.

————. 2003. *Diffusion of Innovations.* 5th ed. New York: Free Press.

Rogers, Everett M., and F. Floyd Shoemaker. 1971. *Communication of Innovations: A Cross-cultural Approach.* 2nd ed. New York: Free Press.

Romney, A. Kimball. 1971. Measuring endogamy. In *Explorations in Mathematical Anthropology,* ed. P. Kay. Cambridge, MA: MIT Press.

————. 1999. Culture consensus as a statistical model. *Current Anthropology* 40: S103–S115.

Romney, A. Kimball, Susan C. Weller, and William H. Batchelder. 1986. Culture as consensus: A theory of culture and informant accuracy. *American Anthropologist* 88 (2): 313–38.

Roscoe, Paul. 1995. The perils of "positivism" in cultural anthropology. *American Anthropologist* 97 (3): 492–504.

Rosenberg, Noah A., Jonathan K. Pritchard, James L. Weber, Howard M. Cann, Kenneth K. Kidd, Lev A. Zhivotovsky, and Marcus W. Feldman. 2002. Genetic structure of human populations. *Science* 298 (5602): 2381–85.

Rosenthal, Ted L., and Barry J. Zimmerman. 1978. *Social Learning and Cognition.* New York: Academic Press.

Ross, Lee, and Richard E. Nisbett. 1991. *The Person and the Situation: Perspectives of Social Psychology.* Philadelphia: Temple University Press.

Roth, Alvin E. 1995. Bargaining experiments. In *The Handbook of Experimental Economics,* ed. J. H. Kagel and A. E. Roth, 253–48. Princeton, NJ: Princeton University Press.

Roth, Alvin E., Vesna Prasnikar, Masahiro Okuno-Fujiwara, and Shmuel Zamir. 1991. Bargaining and market behavior in Jerusalem, Ljubljana, Pittsburgh, and Tokyo: An experimental study. *American Economic Review* 81 (5): 1068–95.

Roughgarden, Joan. 2004. *Evolution's Rainbow: Diversity, Gender, and Sexuality in Nature and People.* Berkeley: University of California Press.

Roughgarden, Joan, Meeko Oishi, and Erol Akcay. 2006. Reproductive social behavior: Cooperative games to replace sexual selection. *Science* 311 (5763): 965–69.

Rushton, J. Philippe. 2000. *Race, Evolution, and Behavior: A Life History Perspective.* 3rd ed. Port Huron, MI: Charles Darwin Research Institute.

Russell, Diana E. H. 1984. The prevalence and seriousness of incestuous abuse: Stepfathers vs. biological fathers. *Child Abuse and Neglect* 8:15–22.

Sahlins, Marshall. 1976a. *Culture and Practical Reason.* Chicago: University of Chicago Press.

———. 1976b. *The Use and Abuse of Biology: An Anthropological Critique of Sociobiology.* London: Tavistock.

———. 1995. *How Natives Think: About Captain Cook, For Example.* Chicago: University of Chicago Press.

Salama, Ibrahim. 1939. *L'Enseignement islamique en Egypte: Son evolution, son influence sur les programmes modernes.* Cairo: Imprimerie Nationale, Boulaq.

Salmoni, Barak. 2001. The miniature society and its surroundings: Schools, students, and homes (manuscript).

Sangren, P. Steven. 1988. Rhetoric and the authority of ethnography. *Current Anthropology* 29 (3): 405–35.

———. 1995. "Power" against ideology: A critique of Focaultian usage. *Cultural Anthropology* 10 (1): 3–40.

Sapir, Edward. 1907. Religious ideas of the Takelma Indians of southwestern Oregon. *Journal of American Folk-lore* 20 (76): 33–49.

———. 1916a. Terms of relationship and the levirate. *American Anthropologist* 18 (3): 327–37.

———. 1916b. Time perspective in aboriginal American culture: A study in method. In *Selected Writings of Edward Sapir in Language, Culture, and Personality*, ed. David G. Mandelbaum, 389–462. Berkeley: University of California Press.

———. 1924. Culture, genuine and spurious. In *Selected Writings of Edward Sapir in Language, Culture, and Personality*, ed. David G. Mandelbaum, 308–31. Berkeley: University of California Press, 1949.

———. 1938. Why cultural anthropology needs the psychiatrist. In *Selected Writings of Edward Sapir in Language, Culture, and Personality*, ed. David G. Mandelbaum, 569–7. Berkeley: University of California Press, 1949.

Sariola, Heikki, and Anntti Uutela. 1996. The prevalence and context of incest abuse in Finland. *Child Abuse and Neglect* 20:843–50.

Savage-Rumbaugh, E. Sue, Rose A. Sevcik, D. M. Rumbaugh, and Elizabeth Rupert. 1985. The capacity of animals to acquire language: Do species differences have anything to say to us? *Philosophical Transcriptions of the Royal Society of London* 308:177–85.

Saxe, Geoffrey B. 1981. Body parts as numerals: A developmental analysis of numeration among the Oksapmin in Papua New Guinea. *Child Development* 52:306–16.

Scheidel, Walter. 1995. Incest revisited: Three notes on the demography of sibling marriage in Roman Egypt. *Bulletin of the American Society of Papyrologists* 32:143–55.

Scheper-Hughes, Nancy. 1995. The primacy of the ethical: Propositions for a militant anthropology. *Current Anthropology* 36 (3): 409–40.

Schmidt, Wilhelm. 1931. *The Origin and Growth of Religion: Facts and Theories.* Trans. H. J. Rose. London: Methuen.

———. 1939. *The Culture Historical Method of Ethnology.* Trans. S. A. Sieber. New York: Fortuny's.

Schneider, David M. 1980. *American Kinship: A Cultural Account*. Chicago: University of Chicago Press.

———. 1984. *A Critique on the Study of Kinship*. Ann Arbor: University of Michigan Press.

Schneider, Jane. 1987. The anthropology of cloth. *Annual Review of Anthropology* 16:409–48.

Scholz, Christopher A. 2005. Lake Malawi Drilling Project. http://malawi drilling.syr.edu (accessed October 2005).

Schwartz, Maxime. 2003. *How the Cows Turned Mad*. Trans. E. Schneider. Berkeley: University of California Press.

Searle, John R. 1995. *The Construction of Social Reality*. New York: Free Press.

Sedra, Paul. 2001. Modernity's mission: Evangelical efforts to discipline the nineteenth-century Coptic community. *Columbia International Affairs Online*, http://www.ciaonet.org/conf/mei01.sep01.html.

Segall, Marshall, Donald Campbell, and Merville Herskovits. 1966. *The Influence of Culture on Visual Perception*. New York: Bobbs-Merrill.

Seielstad, Mark T., Eric Minch, and Luigi L. Cavalli-Sforza. 1998. Genetic evidence for a higher female migration rate in humans. *Nature Genetics* 20: 278–80.

Shapin, Steven. 1994. *A Social History of Truth: Civility and Science in Seventeenth-Century England*. Chicago: University of Chicago Press.

Shennan, Stephen J. 2002. *Genes, Memes, and Human History: Darwinian Archaeology and Cultural Evolution*. London: Thames and Hudson.

Shennan, Stephen J., and James Steele. 1999. Cultural learning in hominids: A behavioral ecological approach. In *Mammalian Social Learning: Comparative and Ecological Perspectives*, ed. H. Box and K. Gibson, 367–88. Cambridge: Cambridge University Press.

Shennan, Stephen J., and J. R. Wilkinson. 2001. Ceramic style change and neutral evolution: a case study from Neolithic Europe. *American Antiquity* 66:577–593.

Shepher, Joseph. 1971. Mate selection among second generation kibbutz adolescents and adults: Incest avoidance and negative imprinting. *Archives of Sexual Behavior* 1:293–307.

———. 1983. *Incest: A Biosocial View*. New York: Academic Press.

Shepherd, John Robert. 1993. *Statecraft and Political Economy on the Taiwan Frontier, 1600–1800.* Stanford, CA: Stanford University Press.

———. 1995. *Marriage and Mandatory Abortion among the Seventeenth Century Siraya,* American Ethnological Society Monograph 6. Arlington, VA: American Anthropological Association.

Silberbauer, George B. 1981. *Hunter and Habitat in the Central Kalahari Desert.* New York: Cambridge University Press.

Skinner, G. William. 1997. Family systems and demographic processes. In *Anthropological Demography: Toward a New Synthesis,* ed. D. I. Kertzer and T. Fricke, 53–95. Chicago: University of Chicago Press.

Skinner, G. William, and Y. Jianhua. n.d. Reproduction in a patrilineal joint family system: Chinese in the lower Yangzi macroregion (manuscript).

Smith, Eric Alden. 1991. *Inujjuamiut Foraging Strategies: Evolutionary Ecology of an Arctic Hunting Economy.* New York: Aldine de Gruyter.

Smith, Eric Alden, and Rebecca L. Bliege Bird. 2000. Turtle hunting and tombstone opening: Public generosity as costly signaling. *Evolution and Human Behavior* 21 (4): 245–61.

Smith, Eric Alden, and Bruce Winterhalder, eds. 1981. *Hunter-gatherer Foraging Strategies: Ethnographic and Archaeological Analyses.* Chicago: University of Chicago Press.

———, eds. 1992. *Evolutionary Ecology and Human Behavior.* New York: Aldine De Gruyter.

Snijders, Tom A. B., Philippa E. Pattison, Garry L. Robins, and Mark S. Handcock. 2004. New specifications for exponential random graph models. Center for Statistics and the Social Sciences Working Paper 42. Seattle: University of Washington.

Snow, Charles Percy. 1998. *The Two Cultures.* Cambridge: Cambridge University Press.

Sober, Elliot. 1991. Models of cultural evolution. In *Trees of Life: Essays in Philosophy of Biology,* ed. P. Griffiths, 17–38. Dordrecht, The Netherlands: Kluwer.

Soltis, Joseph, Robert Boyd, and Peter J. Richerson. 1995. Can group-functional behaviors evolve by cultural group election? An empirical test. *Current Anthropology* 36 (3): 473–94.

Sonbol, Amira al-Azhari. 1988. Egypt. In *The Politics of Islamic Revivalism*, ed. Shireen Hunter, 23–38. Bloomington: Indiana University Press.

Spain, David H. 1987. The Westermarck-Freud incest-theory debate: An evaluation and reformulation. *Current Anthropology* 28 (5): 628–29.

Spelke, Elizabeth S. 1994. Initial knowledge: Six suggestions. *Cognition* 50: 443–47.

Sperber, Dan. 1985. Anthropology and psychology: Toward an epidemiology of representations. *Man*, n.s. 20 (1): 73–89.

———. 1996. *Explaining Culture: A Naturalistic Approach*. Oxford: Blackwell.

Spiro, Melford E. 1958. *Children of the Kibbutz*. Cambridge, MA: Harvard University Press.

———. 1987. *Culture and Human Nature*. Chicago: University of Chicago Press.

———. 1997. *Gender Ideology and Psychological Reality*. New Haven, CT: Yale University Press.

Spradley, James. 1979. *The Ethnographic Interview*. New York: Holt, Rinehart, and Winston.

Starrett, Gregory. 1998. *Putting Islam to Work: Education, Politics, and Religious Transformation in Egypt*. Berkeley: University of California Press.

Steadman, Lyle B., and Charles F. Merbs. 1982. Kuru and cannibalism? *American Anthropologist* 84 (3): 611–27.

Sterling, Sarah. 1999. Mortality profiles as indicators of slowed reproductive rates: Evidence from ancient Eygpt. *Journal of Anthropological Archaeology* 18 (3): 319–343.

Stewart, Kelly J., and Alexander H. Harcourt. 1987. Gorillas: Variations in female relationships. In *Primate Societies*, ed. B. B. Smuts, D. L. Cheney, R. M. Seyfarth, R. W. Wrangham, and T. T. Struhsaker, 155–64. Chicago: University of Chicago Press.

Stocking, George W., Jr. 1968. *Race, Culture, and Evolution: Essays in the History of Anthropology*. Chicago: University of Chicago Press.

———. 1974. *The Shaping of American Anthropology 1881–1911: A Franz Boas Reader*. New York: Basic.

———. 1992. *The Ethnographer's Magic and Other Essays in the History of Anthropology*. Madison: University of Wisconsin Press.

Stoneking, Mark. 2003. Widespread prehistoric cannibalism: Easier to swallow? *Trends in Ecology and Evolution* 18 (10): 489–90.

Stoner, Bradley P., W. L. H. Whitington, S. O. Aral, J. Hughes, H. H. Handsfield, and K. K. Holmes. 2003. Avoiding risky sex partners: Perception of partners' risks vs. partners' self-reported risks. *Sexually Transmitted Infections* 79:197–201.

Strathern, Marilyn. 1988. *The Gender of the Gift*. Berkeley: University of California Press.

Strauss, Claudia. 1992a. Models and motives. In D'Andrade and Strauss 1992, 1–20.

———. 1992b. What makes Tony run? Schemas as motives reconsidered. In D'Andrade and Strauss 1992, 197–224.

Strauss, Claudia, and Naomi Quinn. 1997. *A Cognitive Theory of Cultural Meaning*. Cambridge: Cambridge University Press.

Sugiyama, Yukimaru. 1981. Observations on the population dynamics and behavior of wild chimpanzees of Bossou, Guinea, 1979–1980. *Primates* 22 (4): 435–44.

———. 1994. Tool use by wild chimpanzees. *Nature* 367: 327.

Sugiyama, Yukimaru, and Jeremy Koman. 1979. Social structure and dynamics of wild chimpanzees at Bossou, Guinea. *Primates* 20 (3): 323–39.

Supreme Court of the Republic of China. 1918. See MXDB

Tagg, Stephen K. 1985. Life story interviews and their interpretation. In *The Research Interview: Uses and Approaches*, ed. M. Brenner, J. Brown, and D. Canter, 163–99. Orlando, FL: Academic Press.

Talmon, Yonina. 1964. Mate selection on collective settlements. *American Sociological Review* 29 (4): 491–508.

Tanaka, Mark M., Jochen Kumm, and Marcus W. Feldman. 2002. Coevolution of pathogens and cultural practices: A new look at behavioral heterogeneity in epidemics. *Theoretical Population Biology* 62:111–29.

Taussig, Michael. 1980. *The Devil and Commodity Fetishism in South America*. Chapel Hill: University of North Carolina Press.

———. 1989. History as commodity: In some recent (American) anthropological literature. *Critique of Anthropology* 9 (1): 25–31.

Temerlin, Maurice K. 1975. *Lucy: Growing Up Human, a Chimpanzee Daughter in a Psychotherapist's Family*. Palo Alto, CA: Science and Behavior Books.

Thomas, Keith. 1971. *Religion and the Decline of Magic*. Oxford: Oxford University Press.

Thomason, Sarah Grey. 2001. *Language Contact*. Washington, DC: Georgetown University Press.

Thornhill, Nancy Wilmsen. 1991. An evolutionary analysis of rules regulating human inbreeding and marriage. *Behavioral and Brain Sciences* 14:247–93.

———. 1993. *The Natural History of Inbreeding and Outbreeding*. Chicago: University of Chicago Press.

Tierney, Patrick. 2000a. *Darkness in El Dorado: How Scientists and Journalists Devastated the Amazon*. New York: W. W. Norton.

———. 2000b. The fierce anthropologist. *New Yorker* October 9: 50–61.

Tishkoff, Sarah A., Floyd A. Reed, Alessia Ranciaro, Benjamin F. Voight, Courtney C. Babbitt, Jesse S. Silverman, Kweli Powell, Holly M. Mortensen, Jibril B. Hirbo, Maha Osman, Muntaser Ibrahim, Sabah A. Omar, Godfrey Lema, Thomas B. Nyambo, Jilur Ghori, Suzannah Bumpstead, Jonathan K. Pritchard, Gregory A. Wray & Panos Deloukas. 2007. Convergent adaptation of human lactase persistence in Africa and Europe. *Nature Genetics* 39:31–40.

Tomasello, Michael. 1990. Cultural transmission in tool use and communicatory signaling of chimpanzees? In *Comparative Developmental Psychology of Language and Intelligence in Primates*, ed. S. Parker and K. Gibson, 274–311. Cambridge: Cambridge University Press.

———. 1996. Do apes ape? In *Social Learning in Animals: The Roots of Culture*, ed. C. M. Heyes and B. G. Galef, Jr., 319–346. New York: Academic Press.

———. 1999a. *The Cultural Origins of Human Cognition*. Cambridge, MA: Harvard University Press.

———. 1999b. The human adaptation for culture. *Annual Review of Anthropology* 28:509–29.

———. 2000. Culture and cognitive development. *Current Directions in Psychological Science* 9 (2): 37–40.

Tomasello, Michael, and Josep Call. 1997. *Primate Cognition*. New York: Oxford University Press.

Tooby, John, and Leda Cosmides. 1992. The psychological foundations of culture. In *The Adapted Mind: Evolutionary Psychology and the Generation of Culture*, ed. J. Barkow, L. Cosmides, and J. Tooby, 19–136. New York: Oxford University Press.

Toulmin, Stephen. 1990. *Cosmopolis: The Hidden Agenda of Modernity*. Chicago: University of Chicago Press.

Trigg, Roger. 2001. *Understanding Social Science.* 2nd ed. Oxford: Blackwell.

Trivers, Robert L. 1971. The evolution of reciprocal altruism. *Quarterly Review of Biology* 46:34–57.

Trouillot, Michel-Rolph. 2002. Adieu, culture: A new duty arises. In Fox and King 2002b, 37–60.

Truncer, James. In progress. Harappan cultural elaboration and demographics: An evolutionary perspective (article manuscript).

Turner, Jonathan H. 1995. *Macrodynamics: Toward a Theory on the Organization of Human Populations.* New Brunswick, NJ: Rutgers University Press.

Tylor, Edward B. 1871. *Primitive Culture: Research into the Development of Mythology, Philosophy, Religion, Art and Custom.* London: Murray.

Vandello, Joseph A., and Dov Cohen. 2004. When believing is seeing: Sustaining norms of violence in cultures of honor. In *The Psychological Foundations of Culture,* ed. M. Schaller and C. Crandall, 281–304. New York: Lawrence Erlbaum.

van Schaik, Carel P., M. Ancrenaz, G. Borgen, B. Galdikas, C. D. Knott, I. Singleton, A. Suzuki, S. S. Utami and M. Merrill. 2003. Orangutan cultures and the evolution of material culture. *Science* 299 (5603): 102–5.

van Schaik, Carel P., Robert Deaner, and Michelle Merrill. 1999. The conditions for tool use in primates: Implications for the evolution of material culture. *Journal of Human Evolution* 36 (6): 719–41.

Vayda, Andrew P. 1994. Actions, variations, and change: The emerging anti-essentialist view in anthropology. In Borofsky 1994, 320–29.

Verghese, Abraham. 2002. Wars are made, not born. *New York Times Book Review* September 29: 21.

Vrba, Elisabeth S., George H. Denton, Timothy C. Partridge, and Lloyd H. Burckle. 1995. *Paleoclimate and Evolution, with Emphasis on Human Origins.* New Haven, CT: Yale University Press.

de Waal, Frans B. M. 1999. Cultural primatology comes of age. *Nature* 399: 635–36.

———. 2001. *The Ape and the Sushi Master: Cultural Reflections of a Primatologist.* New York: Basic.

Wallace, Anthony F. C. 1952. *The Modal Personality Structure of the Tuscarora Indians, as Revealed by the Rorschach Test.* Smithsonian Institution, Bureau of American Ethnology, Bulletin 150. Washington, DC: US Government Printing Office.

————. 1961. *Culture and Personality.* New York: Random House.

————. 1966. *Religion: An Anthropological View.* New York: Random House.

————. 1969. *The Death and Rebirth of the Seneca.* New York: Vintage.

————. 1972. *Rockdale: The Growth of an American Village in the Early Industrial Revolution.* New York: Knopf.

————. 1981. *St. Clair: A Nineteenth-Century Coal Town's Experience with a Disaster-Prone Industry.* New York: Knopf.

Walters, S. A. 1942. A genetic study of geometrical-optical illusions. *Genetic Psychology Monograph* 25:101–55.

Wang Shih-ch'ing [Wang Shiqing]. 1996. *Danshui he liuyu hegang shuiyun shi* (History of Water Transportation among Ports on the Tamsui River). Taipei: Sun Yat-sen Institute for Social Sciences and Philosophy.

Wapner, Seymour H., and Heinz Werner. 1957. *Perceptual Development: An Investigation within the Framework of Sensory-Tonic Field Theory.* Worcester, MA: Clark University Press.

Wapner, Seymour H., Heinz Werner, and Paul Comali. 1960. Perception of part-whole relationship in middle and old age. *Journal of Gerontology* 15: 412–16.

Wasserman, Stanley, and Katherine Faust. 1994. *Social Network Analysis: Methods and Applications.* Cambridge: Cambridge University Press.

Wasserman, Stanley, and Philippa Pattison. 1996. Logit models and logistic regressions for social networks: An introduction to Markov graphs and p*. *Psychometrika* 61 (3): 401–25.

Watson, James L. 1988. The structure of Chinese funerary rites: Elementary forms, ritual sequence, and the primacy of performance. In *Death Ritual in Late Imperial and Modern China,* ed. James L. Watson and Evelyn S. Rawski, 3–19. Berkeley: University of California Press.

————. 1993. Rites or beliefs? The construction of a unified culture in late imperial China. In *China's Quest for National Identity,* ed. Lowell Dittmer and Samuel S. Kim, 80–103. Ithaca, NY: Cornell University Press.

Weber, Max. 1922a. *The Sociology of Religion.* Trans. Ephraim Fischoff. Boston: Beacon, 1963.

————. 1922b. *Economy and Society.* Ed. Guenther Roth and Claus Wittich, trans. Ephraim Fischoff et al. Berkeley: University of California Press, 1978.

Weiner, Jonathan. 1994. *The Beak of the Finch: A Story of Evolution in Our Time*. New York: Knopf.

Westermarck, Edward. 1891. *The History of Human Marriage*. London: Macmillan.

———. 1926. *A Short History of Human Marriage*. New York: Macmillan.

Western, Bruce. 2002. The impact of incarceration on wage mobility and inequality. *American Sociological Review* 67 (4): 526–46.

White, Leslie. 1949. *The Science of Culture: A Study of Man and Civilization*. New York: Farrar, Straus.

Whiten, Andrew, and Christophe Boesch. 2001. The cultures of chimpanzees. *Scientific American* 284 (1): 48–55.

Whiten, Andrew, and Richard W. Byrne. 1988. *Machiavellian Intelligence: Social Expertise and the Evolution of Intellect in Monkeys, Apes, and Humans*. Oxford: Oxford University Press.

Whiten, Andrew, and Deborah M. Custance. 1996. Studies of imitation in chimpanzees and children. In Heyes and Galef 1996, 291–318.

Whiten, Andrew, Deborah M. Custance, Juan-Carlos Gomez, Patrica Teixidor, and Kim A. Bard. 1996. Imitative learning of artificial fruit processing in children (*Homo sapiens*) and chimpanzees (*Pan troglodytes*). *Journal of Comparative Psychology* 110 (1): 3–14.

Whiten, Andrew, J. Goodall, W.C. McGrew, T. Nishida, V. Reynolds, Y. Sugiyama, C.E.G. Tutin, R.W. Wrangham, and C. Boesch. 1999. Cultures in chimpanzees. *Nature* 399 (6737): 682–85.

———. 2001. Charting cultural variation in chimpanzees. *Behaviour* 138 (11/12): 1489–1525.

Whiting, Beatrice, and John Whiting. 1975. *Children of Six Cultures: A Psycho-Cultural Analysis*. Cambridge, MA: Harvard University Press.

Wilson, Edward O. 1975. *Sociobiology*. Cambridge, MA: Harvard University Press.

———. 1998. *Consilience: The Unity of Knowledge*. London: Abacus.

Wilson, Richard A. 2002. The politics of culture in post-apartheid South Africa. In Fox and King 2002b, 209–34.

Wissler, Clark. 1923. *Man and Culture*. New York: Thomas W. Crow.

Wittgenstein, Ludwig. 1953. *Philosophical Investigations*. 3rd ed. Trans. G. E. M. Anscombe. Oxford: Blackwell, 2001.

Wohlwill, Joachim F. 1960. Developmental studies of perception. *Psychological Bulletin* 57:249–88.

Wolf, Arthur P. 1966. Childhood association, sexual attraction, and the incest taboo: a Chinese case. *American Anthropologist* 68:883–98.

———. 1968. Adopt a daughter-in-law, marry a sister: A Chinese solution to the problem of the incest taboo. *American Anthropologist* 70:864–74.

———. 1970. Childhood association and sexual attraction: A further test of the Westermarck hypothesis. *American Anthropologist* 72:503–15.

———. 1989. The origins and development of the variation in the Chinese kinship system. In *Anthropological Studies of the Taiwan Area*, ed. Kwang-chih Chang, Kuang-chou Li, Arthur P. Wolf, and Alexander Yin, 241–60. Taipei: National Taiwan University.

———. 1995. *Sexual Attraction and Childhood Association: A Chinese Brief for Edward Westermarck.* Stanford, CA: Stanford University Press.

———. 2001. Culture, culture, culture. *Nederlandsch Economisch-Historisch Archief Jaarboek* 64:75–85.

———. 2004. Introduction. In Wolf and Durham 2004, 1–23.

Wolf, Arthur P., and William H. Durham. 2004. *Inbreeding, Incest, and the Incest Taboo: The State of Knowledge at the Turn of the Century.* Stanford, CA: Stanford University Press.

Wolf, Arthur P., and Chieh-shan Huang. 1980. *Marriage and Adoption in China, 1845–1945.* Stanford, CA: Stanford University Press.

Wolf, Eric. 1982. *Europe and the People without History.* Berkeley: University of California Press.

Wolf, Margery. 1992. *A Thrice-Told Tale: Feminism, Postmodernism, and Ethnographic Responsibility.* Stanford, CA: Stanford University Press.

Woodburn, James. 1970. *Hunters and Gatherers: The Material Culture of the Nomadic Hadza.* London: British Museum.

Worden, Robert P. 1996. Primate social intelligence. *Cognitive Science* 20 (4): 579–616.

Wright, Sewell. 1921. Systems of mating. I. The biometric relations between parent and offspring. *Genetics* 6:111–23.

Yunis, Fathi 'Ali, et al. 1987. *Al-tarbiya al-islamiyya.* Fourth grade. Cairo: Al-Jihaz al-Markazi lil-Kutub al-Jami'iyya wa al-Madrasiyya wa al-Wasa'il al-Ta'limiyya.

———. 1987–88. *Al-tarbiya al-islamiyya.* Fifth grade. Cairo: Al-Jihaz al-Markazi lil-Kutub al-Jami'iyya wa al-Madrasiyya wa al-Wasa'il al-Ta'limiyya.

————. 1988–89. *Al-tarbiya al-islamiyya.* First grade. Cairo: Al-Jihaz al-Markazi lil-Kutub al-Jami'iyya wa al-Madrasiyya wa al-Wasa'il al-Ta'limiyya.

Yusuf, 'Abd al-Tuwab, and Yahya 'Abduh. 1988. *Al-sufuf al-munadhdhama.* Cairo: Safeer.

Zaslavsky, Claudia. 1973. *Africa Counts: Number and Pattern in African Culture.* Chicago: Lawrence Hill.

Zigas, Vincent. 1990. *Laughing Death: The Untold Story of Kuru.* Clifton, NJ: Humana.

# Contributors

**KENICHI AOKI** is Professor of Biological Sciences at the University of Tokyo, and coauthor of "The emergence of social learning in a temporally changing environment: A theoretical model" (*Current Anthropology*).

**CHRISTOPHE BOESCH** is Professor of Primatology and Director of the Max Planck Institute for Evolutionary Anthropology, and author of "Male competition and paternity in wild chimpanzees of the Taï forest" (*American Journal of Physical Anthropology*).

**ROBERT BOROFSKY** is Professor of Anthropology at Hawaii Pacific University and coauthor of *Yanomami: The Fierce Controversy and What We Can Learn from It*.

**SAMUEL BOWLES** is Research Professor and Director of the Behavioral Sciences Program at the Santa Fe Institute, and Professor of Economics at the University of Siena. He is the author of *Microeconomics: Behavior, Institutions, and Evolution*.

**ROBERT BOYD** is Professor of Anthropology at the University of California, Los Angeles, and coauthor of *Modeling the Evolution of Social Behavior*.

**MELISSA J. BROWN** is Assistant Professor of Anthropological Sciences at Stanford University, and author of *Is Taiwan Chinese? The Impact of Culture, Power, and Migration on Changing Identities*. ·

**ROY D'ANDRADE** is Professor of Anthropology at the University of Connecticut and author of *The Development of Cognitive Anthropology*.

**WILLIAM H. DURHAM** is Professor of Anthropological Sciences and Bing Professor of Human Biology at Stanford University, and author of *Coevolution: Genes, Culture, and Human Diversity*.

**MARCUS W. FELDMAN** is Professor of Biological Sciences and Burnet C. and Mildred Finley Wohlford Professor in the School of Humanities and Sciences at Stanford University. He is coauthor of *Niche Construction: The Neglected Process in Evolution*.

**HERBERT GINTIS** is an external faculty member of the Santa Fe Institute and Professor of Economics at the Central European University in Budapest. He is the author of *Game Theory Evolving: A Problem-Centered Introduction to Modeling Strategic Behavior*.

**JOSEPH HENRICH** is Associate Professor of Psychology and Economics at the University of British Columbia, where he holds the Canada Research Chair in Culture, Cognition and Coevolution. He is coauthor of *Why Humans Cooperate: A Cultural and Evolutionary Explanation*.

**YASUO IHARA** is Lecturer in Biological Sciences at the University of Tokyo and coauthor of "Cultural niche construction and the evolution of small family size" (*Theoretical Population Biology*).

**JAMES HOLLAND JONES** is Assistant Professor of Anthropological Sciences at Stanford University and coauthor of "Interval estimates for epidemic thresholds in two-sex network models" (*Theoretical Population Biology*).

**PETER J. RICHERSON** is Distinguished Professor of Environmental Science and Policy at the University of California, Davis, and coauthor of *Not By Genes Alone: How Culture Transformed Human Evolution*.

**GREGORY STARRETT** is Associate Professor of Anthropology at the University of North Carolina at Charlotte and author of *Putting Islam to Work: Education, Politics, and Religious Transformation in Egypt*.

**ARTHUR P. WOLF** is Professor of Anthropological Sciences and David and Lucile Packard Foundation Professor in Human Biology at Stanford University. He is the author of *Sexual Attraction and Childhood Association: A Chinese Brief for Edward Westermarck*.

# Index

'Abduh, Muhammad, 268

Abu-Lughod, Lila, 271–72

Aborigines (in Taiwan): classification
of, 164–65; customs of, 167, 174–78,
180; and identity, 166–68, 169, 170,
171, 173, 174–78, 180; and social
selection, 181

Aché (of Paraguay), 190, 201, 217–19,
220, 224. See also small-scale societies

Achuar (of the Peruvian-Ecuadorian
Amazon), 201, 203, 204, 217, 223.
See also small-scale societies

adaptation, 17–18, 56, 91–93, 206, 305–
6. See also cultural traits: adaptive;
maladaptations; natural selection

African Americans: and racism, 14n6,
294; rates of STIs among, 118, 132;
selective sexual mixing among, 127–
31, 132

agriculturalists, small-scale, 191, 194,
201, 217, 222, 223. See also by name
of group

agriculture: origins of, 89, 90–91; commercial 166, 170, 241; innovation in,
93; and niche construction, 157, 160,
183n4

AIDS epidemic, 111, 119, 133

Aka Pygmies (of the Congo), 61, 83–84

Allen's rule, 66

Allport, Floyd, 237

Ammar, Hamed, 262

Anabaptists, 95, 96

ancestry, 14n6, 182n1; in relation to
identity, 162, 165–66, 169, 170, 173,
207; and transmission, 232

Anderson, Benedict, 22, 292

anthropology: American, 3, 4, 5–6, 7,
8, 9, 13n2, 14n3, 15n14, 22–23, 58,
254–56, 281–83, 291–92; biological
(see evolutionary anthropology);
British, 4, 7, 8, 13n2, 15n14, 74, 76,
190; French, 8, 15n14; German, 8,
15n14, 292; models in, 55, 58, 71,
78–79, 120–21, 124, 302; and the
nation-state, 291–92, 295; poststructuralism in, 6, 7, 13n2, 14n3; quantitative, 58, 154; and recursion, 78–79,
97, 105, 120–21, 301; and reflexivity,
230, 257, 270–74; and science, 8–9,
11–13, 15n15, 16n16, 16n18, 18, 55–56,
71, 142–43, 190–91, 231, 253, 295,
298, 306; and studies of variation/
diversity, 6, 9, 12, 58, 71, 206, 231,
283, 284–91, 295, 297, 302, 303.

anthropology (continued)
   See also archaeology; biology: compared to anthropology; cognitive
   anthropology; culture, concept of:
   anthropological; ethnography; evolutionary anthropology; interpretive
   anthropology; paleoanthropology;
   paradigm; postmodernism; social
   science: anthropology as; variation
antievolutionists, twentieth-century, 232.
   See also Boas, Franz; evolution
Aoki, Kenichi, 4, 7, 12, 68, 74, 78, 178,
   230, 303
archaeology, 9, 37, 87–88, 302, 304, 305
Arens, William, 142
artifacts, 88, 291, 304
asymmetric transmission, 96, 109, 125,
   135, 157
asymmetries: between inheritance systems, 93–94, 96, 98 (see also inheritance tracks); in preferences for social
   interaction, 117, 124 (see also bias;
   homophily). See also symmetry
atomic element number, 118; controversy of, 277–79; data fabrication
   of, 278–79
attention (cognitive), 46, 80, 109, 189,
   207, 208, 209
Au (of New Guinea), 201, 203, 217, 218,
   219. See also small-scale societies
Aunger, Robert, 10, 16n18

Bacon, Margaret K., 222
Baldwin, Mark, 98–99
Banna, Hasan al-, 263
Barry, Herbert, III, 222
Bateson, Gregory, 79
Baum, William M., 81
Becker, Gary, 211, 212
behavior: defined, 14n5; as units, 59, 62.
   See also cultural beliefs: shaping human

behavior; cultural traits: as behavioral;
   cultural units: as behaviors; culture's
   influence on behavior; experimental
   behavior; group-specific behaviors;
   ideas: influencing behavior; sharing;
   social behavior; social structure: influencing individual behavior; variation:
   behavioral
behavioral ecology. See ecology: behavioral
behavioral genetics, 77
behavioral traits. See cultural traits: as
   behavioral
beliefs: adaptive, 138, 206; adoption
   of, 210, 221; between-group differences in, 138, 212, 221–22, 224, 226;
   changes in, 264; collectivizing, 28;
   cultural evolution of, 143, 224–25;
   as culture, 13n2, 139, 212, 256, 275,
   282 (see also cultural beliefs); defined,
   215; in decision theory, 196, 205, 214–
   15; evolution of, 143; false, 42, 215;
   Fore, 27, 143, 144, 146–48; ideas
   and, 14n7, 28, 42, 139, 149, 273, 282;
   importance of, 226, 267; imposition
   of, 159; inferred from behavior, 42;
   and institutions, 25; and meaning,
   148, 178, 264; not important, 166;
   persistence of, 221, 287; political,
   27, 256, 267; religious, 23, 27, 36,
   42, 230, 250, 253, 254; shared, 28,
   267, 275, 282, 287, 291; and soul of
   a nation, 267; transmission of, 178,
   221–22, 224–25, 267; and utility/
   function, 264, 267, 273; variation in,
   138, 215, 282–83, 287, 291; in witchcraft, 27, 146–48, 253
Bergmann's rule, 66
bias: cognitive/learning, 207, 210; in
   imitation, 80; in-group, 210; memory, 300; methodological, 229, 297;
   by race or ethnicity, 117, 120, 124, 126,

131–33, 163, 169, 170, 210, 299; by researcher, 118; in sampling, 60. See also asymmetries; homophily; patriarchal family system; prestige-biased transmission
biological determinism 5, 58, 99
biological evolution. See cultural evolution and biological evolution: comparison of
biological variation, 56, 57, 60, 78, 102; caused by cultural traits, 58; as stable polymorphisms, 64. See also genetic variation
biologists: attracted to cultural explanations, 38, 232; and concept of life, 257; defining cultural traits, 41, 48
biology: as a causal factor, 36; compared to anthropology, 5, 71, 206, 257; distinguished from human institutions, 24, 36, 126; prejudice against, 5, 14n6, 58, 97, 98; and race, 126. See also biological variation; cognitive anthropology; cultural evolution and biological evolution; cultural traits: influence of, on biological variation; cultural transmission: compared to biological inheritance; ecology; evolutionary anthropology; genetic evolution; genetics; neuroscience; sociobiology; traits: biological basis of
birds, 39, 45, 49, 102, 183n2
Bittles, Alan H., 68, 104
Boas, Franz, 8, 58, 232, 255, 283
Boesch, Christophe, 4, 12, 17–18, 19, 62, 82, 92, 185, 230, 305, 306
Borofsky, Robert, 3, 5, 6, 7, 12, 15n11, 23, 231, 255, 271, 297, 302, 303
Boserup, Ester, 93
Bowles, Samuel, 4, 12, 27, 71, 138, 178, 205, 211–27, 230, 300, 305

Boyd, Robert, 4, 7, 12, 41, 73, 111, 119, 210–12, 230, 298, 303–5
brain: and disease, 140 (see also kuru); environmental effects on, 189–90; evolution, 89, 305; as holder of cultural unit, 105, 184; models of, 78; ontogenetic development of, 190, 195, 196, 209; states, 28; shaped by learning, 184 (see also cognition, human: and cultural learning; cognition, nonhuman: and learning). See also endocannibalism; cognition, human; cognition, nonhuman; internalization
brother-sister mating. See consanguineous marriages; inbreeding, human; sib mating
Brown, Melissa J., 5, 7, 12, 15n15, 19, 22, 25, 75, 94, 137–38, 150, 156, 159, 230, 254, 275, 283, 298, 299, 302–4

cannibalism, 141–42, 143; end of, among the Fore, 148–50; gender differences in, 153–54, meaning of, 144–46; and mortuary feasts, 145, 148, 154; and niche construction, 160. See also Fore; kuru
capuchin monkeys, 48, 49, 52; group-specific traditions in, 40. See also primates, nonhuman
Cavalli-Sforza, Luigi L., 14n6, 55, 57, 59, 97, 120, 143, 233, 235, 247, 250
cetaceans, 18, 48, 49
Chen Shui-bian, 182
Chiang Ching-kuo, 173, 174, 182
Child, Irvin L., 222
chimpanzees: adaptive cultural traits in, 43–50 (see also cultural traits); ant-dipping methods of, 50; behavior patterns of, 37, 62; cognition and learning of, 82; cultural complexity

chimpanzees (continued)
in, 41, 43–52; cultural traits in, 18,
46–47, 61–62, 306; diet of, 45; and
kuru epidemic, 140, 141; leaf-clipping
behavior in, 51, 306; nonadaptive
cultural traits in, 45–52, 306 (see also
cultural traits); nut cracking behavior
in, 43–45; tool used among, 44, 50.
See also primates, non-human
China, 91, 171–73, 229, 241, 298; children
in, 94, 96; marriage in, 233–34, 236–
38; regional differences in, 164–65.
See also footbinding; Taiwan; uxorilocal marriages
Chinese, Han, 94, 164–68 passim, 172,
176–77, 298, 304; identity, 165–67,
168–70; and marriage, 229, 234, 236;
and social selection, 180–81. See also
Hoklo
chlamydial infections, 129; hypothetical,
130, 131. See also sexually transmitted
infections
civility, American code of, 32–35
class, 171, 172, 255–56, 274, 300; by
land tax, 241, 242; and language,
84–85
classification. See ethnic identity
climate, influence of: on genetic evolution, 66; on cultural evolution, 89–
91, 97
code of honor. See honor code
coevolutionary theory, four-tracks
of, 156–60. See also evolutionary
approaches to culture; gene-culture
coevolution; inheritance tracks
cognition, human, 80–81, 229, 305;
and cultural evolution, 97, 119, 195,
207–9, 305; and cultural learning,
184, 189, 206–10; environmental
effects on, 189–90, 195, 196; evolved,
in ultimatum game, 199–201 (see also

ultimatum game); and maladaptation, 207, 305; natural selection on,
206, 207, 305 (see also prestige-biased
transmission); and perception, 181;
shared, 19; and social selection, 181.
See also brain
cognition, nonhuman: and learning, 40,
52, 53, 80, 82
cognitive anthropology, 5, 9, 11, 12–13,
196, 303
cognitive architecture, 189–90, 199.
See also brain
cognitive experience, 27, 28, 34, 80,
190, 195, 221, 288 See also cultural
models: as schema; internalization
cognitive mechanisms, adaptive, 138,
181, 196, 199, 206, 207, 209–10, 305.
See also brain
cognitive variation, 6, 20, 21, 27, 33,
181, 192–95; adolescence reduces,
192, 194; and childhood socialization,
196–97, 204–6, 207; and fitness, 207.
See also learning: and variation
Cohen, Dov, 29–30, 32–33, 35
collaboration, between field researchers
and modelers, 12, 297, 300–301
Committee of Union and Progress (CUP),
265–66
compartmental modeling, 74, 120, 121,
122, 125, 130, 131, 134–35. See also
infectious diseases: models for
complexity: environmental, 90, 209;
social, 52, 62, 80, 83, 91, 98, 210.
See also cultural complexity; technological complexity
Comte, Auguste, 8, 253, 265, 268
conformity, 28, 41, 67, 82, 198, 212, 217;
and transmission, 86–87, 109–11, 115.
See also cultural transmission: horizontal; cultural transmission models;
imitation; norms; transmission bias

consanguineous marriages: genetic
consequences of, 68, 104. *See also*
inbreeding, human
consilience, 73, 75–76, 97
constraints, 26, 77, 214–15, 299; population, 125, 131
cooperation, 138, 261, 274; experimental, payoffs to, 219–20 (*see also* ultimatum game); evolution of, 92, 111, 120, 184, 210; and fitness, 96; and group selection, 86, 92; household, 176; large-scale, among humans, 92, 96, 184, 210; and punishment, 111; small-scale, 201. *See also* collaboration, between field researchers and modelers
cortisol response, 35; levels of, 29, 34
credibility. *See* science
cross-cultural studies, 138, 221–22; on childrearing, 283; and evidence of developmental learning patterns, 196–97, 204–5; on the perception of illusions, 190–95; on public-goods game, 216, 219 (see also public-goods game); on the ultimatum game, 201–5, 216–20, 226–27 (*see also* ultimatum game). *See also* cross-cultural variation
cross-cultural variation, 9, 59, 76, 101, 102–3, 138, 196–97, 197–206, 216–17, 283, 303; in hunting, 184–85, 188, 190; among nonhuman primates, 41, 48, 185; in perception, 190–95; in preferences, 138, 197; and socialization, 197, 212, 283; in ultimatum game, 197–206, 216–17, 219. *See also* cultural variation
cultural anthropology. *See* anthropology: American
cultural beliefs, 143, 178; acquisition of, 196, 205, 224–25; behind cannibalism, 143, 146; and power, 149, 159,

181; shaping human behavior, 148, 149, 196, 224–25. *See also* beliefs
cultural complexity, 38–43, 52–53, 275; in chimpanzees, 38, 39, 41, 43, 45, 62; in humans, 8, 13n2, 18, 26, 36, 42–43, 53, 75, 89, 137, 208, 255; and modeling, 78–79, 121, 124, 158; and population size, 88–89; progressive, and racism, 58. *See also* complexity; technological complexity
cultural drift, 60–61, 83, 304
cultural dynamics, 55, 57, 59–61, 143, 225–26, 305–6; compared to epidemic dynamics, 70, 74, 117, 119, 121, 123, 133–34 (*see also* epidemic dynamics); compared to social dynamics, 138, 156, 178, 290–91; of marriage and sexual partners, 103, 120, 246; modeled, 104–13, 121; run-away, 94; signaling, 94; of tool technology, 87. *See also* evolutionary dynamics; population dynamics; social dynamics; transmission dynamics
cultural evolution, 233; as the basis of cultural differences, 76, 212; challenges for, 98–99; climate influence on, 89–90, 97; as consilient phenomena, 75, 97; criticism of, 55, 57–58, 66, 79, 91, 97, 247–48; decision-making forces in, 84–86, 88, 189; as empirical science, 75–76, 98–99; and human history, 83, 87, 97; and learning, 83, 210; and macroevolution, 87–91; and maladaptations, 94–96, 143–44; and microevolution, 84–86; and niche construction, 55, 65, 66; a population phenomenon, 83; and power, 156; quantitative study of, 73, 83; rates of, 88, 90–91, 97; and social structure, 60, 77, 92, 120, 156, 247–48; and science of culture, 77;

---

cultural evolution (continued)
and technology, 93; terms for, 73;
and variation, 59, 283. See also cognition, human: and cultural evolution;
cultural selection; cultural transmission; cultural units; Darwinian theory
of cultural evolution; gene-culture
coevolution; identity: change; inheritance; kuru; maladaptations; microevolution; modeling; natural selection:
and cultural evolution; niche construction; theoretical frameworks
cultural evolution and biological evolution: comparison of, 56, 60, 67, 71,
76, 83, 87, 97, 98–99, 283; enhancement of, 65; interaction of, 65, 95;
opposition between, 143–44. See also
genetic evolution: and cultural
evolution
cultural evolution models, 10, 18, 73,
97–98, 100, 104–12, 212, 233, 304–
5; compartmental models and, 133–
34; criticism of, 55, 57–58; and
cultural context/ history, 18, 55–56,
66; formal, and logical consistency,
78–79, 97; and individual psychology, 78–79; origins of, 14n6, 55, 97;
population genetics-style of, 78;
recursion equations of, 78–79, 97;
requirements for, 59–61, 134, 233–34
cultural homogeneity, 134, 271–72, 284.
See also cultural variation; culture, concept of: as shared; heterogeneity
cultural ideas, 5–6, 19–22, 229, 300;
competition between and mutual
influence of, 21, 66, 71, 94–95, 144–
48, 153, 234, 303 (see also niche construction, cultural); frequency or
distribution of, within a population,
6–7, 58–59, 60, 215–16, 284–87;
300, 302; function of, 257, 259, 264,

273; and hegemonic power, 139, 150,
156, 157; and identity, 163, 166–67,
178–79, 181, 182n1, 299; as meaningful, 6, 14n5, 27, 139, 156, 178, 230,
264, 273; and memory, 300; as motivational, 5, 6, 19–22, 24, 28, 137, 138,
139, 144, 150, 156, 179, 182, 183n6,
230, 303; and reflexivity problem, 5,
230–31, 257, 303; as shared, 5, 6–7,
14n5, 19, 21, 42, 143, 212, 215–16,
267, 273–74, 282, 300, 302 (see also
culture, concept of: ideational; culture, definitions of: ideational); and
social structure, 157, 178–79, 181–82,
183n4, 183n6, 229–30, 273–74, 290,
300, 303; variation in, within a population, 5, 6–7, 150, 215–16, 231, 302,
303 (see also cultural variation)
cultural inertia, 85, 93, 148. See also
persistence
cultural inheritance, 60–61, 76, 79, 93,
98, 138, 157–59. See also coevolutionary theory, four-tracks of; cultural
transmission; inheritance; social
inheritance
cultural models (cognitive), 19–29, 34–
36, 37–38, 288; as schema, 6, 19,
28, 61; standard, in social science,
22. See also modeling; niche construction: models of
cultural niche construction. See niche
construction: cultural
cultural practices, 21, 48, 61, 137, 163,
174, 189; as adaptive, 138, 189, 206;
adoption of, 178, 210, 234; affected
by structure, 160, 176, 179, 221–22,
256; defined, 14n5; functionalization
of, 264, 273; versus ideas, 5, 7, 166,
168, 179, 264, 302; practices: and
learning, 189, 207; maladaptive, 137
(see also cannibalism; footbinding);

as not part of culture, 5, 21–22; and
social evolution, 178; scientific practices as, 276; and transmission, 63.
See also cultural traits: as behavioral;
cultural units; customs
cultural selection, 21–22, 25, 41, 60–61,
83, 84, 88, 92, 93, 160, 181, 221; acting on groups, 86–87, 92, 111, 138,
210, 221, 225; as adaptive, 88, 92;
in chimps, 44; degradation of, 88;
demic, 60; and maladaptation, 93–
95; by primary and secondary values,
63, 92, 148, 158–59. See also decision-
making forces; natural selection: and
cultural evolution; natural selection
(on cultural variation); preferences
cultural structure. See social structure
cultural traits (or variants), 37–38, 55,
83, 230, 233, 237, 281–82; adaptive,
37, 42–43, 45–50, 52, 207; arbitrariness in, 51; as behavioral, 37–43,
59–61, 62, 69–71, 83, 143, 230, 282;
in birds, 45, 49; in cetaceans, 48–49;
changing within a lifetime (revert-
ing), 134; conformity in, 41; continuously varying, 57, 59; discrete, 59;
evolution of, 233, 247, 250; examples
of, in humans, 63, 65, 68, 69–70, 74,
83–84, 233, 234, 237; example of, in
humans, challenged, 237, 247, 250;
frequency of, 60, 66–67, 120–21, 134,
233 (see also cultural ideas: frequency
or distribution of, within a population); as ideational, 58–59, 143, 230,
281; imitation of, 42, 207–9; influence of, on biological variation, 58,
63, 69–70; influenced by other cultural traits, 66–68, 71 (see also niche
construction); maladaptive, 17–18,
37–38, 94, 207; nonadaptive or neutral, 38, 43, 45–52, 207, 304; in non-

human primates, 45, 48–50, 52,
62 (see also chimpanzees); quality of,
94; in sea otters, 49; as semotypes,
66–67, 71; shared, 51–51, 281; shared
meanings of, 18, 41–42, 51–52, 53,
281–82; symbolic means of dissemination of, 42–43, 52–53
cultural transmission, 5, 7, 18, 38, 42,
55, 59–61, 71, 178, 210, 232–33,
303; affected by environment, 68–69;
biased, 65, 105–7, 120, 121; compared
to biological inheritance, 14n6, 38,
56, 59–61, 63, 65, 79, 93–94, 98;
dynamics, 14n6, 55, 117, 119, 120–23,
127, 134; examples of, 67–68, 117;
and founder effects, 61; function as,
273 (see also functionalization); horizontal, 59–60, 61, 67–68, 87, 95,
109–11, 115, 121, 233; of ideas, 16n17,
143, 229–30, 257, 273 (see also ideas:
transmission of); modes of, 59–60,
95, 233–34; in niche construction,
63, 226; oblique, 67, 68, 69, 87, 95,
233; and persistent between-group
differences, 224–26 (see also strong
hypothesis that culture matters);
positive feedbacks in, 224; prestige-
biased, 94, 207–9 (see also prestige-
biased transmission); and social
networks, 134–35; of uxorilocal marriages, 234, 235, 237, 247–48; vertical, 56–57, 59–60, 61, 68–69, 84, 95,
104–9, 114–15, 233. See also social
transmission; transmission bias
cultural transmission models: confounding social conditions of, 229–30,
247–50; epidemic, 120–23, 128–30;
heterogeneity in, 55, 119–20, 123–26,
135–36; niche construction, 63–71,
226; punishment, 100, 104–15; recursion, 60, 120. See also modeling

cultural units, 58–59, 61–62, 229, 232–
33, 301; as behaviors, 18, 59, 61–62,
302–3; compared to genes, 26, 58, 79;
cultural-linguistic, 290, 292; defining, 17, 58; detecting, 17, 58–59, 61,
302; discrete, 79; examples of, 61–62,
232–33, 284–87; fish names, 286–
87; as ideas, 18, 58–59, 299, 302–3,
306; as information, 105; as learned,
233; and modeling, 18, 58–59, 79,
105, 120, 233, 299, 302–3; named
meme, 26, 58, 105; named seme, 26,
61; as nation-states, 292, 294; nonhuman, 62; quantifying, 17, 18, 58–
59, 154; relation of, to each other,
232; as replicators, 26, 79, 120, 257,
288; as socially transmitted, 18, 58–
59, 233, 301, 302–3; story units, 284–
85; unitless, 79; as variable, 18, 59.
See also culture: as single, holistic unit
cultural variants. See cultural traits
cultural variation, 57, 59–60, 66, 71, 79,
98, 229, 282; acquisition of, 60, 76,
79, 197; and childhood socialization,
196–97, 204, 212, 222, 283, 288;
between children and adults, 192–95,
198–200, 203, 204–6; and climate/
environmental variation, 89–91, 209,
304; comparison to genetic variation,
71; and cultural background of carriers of cultural variants, 66–68, 71;
discrete, 134; and fitness, 207, 213,
283; and frequency of a variant, 60,
68, 102–3, 105–15, 120, 154, 284–88;
302; and group selection, 86–87; in
hunting, 84, 185–89, 190, 207; and
intentionality, 60, 303–4; and maladaptation, 95; in mythology, 289; in
nonhuman primates, 82; qualitative
knowledge of, 98, 154; and personality/
temperament, 183n6, 288; in punish-

ment, 101, 190, 196, 199–200, 203–4;
and science, 302; and selection, 60,
95, 221; and sharing, 302; and social
preferences, 196–97; and universals,
9, 58. See also cross-cultural variation;
intracultural diversity; intracultural
variation; natural selection (on cultural variation); variation
culture: as adaptive, 17–18, 37–38, 42–
43, 206–7, 212, 305–6 (see also adaptation; cultural traits: adaptive); and
ethnic identity, 162–63; in nonhumans, 37, 49, 51–52, 306 (see also
cetaceans; chimpanzees; primates,
nonhuman); matters, 212–14 (see also
culture, concept of: importance of;
culture's influence on behavior); as
single, holistic unit, 281, 284, 292
(see also cultural units)
culture, boundedness of, 6, 7–8, 302;
as binding people to the group, 269,
294; as conflating bundled variables,
36, 254; as limited rational choice,
77; and problematic notion of boundaries as clear, 15n11, 272, 290–92;
and relation to national boundaries,
282, 292–93
culture, concept of, 3, 4–8, 28–29; abandoning, 6, 7, 14n4, 230–31, 253–58,
271–72, 274; anthropological, 4–8,
22–23, 41, 42, 154, 206, 212, 230–31,
253–58, 271–72, 274, 281–83, 288,
291–92, 295, 302, 306; American
anthropology's, 3, 5–6, 7, 13n2,
14n9, 22–23; British anthropology
and, 4, 7, 13n2, 76; challenging, 230–
31, 275, 288, 291; as consilient, 75;
as a construct, 281; debates about,
4–7, 13n2, 14n9, 22–23, 37–38, 76–
77, 271–72; frequency-based, 6–7, 62,
302; German, 292; as hegemon, 255,

275, 280–81, 291, 295; ideational, 5–7, 17, 22–23, 42, 62, 302, 306 (see also cultural ideas: as shared; culture, definitions of: ideational); importance of (culture matters), 3, 4–5, 7–8, 13, 62, 77, 117, 212–14, 257, 272, 274; as incoherent, 253–55, 272; as incomplete, 253–54, 255–56 (see also culture, concept of: poststructural criticism); and institutions, 17, 22–23, 230; interpretive, 5–6, 14n8, 22–23, 302; based on learning, 5, 14n4, 38, 41, 206; and nation-states, 231, 256–57, 291–92, 295; as the negation of race, 126, 255; and non-humans, 17, 37–38, 41, 306; as politically contaminated/spoiled, 253–54, 256–58; poststructural criticism of, 4, 6 (see also culture, concept of: as incomplete); and rational choice, 77; and reflexivity problem, 5, 230–31, 257–58, 272–73, 274, 295; a scientific, 8–9, 13, 16n17 (see also science: of culture); as shared, 5–7, 19, 21, 22, 29, 38, 41–42, 48, 53, 119, 143, 166, 215, 231, 275, 281–82, 285, 287–93, 295, 302; and status quo, 275; and tipping point, 302; as trivial, 253–54, 274; and variation, 6–7, 271–72, 302; versus traditions, 48. See also anthropology; cultural complexity; culture: boundedness of; culture, definitions of; meaning

culture, definitions of, 5, 14n8, 17, 19–29, 38–43, 297 (see also social transmission: as basis for defining culture); anthropological, 5–7, 13n2, 19, 36, 42, 212, 215, 281–82, 292; as behavioral, 7, 39, 48, 58–59, 282, 303; as depending on what is included, 17, 19, 25, 28–29, 35–36,

48, 254 (see also institutions); disagreement over, 14–15n8, 36, 37–38, 57–59, 212–13, 282; in economics, 77, 212–13, 215; by frequency/distribution, 7, 215–16, 302; as cognitive, 19, 77, 119, 206–7 (see also cultural models); ideational, 5–6, 13n2, 19, 28–29, 58–59, 75, 139, 212–13, 215, 282 (see also culture, concept of: ideational); as information, 5, 14n8, 19, 105, 119, 184; in terms of learning mechanisms, 38, 41, 48, 105, 184, 215, 282; as meaningful, 5–6 (see also meaning); with national identity, 292; as shared, 5–7, 19, 41, 48, 52–53, 119, 281–82, 302; as socially transmitted, 119, 282, 303; limiting, 7, 19 (see also culture: boundedness of); multilevel/polysemous, 18, 19, 38–43, 50, 52–53; origins of, 14n8, 292; theoretical reasons for specific definitions, 25, 38; unbounded (inability to define), 6, 7, 15n11, 55, 57–58. See also cultural units: defining; culture: boundedness of; motivation

culture's influence on behavior, 3–4, 7–8, 19–29 passim, 139, 156, 211–14, 274–75; challenges to, 17, 28–29, 35–36, 274, 275; and competition between ideas, 21, 27, 147, 167, 230, 303, 305; depending on assumptions about motivation, 19–21, 27, 28–29, 162, 179, 303 (see also motivation); depending on how culture is defined, 17, 19, 28–29, 35–36; difficulty in assessing, 5, 19, 274; through extrinsic power, 137, 139, 149–50, 156, 157–59; and fitness/adaptation, 22, 27, 138, 212, 214, 305; through hegemonic power, 137, 139, 150–54, 156, 157–59; through ideas, 19–20, 137, culture's influence on behavior (*continued*) 156, 178, 179, 183n6, 303; and internalization, 27, 28–29, 35, 183n6, 212, 224; and learning, 138, 184, 224, 283, 305; and maladaptation, 96, 137; through meaning, 137, 139, 147–48, 156, 178; mediated by cognition, 138, 184, 305; mediated by structure, 5, 7, 60, 71, 137, 138, 149–50, 156, 158–59, 179, 181, 211, 212, 215, 226–27, 229–30, 283, 305; and predicting behavior, 20, 21, 213; scientific explanations of, 8, 275; and self-interest, 4, 22, 211–12; versus social influences, 21–22, 24–25, 28–29, 178–79, 281, 231, 253–54, 264, 305 (*see also* institutions); through socialization, 138, 212, 224, 283; and transmission, 71, 224, 303; variation in, across individuals, 20, 183n6, 303. *See also* ideas: causal powers of; institutions; modeling; motivation

customs, 13n2, 18, 59, 74, 81, 215, 257, 264, 283, 302–4. *See also* Chinese, Han: customs; incest; marriage; religion

Dagg, Anne, 10
dairying, evolution of, 63
D'Andrade, Roy, 4, 7, 12, 17, 30–32, 34, 42, 65, 75, 77, 101, 230, 255, 275, 281, 282, 303, 305
Darwin, Charles, 75, 82, 83, 86, 95, 98, 264
Darwinian theory of cultural evolution (sometimes called Darwinism), 73, 75–76, 79, 83–87, 95–96, 143, 159; criticism of, 91, 95, 96, 97, 98; as social science, 76, 84, 92, 95, 98–99. *See also* cultural evolution; social science: and evolutionary theory

Darwinism, social, 8, 267. *See also* Galton, Francis
decision-making forces, 83–87, 88, 91, 92, 93, 214, 215, 225, 234. *See also* cultural selection; cultural transmission; rational-choice theory
decision theory. *See* beliefs; constraints; individual behavior: terms for explaining; preferences
De Lepervanche, Marie, 290
demic diffusion, 60, 61, 86
demographic transition. *See* fertility: control
Demolins, Edmond, 264, 265, 266, 267, 268
descent with modification, 56, 83, 91, 232
developmental linguistics, 80–81
Diamond, Jared, 91
differential homophily, 127–30; effects on STI, 127; measuring, 129; model, 128
diffusion: cultural, 60, 61, 81, 83, 84, 87, 109, 121, 130, 229, 233–34, 257, 297, 302; demic, 60, 61, 86
dissemination of cultural traits, 42
disease. *See* infectious diseases; kuru; STIs
diversity. *See* intracultural diversity
DNA, 56, 57, 63
Dore, Ronald Philip, 270, 272
dual inheritance. *See* inheritance: gene-culture (dual)
Durham, William H., 5, 7, 12, 25, 27, 84, 92, 96, 101, 137, 178, 179, 229, 230, 274, 303, 305
Durkheim, Emile, 14n9, 264–69 passim

ecology, 5, 10, 12, 301; behavioral, 77, 92–93; and cultural evolution, 59; in niche construction, 18, 179; of primates, 44, 50–52, 62; of social structures, 179

ecological inheritance, 157–60. *See also* inheritance; niche construction

economics, 7, 98–99, 206, 211–12; and concept of culture, 255–56; experimental, 137, 138, 197, 216; of marriage, 235; and niche construction, 55; rational-choice theory in, 77; and religion, 261; of small-scale societies, 85, 222; and ultimatum game, 197–99, 216–17, 226

economic structures: influencing institutions, 26; influencing preferences and beliefs, 138, 211, 215, 221–22, 226, 300; and variation within a population, 169–71, 221, 283, 284, 303

Edgerton, Robert B., 85, 222

education: as cultural trait, 68, 95, 96, 250; and effect on behavior, 81, 197, 201, 203, 204, 270–71; language in, 172–73; and mass schooling, 257, 263–65, 273–74; role of religion in, 258–60, 269; theories in, 266–68. *See also* imitation; learning

Eerkens, Jelmer W., 87

effective contact rate, 122, 125, 135–36

Egypt: evaluation of schools in, 260; European colonial administrators in, 262; expansion of religious education in, 264; functionalization in, 264, 269 (*see also* functionalization); institutionalized development of religion in, 262, 266–70; Islam in, 258–65 passim; Muslim reformers in, 262, 263; nationalism in, 267; purpose of elementary education in, 269; religious discourses in, 264–65; secular intellectuals in, 258, 263

Egyptian intellectualism, 266–70

Egyptian religious studies textbooks,

259–62; hygiene in, 259–60; morality in, 259–60, 261

empiricalism, 3, 7–13, 13n1, 29, 73–74, 79, 117, 137, 229–30, 298–99, 301–5. *See also* ethnography; methodological issues

endocannibalism, 141, 142, 148; cultural evolution of, 143–44; and cultural fitness, 148; male and female participation in, 153–54. *See also* cannibalism

endogamy, 117, 132, 134; measuring, 123–24

enlightenment scholars, 8

environment. *See* cultural variation: and climate/environmental variation; ecological inheritance; social structure: compared to ecological environment

epidemic dynamics, 70, 74, 117, 118–19, 132, 133; modeled, 121–22. *See also* cultural dynamics: compared to epidemic dynamics

epidemic models, 69–71, 120–26; advantages in, 121; differential equations in, 121, 122–23; incorporating heterogeneity in, 70, 123–26; loglinear models in, 124–25; mixing in, 131–32; in modeling STIs, 118; simple, 118; structured, 118. *See also* infectious diseases

epidemiology: of infectious diseases, 117–19; of representations, 119

ethnic identity, 162–63, 184, 210; changes in, 178–79, 182; classification based on, in Taiwan, 163–65; defined, 163, 172; and disjuncture with culture, 162, 165–68, 299; emerging from regime change, 172 (*see also* identity: change; regime changes); influencing behavior, 74, 163; and nation-states, 192–93; and prevalence of STIs, 74, 117–18, 120, 127–28, 298–99 (*see also*

ethnic identity *(continued)*
    social interactions: of ethnic groups
    and health); in South Africa, 256. *See
    also* bias: by race or ethnicity; mark-
    ers: and ethnic identity; preferences:
    ethnic; social structure: affecting eth-
    nic identity; uxorilocal marriage:
    ethnic groups participating in; varia-
    tion: ethnic
ethnic stereotypes, 163, 299. *See also*
    bias: by race or ethnicity; prejudice,
    power of
ethnography, 3, 9, 58, 137–38, 174, 178,
    229–31, 284; caution in, 270–74; and
    formal modeling, 156–60; and his-
    torical data, 30, 162–63, 238. *See also*
    empiricalism; intracultural variation;
    methodological issues
European sociology, 265–66, 273
evolution: macro-, 76, 87–91, 97–98;
    micro-, 76, 83–87, 97–98, 179;
    process of, 59–61. *See also* cultural
    evolution; cultural evolution and bio-
    logical evolution; cultural evolution
    models; Darwinian theory of cultural
    evolution; natural selection; social
    evolution
evolution, Darwinian variational.
    *See* Darwinian theory of cultural
    evolution
evolutionary anthropology, 5, 6, 9–12,
    14n6, 16n16, 55, 58, 71, 74, 77, 79,
    142–43, 206, 233, 299. *See also*
    anthropology
evolutionary approaches to culture, 73.
    *See also* cultural evolution; cultural
    evolution models; social science:
    and evolutionary theory
evolutionary archaeology. *See* archaeology
evolutionary culture theory: and kuru,
    73, 142–44. *See also* cultural evolution

evolutionary dynamics, 87, 138, 178–82.
    *See also* cultural dynamics; population
    dynamics; social dynamics; transmis-
    sion dynamics
evolutionary feedback, 55, 62–63, 65.
    *See also* niche construction; social
    structure: and cultural feedback
evolutionary psychology, 10, 58, 71, 92,
    93, 101
exocannibalism, 141. *See also* cannibalism
exogenous mating preference: and
    biased vertical transmission, 104–9;
    and horizontal conformist transmis-
    sion, 109–11
experimental behavior, 197–206, 211–12
experimental economists, 216
experimental games, 197–206, 216–20,
    226–27; mirror patterns of interac-
    tion in, 219
experimental methods and data, 30–
    33, 55; opposition to, 58. *See also*
    methodological issues

Feldman, Marcus W., 4, 7, 12, 18, 73,
    74, 75, 78, 83, 97, 120, 143, 157, 178,
    179, 225, 230, 233–35, 247, 250, 274,
    291, 298, 302–3
fertility: control, 68; and growth, 86;
    and sib mating, 103, 105
fitness, 22, 213, 306; cultural, 148; and
    goals, 183n5; inclusive, 95–96; and
    niche construction, 64–65, 71; and
    sib mating, 104
fixation, 64, 134. *See also* stability
footbinding, 304; banning of, 170–71,
    174, 175; and coadaptation, 96; effect
    of, on fitness, 305–6; as marker of
    identity, 166–67, 174; and prestige-
    biased transmission, 94, 298
foraging, 21, 88, 138, 184, 201, 299, 300;
    bow-and-arrow hunting technologies

in, 185, 186–87, 188–89; and effect on behavior, 203, 217, 220, 222; and effect on cognition, 190–95
Fore, 27, 137, 139–41, 159–60; neighbors of, 141, 145, 153; males as warrior-protectors, 150–54; mortuary customs of, 142; shared idea system of, 143–48; sorcerers, 146–47; warfare among, 145. See also cannibalism; kuru
Foster, George M., 84
Foucault, Michel, 9
Fracchia, Joseph, 57, 58
Frazer, Sir James, 74, 100
Freud, Sigmund, 74
function, 92, 257, 272–74, 299; cognitive, 221
functionalism: and Islamic beliefs, 230, 258–62; variation within, 265, 268. See also Durkheim, Emile
functionalization: definition of, 264; process of, 260–61
funerals. See mortuary practices
Furubotn, E. G., 23

Gajdusek, D. Carleton, 140, 141–42
Galison, Peter, 276–77
Galton, Francis, 57, 58
Gates, Hill, 102
Geertz, Clifford, 5, 6, 14n9, 23, 25, 254
gender: differences in cognition, 195; differences in insult scenarios, 33 (see also honor code); differences in marriage dynamics, 246; differences in STI transmission, 125, 129, 132; and endogamy, 124; and maladaptations, 96; no differences by, in game behavior, 198; relations among the Fore, 150–54 (see also Fore; kuru); role of, in society, 182, 183n6. See also sex-selective abortion
gene-culture coevolution, 57, 73, 87,

210, 214, 304; enhancement feedback loop, 65; modeling of sib mating, 74, 100, 102, 104–12, 113–15; opposition in, 143–44. See also coevolutionary theory, four-tracks of; cultural evolution; inheritance: gene-culture
generalizations: of behavioral rules, 222, 226–27; empirical, 9, 11, 12, 13n1, 162, 271, 289 (see also empiricalism); scientific, 97; theoretical, 14–15n9, 42
genetic evolution, 18, 56–57, 63, 83, 88; and cultural evolution, 38, 65, 76, 78, 83, 85, 88, 91, 94, 97, 143, 195, 210; and diseases, 143; effect from environmental change, 63; and group selection, 86; and modeling niche construction in, 63–67
genetic inheritance, 76, 98, 157–58, 160. See also inheritance
genetic variation, 57, 71, 87, 126, 185; at the level of gametes, 56. See also biological variation; variation
genetics, 5, 7, 14n6, 18, 38, 39, 71; of ancestry, 182n1; behavioral, 77, 78; criticism of, 18, 58; and group differences, 87, 213–14; modeling, 57, 74, 78; molecular, 126; population, 61, 78, 120, 122; in preferences, 221; and race, 126; reductionism in, 98, 99; and social learning, 185, 195, 210. See also genetic evolution; genetic inheritance
gesturing, 53. See also language
Gil-White, Francisco J., 207, 209
Gimi, the, 145, 148, 149, 154
Gintis, Herbert, 4, 12, 27, 71, 138, 178, 205, 230, 300, 305
Gnau (of New Guinea), 203, 219. See also small-scale societies
Goldschmidt, Walter, 85

Gombe chimpanzees, 50–51. See also
    chimpanzees
Graebner, Fritz, 232
great apes, 52
Great Basin of North America, 87
group selection. See cultural selection:
    acting on groups; natural selection:
    acting on groups
group-specific behaviors, 39–43, 212–
    13, 219, 224–26; among the Fore,
    144, 147 (see also Fore); economic
    influences on, 221–22; environmental
    effects on, 196, 197; predicting, 205–
    6, 213–14, 223; in primates, 37, 43–
    44, 50–52, 62; in STIs, 118–19. See
    also behavior; culture; ultimatum
    game
guided variation, 83. See also decision-
    making forces; learning

Hadza (of Tanzania), 21, 185, 188, 201,
    203, 217, 220, 224; hunting technol-
    ogy of, 186–87. See also foraging;
    small-scale societies
Haishan (Taiwan), 238–52. See also uxo-
    rilocal marriages; virilocal marriages
Hakka, 164, 171–73, 236. See also
    Chinese, Han: identity; Taiwan
Han ethnic identity. See Chinese, Han:
    identity
Harbaugh, William T., 198, 204
hegemony, 96, 137, 139; and kuru, 150,
    154, 156. See also nation-states
Henrich, Joseph, 4, 12, 21, 75, 77, 88,
    138, 211, 216, 217, 230, 300, 305
herding communities, 85
heterogeneity: behavioral, in STI trans-
    mission, 70, 118, 123, 126–36; cul-
    tural, 119–20, 134, 291; dynamical
    implications of, 70, 117, 123, 125;
    ethnicity as cultural marker of, 120;

population, 117, 126–34; spatial
    or temporal, 119; transmission, 70,
    125–26. See also cultural variation;
    epidemic models: incorporating
    heterogeneity in; variation: ethnic;
    variation within a population
Hewlett, Barry S., 61
hierarchical social relations. See prestige-
    biased transmission; social structure
Highland New Guinea: warfare in, 86
Hispanic Americans: rates of STIs
    among, 118; selective sexual mixing
    among, 127
history, 223; and contingent processes,
    157, 180–81, 225, 304; contrasted
    with evolution, 18, 55, 56–59, 60–61,
    71; contrasted with culture, 255, 257,
    272, 274, 289–90; as revealing evolu-
    tion, 76, 83, 206; and spread of evo-
    lutionary studies, 98–99
Hodgson, Geoffrey M., 98–99
Hoklo, 164, 166–67, 171–73, 174–78. See
    also Chinese, Han: identity; Taiwan
Holocene, the, 88–89, 91, 93; climate
    change in, 90
homophily, 123–24, 127–29, 131–32.
    See also endogamy
honor code, 29–32, 35; of American
    South, 30–32; disappearance of,
    36; European gentleman's, 30; gen-
    der differences in, 33; shared by
    Americans, 33
Hooper, Judith, 279–80
horticulturalists, 85, 185, 192, 201, 217,
    223. See also Fore
households, 100; and marriage patterns,
    175–76, 237–38; structure and varia-
    tion in, within a population, 219–20,
    283–84
Hua, the, 145
Hull, David, 79

human cognition. See cognition, human

human goals. See motivation

human psychology. See cognition, human

hunter-gatherers. See foraging

hunting, 83–84, 138, 160, 190, 207; and effect on behavior, 219–20, 221–23; technologies, 185–89, 201, 203

Husayn, Taha, 262, 268–69

idealism: in evolutionary and cognitive anthropology, 5, 6, 14n6

ideas, 6; causal powers of, 20, 23, 24, 28–29, 137, 179, 230, 303; collectivizing, 28; difficulty in quantifying, 18, 58–59, 299, 302–3; functionalization of religious, 258–59, 261–62, 264, 273; influencing behavior, 19–20, 28, 62, 137, 138, 139, 150, 179, 181, 230, 265, 299, 302, 303; and institutions, 22–23, 24, 25; internalization of, 27–8, 167; and selection processes, 25; and soul of a nation, 267; transmission of, 14n6, 18, 273, 157–58, 178–79, 229–30, 232, 257, 265, 302; as unit of culture, 5, 18, 58–59, 60, 62, 299, 302–3, 306. See also cultural ideas

ideational system, 17, 22, 143, 230. See also culture, concept of: ideational

identity: change, 162–63, 165, 167, 169–74; group, 126, 292; and social selection, 22, 180–82. See also ethnic identity

Ihara, Yasuo, 4, 7, 12, 68, 74, 78, 178, 230, 303

illusion experiments, 190–95

imitation: and adaptation, 80, 184, 189, 196, 206–9, 230; and concept of culture, 38, 39, 40, 42, 62, 76, 80, 233, 237; and cultural evolution, 83, 188–89; and enculturation, 76, 184,

196; and innateness, 80–82, 196, 230, 237; by nonhuman primates, 38, 39, 40, 42, 62, 82; and prestige, 81, 206–9; and uniqueness of humans, 38, 76, 82, 196; variation in use of, 230, 233–34, 237. See also learning; learning, cultural; learning, social

imposition. See power

inbreeding, human: cases of, 102–3; and depression, 104; in first-cousin marriages, 104; variation in propensity for, 102. See also sib mating

incest, 74; punishment of, 101–2, 109, 112; in simple societies, 101

income-maximizing offer (IMO), 199, 200. See also ultimatum game

individual behavior: affected by social structure, 221; terms for explaining, 214–16. See also behavior

infectious diseases, 69–70; transmission of, 69, 70, 117–19, 125, 127, 128–31, 132, 142–43, 150, 152, 154, 298; epidemiology of, 117–19; models for, 69–70, 120–23; variation in latency of, 140, 144, 150. See also kuru; STIs

inheritance: gene-culture (dual), 76, 93–94, 156–57 (see also gene-culture coevolution); multiple, 157–58, 163, 304; and natural selection, 93; property, 166, 241–42; theory of, 57, 60–61, 98. See also cultural inheritance; ecological inheritance; genetic inheritance; social inheritance

inheritance tracks (or systems), 76, 79, 93–94, 96, 98, 157–60, 163, 304–5

innateness: in aversions, 100–3, 105–6, 112; in preferences, 74, 230, 237

innovation, 59, 60; diffusion of, 83–84, 109, 120–21, 134

institutions, 15n15, 17, 22–24, 210, 212, 221, 230–31, 257, 305; cooperative, 138, 210; honor code as, 30, 35–36; power of, 17, 23, 24, 28–29, 36, 293; schools as, 260, 262–64, 273–74; selection processes of, 25–26

insult experiments, 30–35. See also honor code

integration of theory and data, 12

intellectual developments, Middle Eastern, 265–70; impact from European sociology on, 265–66

intentionality, 60, 303–4

internalization, 6, 7, 27–29, 212, 299, 303; of behaviors, 224–25, 227; and ethnic identity, 166–67; of honor code, 31–33; models of, 36, 121

interpretive anthropology, 5, 9, 13n2, 14n3, 14n8, 14n9, 41, 52; criticism of, 6–7; formulations of culture in, 14n9. See also Geertz, Clifford

intracultural diversity: anthropological treatment of, 231, 271–72, 287–91; and culture as shared, 281–83; data documenting, 284–91; and nation-states, 282, 293–94; usefulness of, in future research, 302. See also cross-cultural variation; cultural ideas: variation in, within a population; cultural variation; variation within a population

intracultural variation: at level of classificatory groups (gender, age, social role, occupation, etc), 76, 183n6, 185, 197, 134; at level of communities, 63, 153; at level of economic structure/market integration, 219, 221, 284; at level of foraging groups, 188–89, 202–3, 221; at regional level, 33, 164; at level of tribes, 85; at level of peoples and nation-states, 231, 283, 288–91, 293; at level of traits, 58.

See also cross-cultural variation; cultural ideas: variation in, within a population; cultural variation; variation within a population

intracultural variation at level of individuals, 6, 33, 59–60, 102–3, 188, 207, 297, 302, 303; data on, 284–91, 302; and fitness, 207, 283 (see also cultural variation: and selection); in ultimatum game, 201–2, 215–16, 217. See also cross-cultural variation; cultural ideas: variation in, within a population; cultural variation; heterogeneity, cultural; variation within a population

Islam, 230, 258–65, 269, 273

Japanese colonial government: household registers of, 238; land-tax registers from, 241–42; social and economic initiatives of, 250; in Taiwan, 170–71 (see also Taiwan)

joint attention, 80

Jones, James H., 5, 7, 12, 74, 95, 163, 179, 298, 299

Keesing, Roger, 291

Kettlewell, Bernard, 280

kinship, 4, 84, 92, 153, 172, 250; Chinese (patrilineal), 234, 238, 252n4. See also marriage; patriarchal family system

Klitzman, Robert, 142, 149

Kohn, Melvin, 221, 226

Koran, the, 263

Kroeber, Alfred L., 58

Kultur, German sense of, 292. See also culture

Kuper, Adam, 14n9, 36, 254–55, 292

kuru, 27, 121, 137, 139–60; cultural evolution of, 142–44, 156–57; deaths, 150–52; Fore's own theory about cause of, 146–48; forms of imposition in, 157–58; gender differences in, 153–55; gene resistance in, 159–60; transmission of, 140–42. *See also* infectious diseases; Fore

Laland, Kevin, 56, 67, 179
Lamalera (of Indonesia), 201, 203, 219, 221, 223, 227. *See also* small-scale societies
language, 42, 53, 61, 65, 80–81, 171; and identity, 84–85, 164–67, 171–73, 292, 294
learning, 38–41, 60; and cultural evolution, 83, 210; through group members, 40, 48, 210; imitative, 39, 41, 188, 196 (*see also* imitation; learning, social); individual, 83, 95; practicing in, process, 184, 189, 196; in primates, 52, 62; strategies, 82; trajectories, 199–201; and variation, 82, 185, 188–89, 197, 204, 207. *See also* education
learning, cultural, 20, 184–89, 204, 206–10; adaptive, 138, 184–89, 206
learning, social, 18, 39–41, 45, 48, 80–82, 196–97, 207, 209, 303; in birds, 39; difference in, between humans and primates, 38, 62, 81, 185; in humans, 80–81, 196–97; mathematical modeling of, 82; and reinforcement, 81
Le Bon, Gustave, 264–67
Lee Teng-Hui, 173
Lévi-Strauss, Claude, 284
Lewontin, Richard C., 57, 58, 62, 157
life, concept of, 257
life force, belief in, 144–46; passing on, 145. *See also* cannibalism
Li Jinghan, 236
Li Nan, 230, 234, 235, 298

Lindenbaum, Shirley, 143, 160
linkage: cultural and biological, 71, 112, 160; disequilibrium, 67, 71
Li Shuzhuo, 234, 235, 236, 237, 238, 239, 298
loglinear modeling, 124–25, 128, 135; parameters of, 129
Lueyang (Shaanxi, China), 234; marriage in, 235, 237, 239. *See also* uxorilocal marriages

Machiguenga (of the Peruvian Amazon), 201, 202, 219, 220, 223. *See also* small-scale societies
macroevolution, 87–91, 98; approaches to studying, 87–88; environmental stages in, 89
macro-micro problem, 79
Mainlanders (in Taiwan), 164–65; formation of identity for, 172
maladaptations, 76, 91–96, 97, 137, 144, 207, 225, 305–6. *See also* adaptation; cultural traits: maladaptive
male-biased hegemony. *See* patriarchal family system
Malinowski, Bronislaw, 4, 13n2
Manchu ethnic group, 169
Mapuche, 201, 219, 220. *See also* small-scale societies
markers: and cultural innovation, 134; of culture, 59, 117, 120; and ethnic identity, 117, 120, 166–67, 172, 174, 177, 208, 210, 298; physical, 29; and religion, 259, 269; and sexual behavior, 126, 133; shared, and adaptive learning, 208, 210
market: integration, 205–6, 219–20, 300 (*see also* ultimatum game); marriage, 21–22, 132, 166, 169–61, 174–76, 230, 135–58, 240–50. *See also* economic structures

marriage: across cultures, 100, 103, 123–24, 132, 166; as cultural trait, 63, 68–69, 95, 117, 134; and cultural transmission, 229, 233–38; forms, 63, 68–69, 100–104, 123–24, 166, 234; market, 21–22, 132, 166, 169–71, 174–77, 230, 235–38, 240–50; in models of, preferences, 104–11. *See also* consanguineous marriages; endogamy; exogamy; incest; sib mating; uxorilocal marriages; virilocal marriages

Marx, Karl, 253, 255

Marxism, cultural, 14n3

Marxist: analysis of religion, 255; materialists, 4, 213; social formations, 232

mathematical modeling. *See* modeling

Mayr, Ernst, 58

McElreath, Richard, 210, 222

Mead, Simone, 160

meaning, 139, 154; in coevolution, 156–57 (*see also* coevolutionary theory, four-tracks of); in concept of culture, 5–6, 9, 14n9, 41, 42, 282, 306; and cultural beliefs and ideas, 139, 147–48, 156, 264; and institutions, 23, 25; shared, 18, 41–42, 51–53, 281–82, 287; variation in, 6, 264. *See also* cultural ideas: as meaningful; cultural traits: shared meanings of; culture's influence on behavior: through meaning

media, mass, 35, 96, 273, 302

Meigs, Anna S., 145

Melanesia, 290–91. *See also* Pukapukans

memes, 58–59, 105; frequency of, 105–7, 109–15

memory, 44, 89, 190, 209, 300

men: Fore, 153–54, 158–59; and honor code, 29–33; and hunting techniques, 83–84; and incest, 100, 112; and marriage, 166, 168–69, 230, 234–35, 238–

52; and STIs, 124–25, 127–32; and tale of Wutu, 285, 296; as tramps, 270–71. *See also* gender; sex ratio; sib mating

Mendel, Gregor, 56, 57

methodological issues, 3, 8, 229–31, 250, 270–72, 299–300; ethnographic, 10, 236–37, 270–72; in developing a paradigm, 11–13; and science, 11–12, 16n16

microevolution, 98; cultural, 85; group selection in, 86–87; language, 84–85; natural selection in, 85–86; and the Spanish Conquest, 84; and transmission of hunting techniques, 83–84

Minnan (Taiwanese), 173

minor marriages, 100, 103. *See also* marriage

mitochondrial DNA, 63

mixing. *See* structured mixing

modeling, 3, 7, 9, 11, 73; approaches, 3, 58, 73, 74, 98, 209–10, 298, 302–3; based on case studies, 73–74; concerns and issues, 9–10, 98, 229–30, 250, 302–5; dynamical versus statistical, 55; gene-culture coevolution, 100, 102, 104–13; importance of, 55, 71, 78–79, 119, 182; heterogeneity in, 123–26 (*see also* cultural transmission models); infectious disease, 117, 128–29 (*see also* epidemic models); integration with empirical data, 10–13, 57–59, 79, 137–38, 182, 210, 297–301 (*see also* paradigm); and logical consistency, 78–79; loglinear, 127–29; microevolution, 87 (*see also* microevolution); niche construction, 62–71; punishment, 111–13; recursion, 60, 78–79, 97, 120–21 (*see also* anthropology: and recursion); and social-network structures, 134–35; variable space in, 107, 110, 113; variation in, 303. *See also* anthropology: models in; coevolutionary theory, four-tracks of; compartmental modeling; cultural evolution models; cultural models; cultural transmission models; epidemic models; prestige-biased transmission; strong hypothesis that culture matters; weak hypothesis that culture matters

Morris, Martina, 127, 135

mortuary practices, 142, 166, 176–78

motivation, 5–6; in cultural models, 20–29 passim; and ideas, 27–28; forces of, 162, 179; in future research, 303; and preferences, 211; sharing 20–21; variation in, 20, 21, 48, 303

Mueller-Lyer illusion, 191–95

multivariate linear regression, 201

National Heath and Social Life Survey (NHSLS), analysis of, 127–30

nationalism: Afrikaner, 256; Egyptian, 267; German, 292; Taiwanese, 165 (*see also* Taiwan); Turkish, 268

nation-states, 22, 231, 256–57, 263–64, 291–95; and colonial regimes, 170–71, 241, 250; fragility of, 293–94; as imagined communities, 22, 292; and morality, 266–69

native categories, separation of, 272

native texts, 271

natural selection, 56, 157–58, 221, 301; acting on groups, 86, 92, 212, 221, 225–26, 267 (*see also* cultural selection, acting on groups); asymmetries in, 93–94 (*see also* symmetry); and cognition, 138, 196, 199–201, 206–9; and cultural evolution, 17–18, 60, 63, 65, 83, 85–86, 88, 94, 111, 196,

206–7, 210, 305 (*see also* decision-making forces); and fraud, 279–80; in the kuru case, 159–60; and niche construction, 62–63, 65, 157, 158, 160, 225–26; in sib mating, 103–4. *See also* coevolutionary theory, four-tracks of; cultural selection; social selection

natural selection (on cultural variation): meaning cultural selection, 60, 83, 95; meaning survival and reproduction of human carriers of specific cultural variants, 63, 64–65, 85–86, 88, 93, 94, 95, 159–60, 207–10, 305

negotiation. *See* social negotiation

networks, sexual, 118, 135. *See also* social networks

neuroscience, 5, 28, 140, 196. *See also* brain

niche construction, 18, 56, 62–71, 156–58, 160, 179, 225; adaptation reconsidered due to, 56; and antibiotic resistance, 70–71; cultural, 55–56, 63, 64–71, 183n4, 225–26, 274; defined, 56; gene-based, 63–64, 157, 160; and language, 65; models of, 63–71, 157–58; and multiple traits, 71; social, 18, 55, 70, 179, 274; and subsistence, 63, 68, 157, 160, 183n4

Ninov, Victor, 231, 277–79

Nisbett, Richard E., 29, 30, 35

nomadic foragers. *See* foraging

norms, 24, 212, 224, 290; of civility, 32, 35; and collectivization, 28; cultural, 74, 133, 177; and fertility control, 68; and institutions, 23–24, 26; and internalization, 27, 212, 224; nonhuman, 40, 50–51; as preferences, 214, 221; prestige, 95; and punishment, 199, 221, 224–25, 258; for sharing, 221, 223; and socialization, 199, 224–25; and violence, 34–35

n-person prisoner's dilemma. See public-goods game
nut cracking, 43–45. See also chimpanzees

Odling-Smee, F. John, 56, 179, 183n4, 225
ontogeny: brain development during, 189–90; cultural learning during, 184, 189, 204; social preferences during, 196–97, 204
optimization, 189
orangutans: behavioral patterns of, 62; cultural traits in, 41; cultural variants in, 48. See also primates, nonhuman
Oregon children in ultimatum game, 203, 204, 205. See also ultimatum game
Orma (of Kenya), 219, 227. See also small-scale societies

paleoanthropology, 88
paradigm, 3, 71, 60, 70, 71, 73, 229; scientific, for anthropology, 4–5, 8, 11–12
parental influence. See uxorilocal marriages: parental influence; social influences
Parsons, Talcott, 22, 23
pastoralism, 68, 85, 138, 185, 191, 201, 217, 299; and effect on behavior, 217, 222; and effect on cognition, 191–95
patriarchal family system, 150–53, 154, 234
patrilineal kinship, ideology of, 238
People's Republic of China (PRC), 173
peppered moths, controversy over, 279–80
persistence: cultural, 138, 224–25, 287, 304; individual, 175, 177–78; of norms, 212, 225. See also cultural inertia

personality, 221–22
phenogenotypes, 65
Pleistocene, the: climate change in, 90; cultural evolution in, 88–89, 93
Polanich, Judith K., 81
population dynamics, 119, 122, 123, 125. See also cultural dynamics; epidemic dynamics; evolutionary dynamics; social dynamics; transmission dynamics
population structure, 5, 131, 162, 210
positivism, 265, 276
postcultural concepts, 6, 256–57
postmodernism, 4, 5, 9, 11, 16n16. See also anthropology, poststructuralism in
power, 139, 156; of culture over human behavior, 144. See also hegemony
predispositions, 112, 221–22, 225–26
preferences, 138, 211, 214, 215; aggregated, 127, 133; economic influences on, 221–22; ethnic, 74, 117, 120, 123–24, 131, 298; in experiments, 32–33; in group differences, 224, 226; in the kuru case, 158–59; marriage, 123, 230; models of mating, 104–13, 124–26, 130, 133, 135; and sex-selective abortion, 67–69 (see also son preference); sexual, 74, 103, 111; social, 196–98, 204; and transmission, 117, 207–9; in ultimatum game, 201. See also beliefs
prejudice, power of, 26–28. See also bias
prestige-biased (or rank-biased) transmission, 22, 81, 84–85, 94, 147, 207–10, 298; avoidance of, 95, 96. See also social rank
primates, nonhuman, 49; compared to humans, 82, 185; existence of culture in, 61–62. See also capuchin monkeys; cognition, nonhuman

prions, 140; PRNP gene in, 159
public-goods game, 216, 219, 227; small-
   scale societies in, 217, 218. *See also*
   ultimatum game
Pukapukans (of Cook Islands), 284–87,
   295–96
punishment: cost of, 106, 109, 211; and
   exogenous mating preference, 104–
   9; memes for, 105, 113–14; modeling
   threat of, and biased vertical transmission, 113; and sib mating preference, 111–13; in ultimatum game, 199,
   203–4 (*see also* income-maximizing
   offer; ultimatum game)
punishment equilibrium, 106, 107, 109, 112

quantitative modeling. *See* modeling
Quichua (of the Ecuadorian Amazon),
   201, 223. *See also* small-scale societies

race: as biological concept, 126; and
   epidemics, 131; and STIs, 126–27.
   *See also* bias: by race or ethnicity
racial stereotypes. *See* bias: by race or
   ethnicity
Radcliffe-Brown, Alfred R., 273
Ramadan, 261
rational-choice theory, 77, 198. *See also*
   decision-making forces
rank. *See* prestige-biased transmission;
   social rank
recursion. *See* anthropology: and recursion; modeling, recursion
reflexivity. *See* anthropology: and reflexivity; cultural ideas: and reflexivity
   problem; culture, concept of: and
   reflexivity problem; social experience:
   construction of
regime changes: in Taiwan, 164–65, 168–
   74 (*see also* Taiwan); in Taiwan and
   China, 165

reinforcement: in cultural and social
   inheritance, 158–59; and social
   learning, 81
relativism, 15n10. *See also* interpretive
   anthropology
religion: and culture, 23, 27, 36, 42, 254,
   255, 289–90; in education, 263–64;
   function of, 258–62, 269, 273; and
   identity, 166, 167, 176
replication, 11–12, 276–80, 305. *See* cultural units: as replicators; imitation
Richards, Robert J., 98–99
Richerson, Peter, J., 4, 7, 12, 41, 73, 111,
   119, 210–12, 230, 298, 303, 304, 305
Riza, Ahmed, 266
Robbins, Joel, 26
Romney, A. Kimball, 15n12, 123, 124
Roughgarden, Joan, 11
run-away forces, 94–95. *See also* cultural
   selection; decision-making forces;
   guided variation; maladapation
Rwanda, 294

Sabahaddin Bey, Prince, 266
Sahlins, Marshall, 254, 292
Saint-Simon, Claude Henri, 265
sampling, 94, 101, 127, 298, 300, 305;
   effects, 60–61, 84. *See also* cultural
   drift; methodological issues
San (of the Kalahari), 185, 188; hunting
   technology of, 186–87; in illusion
   experiments, 191. *See also* foraging
Sander parallelogram, 191, 193–95
Sangu farmers, 203. *See also* small-scale
   societies
Sapir, Edward, 255, 283
science, 8–13, 15–16n16, 275–81; concept of, 5, 11, 16n18, 257, 275, 280
   (*see also* anthropology: and reflexivity;
   cultural ideas: and reflexivity problem);
   of culture: 3–4, 5, 13, 295, 297–300,

382    Index

science *(continued)*
306 (*see also* culture, concept of: a
scientific; social science: and cultural
ideas). *See also* anthropology: and
science; replication; social science;
truth: and science
scientific methods, 3, 11, 300
sea otters, 49
Searle, John, 23
Segall, Marshall, 190–91
selection. *See* cultural selection; natural
selection; sexual selection; social
selection
selective sexual mixing. *See* structured
mixing
self-direction, 221
seme, 61
semotypes, 66–67, 71. *See also* cultural
traits
Seneca Indians, 289. *See also* intracultural
diversity
sex differences. *See* gender differences
sex ratio: biased, 69, 93–94, 241; at birth
(SRBs), 67–69
sex-selective abortion or infanticide, 67–
68, 69, 94
sexual activity, variation in, 118, 298
sexual experience, 102–3
sexually transmitted infections. *See* STIs
sexual partners, variation in, 74, 118, 124,
127–32
sexual selection, 11, 41, 92, 94. *See also*
signaling; structured mixing
shared social experiences, 172. *See also*
structural change
sharing (as a behavior), 211, 220, 221,
223, 225, 226, 281–82, 302. *See also*
beliefs: shared; cognition, human:
shared; cultural ideas: as shared;
cultural traits: shared; cultural traits:
shared meanings in; culture, concept

of: as shared; Fore: shared ideas system of; markers: shared, and adaptive
learning; meaning: shared; motivation: sharing
sib mating, 100, 102; effect on fertility,
103; effect on fitness, 104; error rate
in, 103–4; and marriages, 101, 103;
threat of punishment in, 111–13
signaling, 94
*sim-pua*. *See* minor marriages
Skinner, William, 94
small-scale agriculturalists. *See* agriculturalists, small-scale
small-scale societies, 201–5, 211–12, 217;
adults from industrialized and, 204;
economic structure in, 222; map of,
218. *See also* Aché; ultimatum game
Snow, Charles Percy, 55
social behavior, 184; environmental
effects on, 196–97
social complexity, 52, 62, 80, 83, 91, 98,
210
social criticism, European and Middle
Eastern, 273. *See also* postmodernism
social dynamics, 10, 138, 159, 162, 163,
178–82, 183n2, 304–5; defined, 156;
as power dynamics, 179, 291, 293; as
structure-based, 179. *See also* cultural
dynamics; evolutionary dynamics;
population dynamics; transmission
dynamics
social experience: construction of, 230,
257, 271–72 (*see also* culture: concept
of: and reflexivity problem); of the
environment, 56, 190, 195–97; as
expertise, 81, 270, 286, 287, 296;
of field work, 10, 300; function of,
261; and internalization of identity,
162, 163, 172, 292; of markets, 205–
6; and preferences, 211, 215; of structure, consequences of, 162, 221–22,

254, 273–74; as unnecessary to meaning, 53. *See also* cognitive experience; internalization

social evolution, 92–93, 150, 159, 273–74. *See also* social inheritance; social transformation

social hierarchy, 162, 179, 182, 183n2, 274; as changeable, 179–80; as functional, 259; in institutions, 15n15, 24; as learned by children, 196; as mediating culture, 5. *See also* social structure

social influences, 174, 231, 283; constraining human behavior, 138, 157–58, 179–81, 212, 221-22, 226, 299, 305; of European intellectuals in Egypt, 265–70; on human decisions, 211, 215, 234; of mass media, 96, 273, 302; on nonhuman behavior, 39–40, 50; of poverty, 241; and selection, 60, 138, 179–80, 181; and transmission, 61, 71. *See also* cognitive experience; social structure: influencing individual behavior; uxorilocal marriages: parental influence in; uxorilocal marriages: sibling influences in

social inheritance, 138, 157–59, 179–82, 304–5. *See also* cultural inheritance; inheritance

social interactions, 74, 117, 163, 174, 184, 210; cognitive modules for, 199–201; of ethnic groups and health, 74, 117–18, 120, 127–28, 131, 133 (*see also* ethnic identity: and prevalence of STIs); and institutions, 23, 25, 26; nonhuman, 50. *See also* preferences; social relations

social learning. *See* learning, social

social negotiation, 7, 163, 169, 174–78, 180, 277, 279

social networks, 92, 176, 274; and struc-ture, 15n15, 134–35, 162, 180, 183n2, 185, 299; theory for modeling, 132, 134–35, 298

social organization, 77, 85, 185, 213, 261, 292; and cooperation, 86, 92, 96; and diversity, 231, 288, 295 (*see also* variation within a population: and social structure); and institutions, 230, 257–58; manipulation of, 181; structural-functional theories of, 265; and technology, 83, 92. *See also* social structure

social rank, 30, 181. *See also* prestige-biased transmission

social relations, 81, 220, 299; between individuals and the group, 5, 79; and causation, 22, 223; conceptualization of, 290; and function, 269–70, 274; and institutions, 24–25, 269–70; and networks, 134, 162, 185; and selection, 86, 138; and structure, 156, 157–58, 162, 163, 169, 179–81, 183n2, 185, 274. *See also* gender: relations among the Fore; social interactions

social science, 9, 55, 75, 77, 303; anthropology as, 4, 12, 15–16n16, 71, 253, 271–73, 275; and cultural ideas, 20, 21, 22, 230; as culture, 257–58, 270–74; and evolutionary theory, 14n6, 55, 71, 75–78, 79, 84, 87, 92, 233–34, 274; and functionalism, 92, 230, 273–74; and history, 71, 76; on institutions, 22–23, 24–25; origins of, 253, 265; in pre-science phase, 12, 71; and progress, 253, 266–67. *See also* science

social selection, 25–26, 92, 138, 179–82, 210. *See also* functionalization

social structure (systems), 24–25, 156–59, 162–63, 178–82, 304–5; and adaptation, 92, 138; affecting ethnic

social structure (systems) (continued)
identity, 138, 162–63, 168–74, 178,
182, 210, 299, 305 (see also structured
mixing); compared to ecological environment, 18, 179; and cultural feedback, 55, 62–63, 65, 179, 212, 224–
25; as culture, 274; defined, 156, 162;
distinct from culture, 4, 13n2, 17, 22,
24, 138, 162–63, 178, 182, 183n4,
223, 229–30, 305; and economic
structures, 138, 169–71, 212, 221–23,
226–27, 265–66, 283, 295; evolutionary dynamics of, 178–82; as hierarchical, 15n15, 162, 179–80, 182,
183n2; importance of, 137–38, 156,
179, 231; influencing culture, 60, 138,
156, 157–58, 179, 215, 231 (see also
culture's influence on behavior: mediated by structure); influencing individual behavior, 138, 174–78, 179–
80, 212, 215, 221–22, 226–27, 253–
54, 283; and institutions, 15n15, 17,
22–23, 24–25, 210, 260, 274, 305;
and internalization, 227, 303; and
learning, 210, 283; and marriage,
169, 175–77; and mating, 117–19,
126, 131–32; modeling, 71, 182, 301,
304–5; motivational force in, 23, 138,
162, 179, 303; and niche construction, 160, 179, 274; perception of,
180; and power, 157, 174, 179, 180,
274, 295; quantification of, indirect,
154–56; and religion, 176, 255; and
scientific contributions, 15n15, 295,
297; and selection, 60, 71, 92, 138,
179–81, 306; and socialization, 138,
221–22, 283, 288–89; and social
roles, 22, 24–25, 138, 180–82; and
transmission, 71, 92, 178, 210, 229–
30, 299; and variation, 71, 303. See
also social dynamics; social networks:

and structure; social organization;
structural change
social transformation, 163, 173–74, 178,
230, 232–33, 247–50, 257, 273–74.
See also social evolution
social transmission: as basis for defining
culture, 119, 282; in chimpanzees,
62; of cultural units, 18, 58–59, 233,
301–3; of diseases, 117; of functionalist theory, 262–70; of ideas, 157,
178–79, 303; of structural conditions,
230 (see also uxorilocal marriages).
See also cultural transmission; transmission dynamics
social units, 39, 61, 220, 232; for cooperation, 86; contrasted with ideas, 22;
of cultural sharing, 281, 290; defining, 304–5; population, 86, 222; as
states, 292
socialization (of children), 221–22, 283;
and economic structure, 211–12, 222,
283; processes, 138, 212, 224–25,
288. See also cognitive variation: and
childhood socialization
society, concept of, 36, 265, 266, 268,
270, 290. See also social structure
sociobiology, 10, 58, 71, 213
Sonbol, Amira al-Azhari, 258
son preference, 67–69, 230, 235, 241–42
Spanish Conquest, 84
Spencer, Herbert, 264–68
Spradley, James, 270–72
stability: in cultural traits, 134; genetic,
63, 64; in institutions, 26, 258, 271;
in models, 106–13
Starrett, Gregory, 4, 5, 7, 8, 12, 14, 179,
230, 275, 281, 295, 297, 303
stereotypes, 26–27. See also bias: by race
or ethnicity
Stigler, George, 211, 212
STIs (sexually transmitted infections),

74, 117; core group model in, 133; epidemic models of, 118–19; epidemiology of, 118, 133; importance of social networks in, 298–99; transmitted in heterogeneous populations, 126–30. *See also* ethnic identity: and prevalence of STIs; gender: differences in STI transmission

strategic complements, 224

Strathern, Marilyn, 290

states. *See* nation-states

strong hypothesis that culture matters, 212–14, 226–27; experimental evidence for, 223–24; modeling, 224–26; testing, 213–14. *See also* cultural models

structural change, 162–63, 168–78; affecting cultural practices, 160, 176–77, 179; and kuru, 160; individual-level decisions in, 175; process of, 174–78; and tomb-sweeping, 176–77. *See also* social structure

structural-functionalist theories, 4, 13n2, 265

structural transmission, 229

structured mixing, 74, 117–20, 123–31. *See also* epidemic models

Susceptible-Infected model. *See* infectious diseases: models for

Susceptible-Infected-Removed model. *See* infectious diseases: models for

Susceptible-Infected-Susceptible (SIS) system, 126

symmetry, 55–56, 63, 65, 94. *See also* asymmetries

Taï chimpanzees. *See* chimpanzees

Taiwan, 94, 121, 138, 162–82, 229, 238–50. *See also* Aborigines; Chinese (Han); footbinding; Haishan; Hakka; Hoklo

Tasmania, 88–89, 92

technological complexity: in chimpanzees, 43–45, 50–51; in humans, 81, 83, 88–92, 93. *See also* complexity; cultural complexity; foraging: bow-and-arrow hunting technologies in

theoretical frameworks, 10–12, 23, 55–56, 61, 77, 78, 83, 117, 156–57, 178–79, 184, 210, 215–16, 289–90, 291. *See also* methodological issues; paradigm; science

Tomasello, Michael, 80

traits: biological basis of, 56, 109, 112; morphological (physical, associated with race), 126; personality, 288; skill and knowledge as, 207; variation in, 55, 57, 58, 59, 121. *See also* cultural traits: as behavioral

training, of anthropologists, 71, 270, 274, 297, 300–301

tramps, 270–71; pure, 272

transformations. *See* history: contrasted with evolution; social transformation

transmissible spongiform encephalopathies (TSE), 140. *See also* prions

transmission bias, 65, 80, 83, 86, 87, 94, 134; toward complex culture, 88–89; in punishment, 105–7, 109–10, 112–13, 114, 134. *See also* asymmetric transmission; bias; cultural transmission; prestige-biased transmission

transmission dynamics, 59–61, 64, 119–20, 125–26, 127, 181; modeled, 120, 128–31. *See also* cultural dynamics; epidemic dynamics; evolutionary dynamics; population dynamics; social dynamics

Trobriand Islands, incest among, 102

truth, 5, 13n1, 173, 231, 253, 291, 295; and religion, 263, 266; and science, 8, 9, 16n16, 22, 301

Tsimane (of the Bolivian Amazon), 200, 201, 202. *See also* small-scale societies
Tuscarora Indians, variation among, 288. *See also* intracultural diversity
Tylor, Edward B., 4, 13n2, 76, 253

ultimatum game (UG), 197–206, 216–20, 226–27; cognitive modules, 199–201; different age cohorts in, 198–99; income-maximizing offer (IMO) in, 199, 200; industrialized societies in the, 197–98; market integration in, 205, 219; offer rates in, 217, 220; predictions in, 205–6; regression analysis in, 219, 220; sense of fairness in, 199, 203–4; small-scale societies in, 201–5, 217, 218; strategic behavior in, 199; rejection rates in, 203, 217; taste for punishment in, 199, 200, 203–4 (*see also* punishment); variation in, 197–206, 215–17, 219. *See also* experimental games
uniform homophily model, 128
United States: regional differences in, 32, 34–36; southern, 29–31; ultimatum game (UG) in, 198–205
units of culture. *See* cultural units
Upper Paleolithic Europe, 88–89
Urapmin of New Guinea, 26
uxorilocal marriages, 21–22, 68–69, 234–52; concealing, 236–37; critical conditions for, 241; cultural transmission of, 234, 237; definition of, 234; economics of, 238; effect of poverty on, 238, 241; ethnic groups participating in, 235–36; frequency of, 235, 247, 249; gender difference in, 247; and illegitimate children, 240; inclination toward, 237; by land tax, 241;

242; parental influence in, 234–40 (*see also* social influences); parental transmission in, 242–45, 246–47, 248; and prostitutes, 240, 241; reputation of, in China, 238; shift away from, 176; sibling influences in, 241–48, 250; social conditions of, 247–50; and son preference, 68; vertical transmission of, 69

variation: in actions, 163, 174, 303; behavioral, 55, 70, 71, 123, 185, 219; continuous, 57, 59; discrete, 59, 134; ethnic, 74, 119–20, 123, 164, 293, 298; functional, 51; generation of, 83; heritable, 79; neutral, 95; random, 83; rules of, 57. *See also* biological variation; cross-cultural variation; cultural variation; cognitive variation; genetic variation; heterogeneity; intracultural variation
variation within a population: causes of, 55, 195; and history, 55, 56–59, 60–61, 66, 71; production of, 56, 59; and sampling, 60–61; and social negotiation, 163, 174; and social structure, 221, 288, 303 (*see also* social organization: and diversity). *See also* intracultural diversity; intracultural variation
Veblen, Thorstein, 98–99
vertical transmission. *See* cultural transmission
virilocal marriages, 68, 175; parental influence in, 235, 238–40. *See also* social influences
visual perception, 184
vitality. *See* life force

Waal, Frans B. M. de, 62
Wallace, Anthony, 281, 288–90, 295

weak hypothesis that culture matters, 212–14, 226. *See also* cultural models

Weber, Max, 14n9, 19, 260

Westermarck, Edward, 100–101, 103–4

Westermarck hypothesis, 100, 106, 109

Whewell, William, 75

Whiten, Andrew, 62

Whiting, Beatrice, 283

Whiting, John, 283

Wilson, Edward, 75

Wissler, Clark, 237

Wittgenstein, Ludwig, 14n9

Wolf, Arthur, 5, 7, 10, 12, 21, 77, 137, 156, 178, 229–30, 254, 257, 297, 298, 303

Wolf, Eric, 255

women: and footbinding, 94–96, 166–69, 306; Fore, 148–49, 153–54, 158; and honor code, 29–33; and language, 85; and marriage, 169, 174–76, 232, 236, 238–43, 245–52; and STIs, 102, 124–25, 127–32; tale of Wutu, 285, 296. *See also* fertility; gender; sex ratio; sib mating

Worden, Robert P., 183n5

writing, 42, 89

wudu', ritual purity of, 259. *See also* Egyptian religious studies textbooks

Wutu, tale of, 284–85, 295–96

Y-chromosomal DNA, 63

Zaghlul, Sa'd, 266, 268